Recycling of Organic Wastes in Agriculture: Serving for Sustainable Agriculture

Recycling of Organic Wastes in Agriculture: Serving for Sustainable Agriculture

Guest Editors

Halyna Kominko
Grzegorz Izydorczyk

MDPI

Basel • Beijing • Wuhan • Barcelona • Belgrade • Novi Sad • Cluj • Manchester

Guest Editors
Halyna Kominko
Faculty of Chemical Engineering and Technology
Cracow University of Technology
Cracow
Poland

Grzegorz Izydorczyk
Faculty of Chemistry
Wroclaw University of Science and Technology
Wroclaw
Poland

Editorial Office
MDPI AG
Grosspeteranlage 5
4052 Basel, Switzerland

This is a reprint of the Special Issue, published open access by the journal *Agronomy* (ISSN 2073-4395), freely accessible at: https://www.mdpi.com/journal/agronomy/special_issues/J5JBQP613X.

For citation purposes, cite each article independently as indicated on the article page online and as indicated below:

Lastname, A.A.; Lastname, B.B. Article Title. *Journal Name* **Year**, *Volume Number*, Page Range.

ISBN 978-3-7258-7076-9 (Hbk)
ISBN 978-3-7258-7077-6 (PDF)
https://doi.org/10.3390/books978-3-7258-7077-6

Contents

About the Editors

Halyna Kominko

Halyna Kominko holds a PhD in chemical engineering and is currently an Assistant Professor (Research and Teaching) at the Faculty of Chemical Engineering and Technology, Cracow University of Technology, Poland. Her research focuses on the valorisation of waste materials, particularly sewage sludge and other organic and industrial residues, with an emphasis on their transformation into mineral–organic fertilizers and soil amendments for agricultural use. Her scientific interests include circular economy approaches, nutrient recovery, soil–waste interactions, fertilizer efficiency, and the environmental risks associated with waste recycling in agriculture. Her work combines chemical engineering principles with environmental and agricultural sustainability.

Grzegorz Izydorczyk

Grzegorz Izydorczyk is an Assistant Professor at Wrocław University of Science and Technology, specializing in analytical chemistry and biogeochemistry, with a particular focus on fertilizer technology and the management and valorisation of organic waste. His research concentrates on the identification and quantitative determination of elements, as well as inorganic and organic compounds in complex environmental and material matrices (e.g., soils, sediments, biomass, composts, digestates, and fertilizer feedstocks and products). As Technical Manager of the Chemical Laboratory of Multi-Elemental Analysis, accredited by the Polish Centre for Accreditation, he is responsible for supervision of instrumentation, ensuring the quality of analytical results, and maintaining work standards in line with accreditation requirements. He has extensive experience in analyses using advanced instrumental techniques, including ICP-OES/ICP-MS, ion chromatography (IC), and TOC/TN analyzers.

agronomy

Editorial

Recycling of Organic Wastes in Agriculture: Serving for Sustainable Agriculture

Halyna Kominko [1] and Grzegorz Izydorczyk [2,*]

1 Department of Chemical Technology and Environmental Analytics, Faculty of Chemical Engineering and Technology, Cracow University of Technology, Warszawska Street 24, 31-155 Cracow, Poland; halyna.kominko@pk.edu.pl
2 Department of Advanced Material Technologies, Faculty of Chemistry, Wroclaw University of Science and Technology, 50-370 Wroclaw, Poland
* Correspondence: grzegorz.izydorczyk@pwr.edu.pl

1. Introduction

The transition toward sustainable agriculture is increasingly shaped by two converging pressures: the need to secure food production under climate and resource constraints and the imperative to reduce environmental burdens associated with linear "take–make–dispose" systems [1]. Within this context, the recycling and valorization of organic wastes has emerged as a practical pathway to strengthen circularity in agri-food systems. Organic waste streams—originating from livestock production, municipal services, agro-industrial processing, and wastewater treatment—are commonly rich in organic matter and essential nutrients. When properly managed, their agricultural reuse can support soil organic matter restoration, improve nutrient cycling, and reduce dependence on mineral fertilizers while simultaneously alleviating challenges related to waste storage and disposal [2].

At the same time, the agricultural use of organic wastes raises legitimate concerns that must be addressed with equal rigor. Alongside beneficial constituents, organic residues may introduce undesirable contaminants (e.g., heavy metals, emerging pollutants, microplastics, pathogens) and may contribute to gaseous losses during treatment and application, potentially offsetting environmental gains if not mitigated [3,4]. These trade-offs make it essential to evaluate organic waste recycling through an integrated lens—one that considers not only agronomic performance but also soil quality, environmental risk, and socio-economic feasibility [5,6].

This Special Issue, "Recycling of Organic Wastes in Agriculture: Serving Sustainable Agriculture" (Agronomy, section "Agricultural Biosystem and Biological Engineering"), was conceived to showcase recent advances that help maximize the environmental and agronomic benefits of organic waste utilization while recognizing and managing associated risks. We welcomed contributions spanning the valorization of diverse organic waste streams, assessment of fertilizer potential, impacts on soil properties and crop productivity, identification of potential contaminants and approaches to reduce their occurrence, and broader socio-economic analyses linked to implementation at farm and regional scales.

The nine papers published in this Special Issue collectively reflect the breadth and interdisciplinarity of the field. Several contributions emphasize waste-to-input pathways that convert residues into value-added products capable of supporting soil fertility and plant performance. These include the development of organo-mineral fertilizers from industrial residues (e.g., deinking paper sludge) and soil organic amendments derived from municipal solid waste fractions (biochar and compost), demonstrating the continued innovation in converting heterogeneous waste streams into more standardized agricultural

 https://doi.org/10.3390/agronomy16030398

inputs. The issue also highlights biological and microbiome-oriented approaches—such as circular biostimulant development using winery wastewater as a cultivation substrate and studies on manure-derived inputs affecting soil microbial communities—underscoring the role of biological processes in improving nutrient use efficiency and soil functioning.

Composting and process optimization constitute another important theme. Work on biochar-amended pig manure composting advances the toolbox for assessing compost maturity and immobilizing metals such as Cu and Zn, directly tackling a common barrier to safe land application of manure-derived products. Complementarily, research on sewage sludge composting explores strategies to curb nitrogen losses through targeted amendments (e.g., biochar combined with "garbage enzyme"), linking process control with microbial mechanisms of nitrogen transformation [7].

Risk assessment and environmental safeguards are addressed explicitly in the Special Issue through geospatial modeling of agricultural microplastic hotspots associated with biosolid application [8,9]. Such contributions reinforce that sustainable recycling is not solely about nutrient recovery—robust monitoring, modeling, and preventive management are equally necessary to avoid the unintended spread of pollutants across landscapes. Finally, a state-of-the-art review on biomethane production in the Mediterranean region broadens the perspective from field-scale reuse to system-level circular bioeconomy models, where anaerobic digestion can simultaneously deliver renewable energy and nutrient-rich digestate for agricultural soils [10].

Taken together, the contributions demonstrate that "closing the loop" in agriculture is a multi-criteria challenge: it requires technological solutions for valorization, agronomic evidence of performance, and risk-informed frameworks that account for contaminants, emissions, and implementation barriers. This introduction introduces the rationale and scope of the Special Issue, then synthesizes the published papers into coherent thematic threads before outlining key takeaways and future research directions for advancing organic waste recycling as a pillar of sustainable and resilient agriculture [11].

2. An Overview of Published Articles

Contribution 1 addresses the recycling of animal-derived organic wastes in agriculture by evaluating the short-term effects of cattle manure and poultry litter application on soil chemical properties and bacterial communities. Based on a 60-day field experiment, the study demonstrates that poultry litter induced more pronounced changes in soil chemistry and microbial community structure than cattle manure, particularly through increases in soil organic carbon and phosphorus availability and reductions in soil acidity. The results indicate that the observed shifts in soil microbiota were primarily driven by changes in soil chemical properties rather than by the direct transfer of microorganisms from the applied residues. Overall, this contribution highlights the potential of recycling organic waste as an effective strategy to improve soil fertility while supporting sustainable agriculture.

Contribution 2 explores the valorization of municipal solid waste through the conversion of its combustible and compostable fractions into biochar and compost for agricultural use. The study is based on a controlled pot experiment in which sunflower was cultivated for 75 days in soils amended with municipal solid waste–derived biochar and compost applied individually and in combination at different doses. The results demonstrate that the combined application of biochar and compost, particularly at a medium dose, led to marked improvements in key soil health indicators, including soil organic matter, cation exchange capacity, water-holding capacity, and the availability of essential nutrients. These improvements translated into enhanced sunflower growth, yield, and oil content, with performance comparable to that achieved using conventional chemical fertilizers. Importantly, the study shows that increasing amendment rates did not necessarily result in proportional yield gains, highlighting the importance of optimized application strategies.

Contribution 3 presents an innovative approach to recycling industrial organic waste by valorizing deinking paper sludge from the paper recycling industry for the production of organo-mineral fertilizers. The study combines acid-hydrolyzed deinking sludge with poultry litter ash to formulate PK-type fertilizers enriched with organic matter, secondary nutrients, and micronutrients. The resulting fertilizers were comprehensively characterized in terms of nutrient composition, heavy metal content, and phosphorus speciation, followed by an eight-week soil incubation experiment to assess their behavior under controlled conditions. The results demonstrate that the fertilizers met regulatory requirements for organo-mineral fertilizers and provided a substantial proportion of potentially bioavailable phosphorus, with a gradual release pattern favorable for reducing nutrient losses. Soil incubation studies revealed improvements in soil pH, organic carbon content, phosphorus availability, and microbial activity, as indicated by increased catalase activity.

Contribution 4 presents a GIS-based geospatial modeling approach to assess the risk of agricultural microplastic transport associated with the land application of biosolids. The study combines spatial data on rainfall, soil properties, land use, and proximity to waterways to map the relative risk of microplastic movement through three pathways: surface runoff, retention in soil, and groundwater infiltration. The results show that the risk of microplastic pollution is highest in agricultural areas located near rivers, in regions with higher rainfall, and in soils that favor particle mobility. By clearly identifying high- and low-risk zones, the study demonstrates how spatial risk mapping can support more informed decisions on where biosolids can be safely applied and where their use may increase the risk of microplastic pollution.

Contribution 5 describes the development of a circular, microbe-based biostimulant that combines a plant growth–promoting *Bacillus* strain with winery flotation wastewater as a low-cost growth medium. A bacterial isolate from the grapevine rhizosphere, named *Bacillus* sp. 10/R, was thoroughly analyzed for its biochemical characteristics, enzyme production, and plant growth–promoting traits. At the same time, winery flotation wastewater was assessed as a cultivation substrate by tracking nutrient use and enzyme production during bacterial growth. The resulting bacterial cultivation broths were then tested as seed treatments to examine their effects on barley germination and early seedling growth. The findings show that *Bacillus* sp. 10/R has a broad enzymatic capacity and strong pectin-degrading activity, supporting its potential as a plant growth–promoting microorganism. Winery flotation wastewater has been shown to promote bacterial growth and metabolite production while reducing the wastewater's organic nutrient content. The authors therefore suggest further research to better understand plant responses, optimize application methods and dosage, and evaluate the technical and economic feasibility of this biostimulant approach under greenhouse and field conditions.

Contribution 6 presents a study on *Opuntia* spp., with a particular focus on residue valorization and circular economy practices in Mexico. The article combines a bibliometric analysis of international literature with a critical review of applied Mexican studies to identify research trends, regional gaps, and opportunities for closing production cycles through the reuse of cladode residues. The study shows that three main valorization pathways are composting, anaerobic fermentation, and direct soil application of *Opuntia* residues, with composting being the most popular. The authors argue that these practices fit well with smallholder farming systems and circular economy principles, but their wider use is limited by the lack of standardized methods, long-term field validation, and policy support. Overall, the article highlights composting and residue reuse as practical strategies

https://doi.org/10.3390/agronomy16030398

to close nutrient cycles and improve the sustainability of *Opuntia*-based production systems in Mexico.

Contribution 7 explores strategies to reduce nitrogen losses during sewage sludge composting by using additives such as garbage enzyme, biochar, and pelelith. The research highlights that combining garbage enzyme with biochar or pelelith can improve nitrogen retention, enhance ammonia-oxidizing microbial activity, and promote more efficient nitrogen transformation. These additives also influence the structure and function of ammonia-oxidizing microbial communities, contributing to higher-quality compost and reduced environmental impacts. The findings suggest that integrating enzymatic and carbon-based amendments offers a promising, sustainable approach to improve composting efficiency and nutrient management, while further studies are needed to fully understand the underlying microbial mechanisms.

Contribution 8 presents an innovative approach to evaluating compost maturity and heavy metal passivation in biochar-amended pig manure using self-organizing maps (SOM) integrated with three-dimensional excitation–emission matrix fluorescence spectroscopy. The research examined different biochar dosages and their effects on compost quality, organic matter transformation, and Cu/Zn stabilization. The findings demonstrate that higher biochar additions accelerate compost maturation and enhance heavy metal passivation, though moderate dosages can achieve comparable results while offering better cost-effectiveness. By combining SOM with advanced spectroscopic techniques, the authors successfully tracked organic matter evolution and identified key microbial communities responsible for metal stabilization. This work provides a novel methodology for real-time compost quality assessment and offers practical insights for optimizing biochar application in livestock waste management systems.

Contribution 9 examines biomethane production in the Mediterranean region as a renewable energy solution. The authors evaluate how anaerobic digestion can convert organic waste, such as olive pomace, citrus peel, grape residues, livestock manure, and municipal biowaste, into clean energy and valuable fertilizer. The study highlights the dual benefits of this technology: producing biomethane as a replacement for natural gas while creating digestate that improves soil health in areas affected by degradation and desertification. The review discusses upgrading technologies that transform biogas into high-quality biomethane, along with policy frameworks like the EU Renewable Energy Directive that support sector growth. Case studies from Italy, Spain, France, Greece, and Tunisia demonstrate the practical application of these systems in rural and island communities. Despite the clear potential, challenges remain, including variable feedstock availability, limited infrastructure, and inconsistent policies across countries. The authors conclude that biomethane can play a central role in achieving energy independence, reducing greenhouse gas emissions, and supporting sustainable agriculture in the Mediterranean within a circular economy approach.

3. Conclusions and Future Perspectives

3.1. Conclusions

The contributions collected in this Special Issue confirm that recycling organic waste streams into agriculture can simultaneously support soil functioning, nutrient cycling and productivity—provided that agronomic benefits are pursued together with rigorous risk management. Across the presented studies, a consistent message emerges: organic amendments and derived products (e.g., composts, digestates, organic–mineral fertilizers and biochar-based materials) can improve soil fertility indicators, enhance nutrient availability, and, in several cases, contribute to stabilizing soil organic matter or improving soil physical properties. At the same time, the "circularity dividend" is not automatic. Outcomes depend

https://doi.org/10.3390/agronomy16030398

on feedstock origin, processing technology, product stability, application rate and timing, as well as site-specific conditions such as soil type, climate and management history.

Importantly, the studies also underscore that the safe return of organic residues to land must address multiple, partly interacting hazards. These include the persistence and mobility of potentially toxic elements, the presence of pathogens and antibiotic resistance determinants, the fate of emerging contaminants and microplastics, and climate-relevant gaseous emissions during storage and after field application. Several contributions point toward practical pathways to mitigate these risks, including improved stabilization and sanitation during processing, blending strategies (e.g., organic–mineral formulations), and decision-support approaches that optimize sourcing and allocation of organic resources at the landscape scale. Overall, the Special Issue supports a shift from viewing organic wastes as uniform "fertilizer substitutes" toward treating them as engineered, quality-controlled soil inputs whose performance and safety must be demonstrated within local agroecosystems and governance frameworks.

3.2. Future Perspectives

Building on the evidence and challenges discussed in this Special Issue, several priorities can guide research and implementation:

1. **Long-term, system-level field evidence.** Multi-year field trials across contrasting soils and climates are needed to quantify yield stability, soil carbon trajectories, nutrient balances and resilience outcomes under repeated applications, including interactions with cropping systems and reduced mineral fertilization strategies.
2. **Harmonized quality metrics and fit-for-purpose standards.** Comparable analytical protocols and transparent criteria are needed for product maturity/stability, nutrient release patterns and contaminant thresholds (including emerging pollutants and microplastics). Linking these metrics to intended agronomic functions (fertility building, carbon management, remediation) would strengthen certification and market confidence.
3. **Integrated risk assessment for multiple stressors.** Future work should move beyond single-contaminant approaches toward frameworks that combine chemical hazards (potentially toxic elements, organic pollutants), biological risks (pathogens, antibiotic resistance) and physical contaminants (microplastics), accounting for co-occurrence, transformation during processing and bioavailability after land application.
4. **Emission-aware circularity.** Nutrient-loop closure should be evaluated together with greenhouse gas and reactive nitrogen emissions (CH_4, N_2O, NH_3) along the full chain—from collection and treatment to storage, transport and field application. Mitigation options (e.g., covers, additives, incorporation timing, biochar co-application) deserve standardized testing under realistic conditions.
5. **Design of advanced products and blends.** Organic–mineral fertilizers, biochar-enhanced amendments and tailored compost/digestate formulations offer opportunities to improve nutrient use efficiency and reduce losses. Research should clarify mechanism-based design rules (e.g., sorption, pH buffering, microbial interactions) and quantify trade-offs between performance, cost and safety.
6. **Spatial planning and decision support at the landscape scale.** Because feedstock availability, transport constraints and environmental sensitivity vary across regions, models combining GIS, logistics, soil vulnerability and regulatory limits can support optimal allocation of organic resources. This is crucial to avoid local over-application of phosphorus or contaminants while maximizing benefits where soils are most responsive.

https://doi.org/10.3390/agronomy16030398

7. **Governance, adoption and economics.** Key barriers include stakeholder acceptance, regulatory harmonization, traceability and liability, and stable business models for waste-derived products. Coupling agronomic trials with techno-economic and life-cycle assessments can translate scientific findings into implementable circular strategies.

Taken together, these directions point toward an integrated agenda in which organic waste recycling is treated as a managed, evidence-based intervention—optimizing agronomic value while ensuring environmental protection and societal trust.

Author Contributions: Conceptualization, H.K.; methodology, H.K. and G.I.; formal analysis, H.K. and G.I.; data curation, H.K. and G.I.; writing—original draft preparation, H.K. and G.I.; writing—review and editing, H.K. and G.I. All authors have read and agreed to the published version of the manuscript.

Funding: This research received no external funding.

Data Availability Statement: No new data were created or analyzed in this study. Data sharing is not applicable to this article.

Acknowledgments: We, as Guest Editors of the Special Issue "Recycling of Organic Wastes in Agriculture: Serving Sustainable Agriculture", would like to express our sincere gratitude to all authors whose valuable contributions made this issue a success. We also extend our heartfelt thanks to Kermit Zhang, Section Managing Editor, for his continuous support and cooperation throughout the publication process.

Conflicts of Interest: The authors declare no conflicts of interest.

List of Contributions:

1. Souza, G.; Sant, L.; Amoroso, L.; Carvalho, L. De; Santos, M.; Andrade, R. De; Pinheiro, D.G.; Zonta, E.; Coelho, S. Short-Term Effects of Poultry Litter and Cattle Manure on Soil's Chemical Properties and Bacterial Community. *Agronomy* **2024**, *14*, 1382.
2. Aslam, S.; Nazir, A. Valorizing Combustible and Compostable Fractions of Municipal Solid Waste to Biochar and Compost as an Alternative to Chemical Fertilizer for Improving Soil Health and Sunflower Yield. *Agronomy* **2024**, *14*, 1449.
3. Kominko, H.; Gorazda, K.; Łoś, K.; Wzorek, Z. Valorisation of Deinking Paper Sludge for Fertiliser Purposes: New Perspective in Sustainable Agriculture. *Agronomy* **2024**, *14*, 2788. https://doi.org/10.3390/agronomy14122788.
4. Mcnulty, K.; Nahar, K. Geospatial Modelling Predicts Agricultural Microplastic Hotspots from Biosolid Application Risks. *Agronomy* **2025**, *15*, 47.
5. Dujković, T.; Danilov, I.; Vlajkov, V.; Savić, M.; Šumić, Z.; Jokić, A.; Grahovac, J. Circular Approach in Development of Microbial Biostimulants Using Winery Wastewater. *Agronomy* **2025**, *15*, 2272.
6. Torres-Sandoval, A.J.; Ortiz-hernández, Y.D.; Tavera-cortés, M.E.; Acevedo-ortiz, M.A.; Lugo-espinosa, G. Closing the Loop in Opuntia Cultivation: Opportunities and Challenges in Residue Valorization. *Agronomy* **2025**, *15*, 2311.
7. Jiang, J.; Cui, H.; Bhople, P.; Chater, C.C.C.; Yu, F. Biochar Combined with Garbage Enzyme Enhances Nitrogen Conservation during Sewage Sludge Composting: Evidence from Microbial Community and Enzyme Activities Related to Ammoniation. *Agronomy* **2024**, *14*, 1162.
8. Zhang, H.; Yu, X.; Luo, L.; Sun, Y.; Zhou, L. Self-Organizing Map-Based Assessment of Compost Maturity and Cu/Zn Passivation in Biochar-Amended Pig Manure. *Agronomy* **2025**, *15*, 778.
9. Comparetti, A.; Ciulla, S.; Greco, C.; Santoro, F. State of the Art of Biomethane Production in the Mediterranean Region. *Agronomy* **2025**, *15*, 1702.

https://doi.org/10.3390/agronomy16030398

References

1. Umesha, S.; Manukumar, H.M.G.; Chandrasekhar, B. Sustainable agriculture and food security. In *Biotechnology for Sustainable Agriculture: Emerging Approaches and Strategies*; Elsevier: Amsterdam, The Netherlands, 2017; pp. 67–92. [CrossRef]
2. Sharma, B.; Vaish, B.; Monika; Singh, U.K.; Singh, P.; Singh, R.P. Recycling of Organic Wastes in Agriculture: An Environmental Perspective. *Int. J. Environ. Res.* **2019**, *13*, 409–429. [CrossRef]
3. Semblante, G.U.; Hai, F.I.; Huang, X.; Ball, A.S.; Price, W.E.; Nghiem, L.D. Trace organic contaminants in biosolids: Impact of conventional wastewater and sludge processing technologies and emerging alternatives. *J. Hazard. Mater.* **2015**, *300*, 1–17. [CrossRef] [PubMed]
4. Seruga, P.; Krzywonos, M.; Paluszak, Z.; Urbanowska, A.; Pawlak-Kruczek, H.; Niedźwiecki, Ł.; Pińkowska, H. Pathogen Reduction Potential in Anaerobic Digestion of Organic Fraction of Municipal Solid Waste and Food Waste. *Molecules* **2020**, *25*, 275. [CrossRef] [PubMed]
5. El-Naggar, A.; El-Naggar, A.H.; Shaheen, S.M.; Sarkar, B.; Chang, S.X.; Tsang, D.C.; Rinklebe, J.; Ok, Y.S. Biochar composition-dependent impacts on soil nutrient release, carbon mineralization, and potential environmental risk: A review. *J. Environ. Manag.* **2019**, *241*, 458–467. [CrossRef] [PubMed]
6. Sharma, L.D.; Sarangthem, I.; Thangjam, R.; Sadhukhan, R.; Oinam, N.; Yanglem, B.; Banarjee, L.; Anal, H.J.; Jatav, H.S. Sewage Sludge and Its Health Risk Assessment: Opportunities and Challenges. In *Sustainable Management and Utilization of Sewage Sludge*; Springer: Berlin/Heidelberg, Germany, 2022; pp. 205–225. [CrossRef]
7. Jiang, J.; Wang, Y.; Yu, D.; Yao, X.; Han, J.; Cheng, R.; Cui, H.; Yan, G.; Zhang, X.; Zhu, G. Garbage enzymes effectively regulated the succession of enzymatic activities and the bacterial community during sewage sludge composting. *Bioresour. Technol.* **2021**, *327*, 124792. [CrossRef] [PubMed]
8. Arab, M.; Yu, J.; Nayebi, B. Microplastics in Sludges and Soils: A Comprehensive Review on Distribution, Characteristics, and Effects. *Chemengineering* **2024**, *8*, 86. [CrossRef]
9. Ramage, S.J.F.F.; Coull, M.; Cooper, P.; Campbell, C.D.; Prabhu, R.; Yates, K.; Dawson, L.A.; Devalla, S.; Pagaling, E. Microplastics in agricultural soils following sewage sludge applications: Evidence from a 25-year study. *Chemosphere* **2025**, *376*, 144277. [CrossRef] [PubMed]
10. McNulty, K.; Nahar, K. Geospatial Modelling Predicts Agricultural Microplastic Hotspots from Biosolid Application Risks. *Agronomy* **2024**, *15*, 47. [CrossRef]
11. Ligarda-Samanez, C.A.; Huamán-Carrión, M.L.; Calsina-Ponce, W.C.; De la Cruz, G.; Huamaní, D.F.C.; Cabel-Moscoso, D.J.; Garcia-Espinoza, A.J.; Sucari-León, R.; Aroquipa-Durán, Y.; Muñoz-Saenz, J.C.; et al. Technological Innovations and Circular Economy in the Valorization of Agri-Food By-Products: Advances, Challenges and Perspectives. *Foods* **2025**, *14*, 1950. [CrossRef] [PubMed]

https://doi.org/10.3390/agronomy16030398

agronomy

MDPI

Article

Short-Term Effects of Poultry Litter and Cattle Manure on Soil's Chemical Properties and Bacterial Community

Gustavo Souza Lima Sant'Anna [1], Lucas Amoroso Lopes de Carvalho [2], Maura Santos Reis de Andrade da Silva [1], João Vitor da Silva Gonçalves [1], Daniel Guariz Pinheiro [2], Everaldo Zonta [3] and Irene da Silva Coelho [1,*]

1 Department of Veterinary Microbiology and Immunology, Veterinary Institute, Federal Rural University of Rio de Janeiro, Seropédica 23897-970, RJ, Brazil; gsouzalimas@gmail.com (G.S.L.S.); maura@ufrrj.br (M.S.R.d.A.d.S.); joaovdsgoncalves@gmail.com (J.V.d.S.G.)
2 Department of Agricultural and Environmental Biotechnology, School of Agricultural and Veterinary Sciences (Jaboticabal Campus), Paulista State University, Jaboticabal 14884-900, SP, Brazil; lucas.amoroso@unesp.br (L.A.L.d.C.); daniel.pinheiro@unesp.br (D.G.P.)
3 Department of Soils, Institute of Agronomy, Federal Rural University of Rio de Janeiro, Seropédica 23897-970, RJ, Brazil; ezonta@ufrrj.br
* Correspondence: irenecoelho@ufrrj.br

Abstract: The expansion of animal husbandry for food production has necessitated effective management of livestock waste, including cattle manure and poultry litter. Using these byproducts as organic fertilizers in agriculture represents a sustainable approach to their disposal. While these residues offer known benefits for soil health and crop productivity, further studies are needed to explore the effect of different manure microbiota on soil composition. The objective of this study was to address this gap and contribute to the advancement of this area. A randomized block design experiment was set up in the field with three replications per treatment, including the application of cattle manure, poultry litter, and a control treatment without fertilizer. After a 60 day period, the chemical characteristics and bacterial population composition of the soil were analyzed using 16S rRNA gene sequencing. Organic carbon, phosphorus (P), aluminum (Al), and the pH level played pivotal roles in changing the structure of the soil's bacterial community. Notably, the incorporation of poultry litter induced more pronounced changes in both the chemical properties and bacterial community composition compared with cattle manure. Bacterial groups were enriched in the soils treated with poultry litter, which may indicate enhanced soil fertility. This association may stem from both the chemical alterations resulting from poultry litter application and the direct transfer of microorganisms from this organic fertilizer to the soil.

Keywords: animal waste; organic fertilizer; organic fertilization; bacterial diversity

1. Introduction

In 2023, global production of chicken meat and beef reached 103 million and 59 million metric tons, respectively. During the same year, Brazil emerged as one of the world's main meat producers, contributing approximately 15 million metric tons of chicken meat and 11 million tons of beef. These statistics positioned Brazil as the second largest producer worldwide in both sectors [1].

However, the escalating demand for meat, driven by economic and population growth, has resulted in substantial environmental concerns [2]. In addition to greenhouse gas emissions due to deforestation for pasture expansion, livestock farming increases waste generation, including cattle manure and poultry litter, which often carry pollutant loads [3–5]. If left unaddressed, these waste streams pose significant environmental liabilities.

A sustainable and promising alternative is to use this waste as organic fertilizer, given its rich nutritional composition which includes essential elements that improve the soil and increase plant production, reducing dependence on chemical fertilizers [6]. Organic fertilizers offer macro- and micronutrients and increase organic matter, thereby improving soil fertility and preserving its biodiversity [7,8]. They are also excellent sources of nutrients, increasing microbial diversity and richness [9,10]. Furthermore, microorganisms incorporated into the soil can influence nutrient cycling, affecting their availability to plants and, consequently, crop productivity [11]. This approach, however, has some drawbacks, including increased heavy metal levels [12], incorporation of pathogenic bacteria [13], and the spread of antimicrobial resistance genes in soil [14].

Therefore, investigations into soil microbiota alterations after the application of these residues are essential to understand the potential benefits and risks to soil health. However, data on changes in bacterial composition following the incorporation of animal production residues in tropical soils are limited. Therefore, gathering information from different soils, climates, and production systems is crucial to deepening the understanding of this dynamic. This necessity is further underscored by the scarcity of comparative studies on microbial structures between different waste sources [15]. Additionally, the immediate impact of these fertilizers on the incorporation or enrichment of microbial groups needs to be better understood [16], as these effects can directly influence crops subsequent to this practice. Considering the critical importance of understanding the effects of organic fertilizers on soil, the objective of this study was to evaluate the short-term impact of incorporating cattle manure and poultry litter on the chemical characteristics and bacterial community of soil.

2. Materials and Methods

Cattle manure (CAM-F) was sourced from a corral within the dairy cattle farming sector of UFRRJ, while poultry litter (POL-F) was obtained from a chicken egg production unit. Both fertilizers were used in their fresh states without undergoing a composting process.

Determination of the Ca, Mg, Na, K, P, Mn, Ni, Fe, Cd, Cr, Cu, Pb, and Zn contents followed the methodology proposed in [17]. For this, 0.5 g of the compound samples were weighed, ground, and sieved. Subsequently, they were digested in nitric acid on a heated plate. The P content was determined by colorimetry, while Na and K were analyzed using flame spectrophotometry. Furthermore, Ca, Mg, Cd, Cr, Cu, Pb, Mn, Ni, Fe, and Zn were quantified using atomic absorption spectrometry (Table 1).

Table 1. Chemical composition of cattle manure (CAM-F) and poultry litter (POL-F) incorporated into soil.

	N	C	P	K	Na	Ca	Mg	Fe	Mn	Zn	Cu	Al	Ni	Pb	Cr	Cd
		------------ $g\ kg^{-1}$ ------------							------------ $mg\ kg^{-1}$ ------------							
CAM-F	12.3	227	0.15	13.4	7.2	15.9	4.2	2.6	251.2	59	17.8	0	12.8	3.4	0	0
POL-F	28.6	325	0.54	17.6	50.5	61.3	6.6	1.9	463.6	471.4	91.6	127.7	5.8	12.6	0	0

The experiment was conducted on a farm located at the coordinates 22°49′19.79″ S and 43°44′16.43″ W at Reta dos 800 in Piranema, a rural area of Seropédica in Rio de Janeiro, Brazil. The soil was classified as Ultisol [18], corresponding to Argissolo Vermelho-Amarelo in the Brazilian Soil Classification System [19]. According to the Köppen classification system, the climate in the region is Aw (Figure 1) [20]. The experimental site comprised a bahiagrass (*Paspalum notatum*) pasture in a state of degradation. Prior to commencing the experiment, the soil was plowed and received 2.5 Mg ha^{-1} of dolomitic limestone with a relative total neutralizing power (rTNP) of 85.6%, 40% CaO, and 10% MgO. This limestone was incorporated to a depth of 20 cm using a mini-tractor equipped with a rotating hoe on 25 April 2019. The experiment officially began 60 days after the liming process. Figure 1 shows climatic data, including precipitation and maximum and minimum temperatures, spanning from the beginning to the end of the experiment.

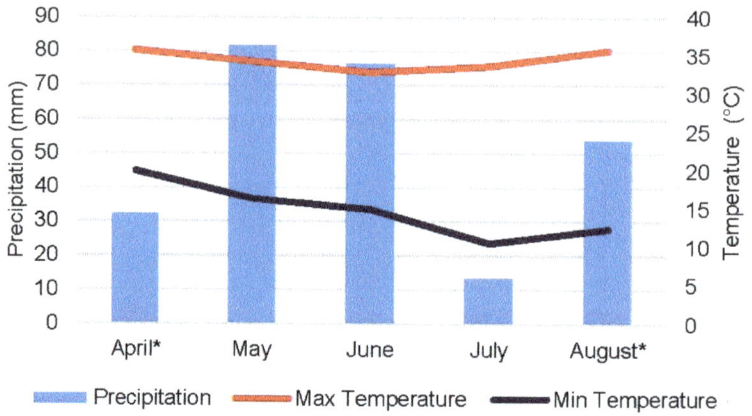

Figure 1. Monthly climate data from the automatic meteorological station of the national meteorology institute (INMET) during the experiment's execution period (INMET, 2024). * Data from 25 April to 26 August 2019.

The randomized block design included three replications per treatment, each characterized using soil with CAM-F (CAM-S) or POL-F (POL-S), and a control treatment (CTL-S) without fertilizers, totaling nine experimental plots of 4 m^2 each (2 $\times$ 2 m). Fertilizers were applied at a dose of 100 kg ha^{-1} of N by incorporating 8.1 Mg ha^{-1} of CAM-F and 3.5 Mg ha^{-1} of POL-F at a depth of 20 cm, using a mini-tractor with a rotating hoe.

A total of five simple samples were collected 60 days post fertilization from the 0–10 cm layer of each experimental plot, resulting in three composite samples per treatment. For the determination of exchangeable nutrient contents, the samples were analyzed as follows. Ca, Mg, and Al were extracted with a KCl solution and quantified via titration. P, K, and Na were extracted with Mehlich 1 solution, with P quantified by colorimetry and K and Na quantified by flame photometry. H + Al was evaluated using a 0.025 mol L^{-1} calcium acetate solution. The soil's pH was determined in water, and the cation exchange capacity (CTC) and base saturation (V) were calculated. The soil organic carbon (SOC) was determined by the wet oxidation method using potassium dichromate, and the total nitrogen was determined through Kjeldahl digestion [21].

A DNA PowerSoil extraction kit (MO Bio Laboratories Inc., Carlsbad, CA, USA) was used for the extraction of both soil and fertilizer DNA, following the manufacturer's protocol. Subsequently, the extracted samples were stored at −20 °C, and the quantity and quality of DNA were evaluated using a spectrophotometer (NanoDrop 1000, Thermo Fisher Scientific Inc., Waltham, MA, USA). The overall integrity of the total DNA was further assessed through electrophoresis on 0.8% agarose gel [22].

For sequencing purposes, a library of amplicons targeting the V3–V4 variable region of the 16S rRNA gene was amplified using Bakt_341F (CCTACGGGNGGCWGCAG) and Bakt_805R (GACTACHVGGGTATCTAATCC) primers. Sequencing was conducted using a paired-end 2 $\times$ 250 system on the MiSeq platform (Illumina Inc., San Diego, CA, USA) at Macrogen Inc. (Seoul, South Korea).

The quality of the sequenced data was initially evaluated using FastQC software v.0.11.9 [23]. The "fastx_info" and "fastq_eestas2" functions of USEARCH software v.11.0.667 were used with the libraries [24], displaying the distribution of qualities, sequence sizes, and expected errors. The "search_oligodb" function of the same software was used to detect the presence and position of primer pair sequences referring to the V3–V4 region of the 16S rRNA gene (341F '5-CCTACGGGNGGCWGCAG-3' and 805R '5-GACTACHVGGGTATCTA ATCC-3'). Subsequently, the primers were removed with Atropos software v.1.1.31 [25], which filtered sequences whose primers were not present ("--discard-untrimmed").

To ensure better quality, the final portion of the sequences was trimmed using Fastp software v.0.23.2 [26] to remove up to 20 ("--max_len1 230") and 30 ("--max_len2 220") low-quality final bases from the forward and reverse libraries, respectively. Additionally, complete reads with a total average Phred score (Q) less than Q20 were removed ("--average_qual 20"). Finally, the library pairs were merged by overlapping using Flash software v.1.2.11 [27], accepting a minimum overlap of 10 bases ("--min-overlap 10").

Merged reads with between 390 and 430 bases were subjected to the "DADA2" pipeline [28] using the "dada2" package v.1.22.0 of the R statistical software v.4.1.2 [29]. Initially, the reads were filtered with the "filterAndTrim" function, considering an expected error of four ("maxEE = 4"). Subsequently, the probability of errors in the bases was estimated ("learnErrors"), and the sequences were corrected according to the obtained model ("dada"). Amplicon variant sequences (ASVs) were designated in each sample, investigated, and filtered for the presence of possible chimeric sequences ("removeBimeraDenovo").

All ASVs were taxonomically annotated against the SILVA v.138 reference sequence database [30], with additional support from the RDP v.18 [31] and GTDB v.202 [32] databases to identify potential contaminants. ASVs not assigned to bacteria or archaea, as well as those assigned to chloroplasts or mitochondria, were filtered out. Furthermore, ASVs prevalent in only a single replicate were excluded from the analysis. The ASV counts and taxonomic annotations were exported in "phyloseq" format using the R "phyloseq" package v.1.38.0 [33]. Subsequently, the data were transformed into compositional data ("method = 'total'") by the "phyloseq_standardize_otu_abundance" function of the R "metagMisc" package v.0.04 [34].

Sampling effectiveness was inferred by rarefaction curves obtained through "amp_rare curve" analysis of the R "ampvis2" package v.2.7.17 [35]. Alpha diversity was estimated by assessing the observed richness and diversity measures (Shannon and Gini-Simpson indices) obtained using the "alpha" function from the R package "microbiome" v.1.16.0 [36]. The obtained measures were statistically compared using the Kruskal–Wallis test (p value ≤ 0.1), and the post hoc Fisher's LSD test was used for pairwise comparison and grouping of the means (p value ≤ 0.1). Both were implemented using the "kruskal" function from the R package "agricolae" v.1.3.5 [37]. A beta diversity analysis was performed by calculating the Bray–Curtis distances with the "distance" function of the "phyloseq" package [33]. The distances were statistically compared between the groups with the "adonis" function from the same package using permutational multivariate analysis of variance (PERMANOVA), considering a p value of 0.05. Furthermore, we used the distances to carry out a canonical analysis of principal coordinates (CAP), where the physical-chemical parameters measured were added to the distances between the compositions and compared for their significance using the "ordinate" function of "phyloseq" with the same significance threshold (p value ≤ 0.05). A graphical representation was obtained with the "plot_ordination" function from the "phyloseq" package. The assessment of differentially abundant taxa involved identifying those present in significantly greater amounts in one treatment compared with the other. This analysis used the "DESeq2" approach v.1.34.0 [38], which compares means based on a negative binomial model and uses the Wald test (adjusted p value < 0.05). Graphical representations of these analyses were prepared using the R "ggplot2" package v.3.3.6 [39].

Additionally, the chemical properties of the soil and the microbiota were correlated while considering the taxonomic level of the family. To filter out low-abundance taxa, we only considered families that corresponded to at least 1% of the abundance of the samples. Pearson coefficients were determined between each taxon and soil property using the "corr.test" function of the R "psych" package v.2.3.9 [40], where we considered a p value of 0.05. A heatmap was created using the "ggplot2" package [39], utilizing the "geom_tile" function to plot correlation values and "geom_text" to indicate significant correlations.

Data on chemical characteristics were initially subjected to analysis of variance using the F test. Then, the Shapiro–Wilk test verified the normality of residuals, and the Bartlett test verified the homoscedasticity. If the F test of the analysis of variance showed significant differences ($p < 0.05$), then the treatment means were compared using the Tukey test at 10% probability.

3. Results

In Table 2, it is observed that POL-S presented higher levels of Ca and P contents compared with CTL-S, along with a higher soil organic C (SOC) content than CAM-S. However, CAM-S and CTL-S presented higher Al concentrations in contrast to POL-S. Differences in bacterial communities were observed in the statistical comparison of Bray–Curtis distances using the PERMANOVA test (p value: 0.004). When relating this information to the environmental variables through the canonical analysis of principal coordinates (CAP), the treatment tended to separate, with POL-S being distinct from the others. However, the inclusion of environmental variables did not result in statistically significant sample separation. However, Figure 2 illustrates that CTL-S and CAM-S were associated with Al and H + Al, whereas POL-S showed correlations with Ca, Mg, SB, V, and CEC. Regarding alpha diversity, there was no significant difference in the observed richness of ASVs between treatments. However, POL-S exhibited lower Shannon and Gini–Simpson indices compared with those of CTL-S and CAM-S (Table 3).

Table 2. Soil chemical composition 60 days after the application of cattle manure (CAM-S) or poultry litter (POL-S) and in the control soil (CTL-S) which did not receive residue application.

	pH *	Ca	Mg	Na	K	Al	H + Al	SB	CEC	V	P	K	N	SOC
	1:2.5				cmol$_c$ dm^{-3}					%	mg dm^{-3}			%
CTL-S	4.90	1.26 ^{b}	1.49	0.03	0.16	0.05 ^{a}	4.37	2.78	7.14	38.09	3.00 ^{b}	63.00	0.13	3.47 ^{ab}
CAM-S	5.25	1.90 ^{ab}	1.71	0.05	0.28	0.05 ^{a}	4.01	3.66	7.68	47.71	8.33 ^{ab}	108.67	0.19	3.43 ^{b}
POL-S	5.36	2.03 ^{a}	1.71	0.05	0.23	0.02 ^{b}	3.72	3.78	7.51	50.56	17.33 ^{a}	89.00	0.16	4.86 ^{a}

* pH in water. SB = sum of bases; CEC = cation exchange capacity; V = base saturation; and SOC = soil organic carbon. Different letters denote significant differences according to Tukey test at 10% significance level ($p < 0.10$).

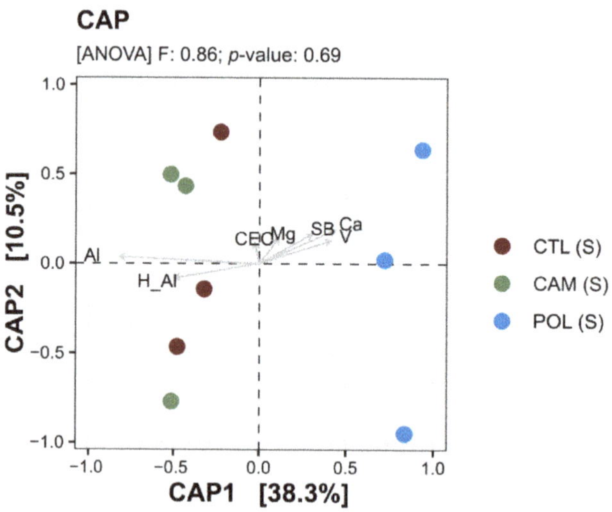

Figure 2. Structure of bacterial communities in soils that received cattle manure (CAM-S) or poultry litter (POL-S) and in the control (CTL-S) without residues application. Distance-based constrained analysis of principal coordinates (CAP) using the Bray–Curtis distance between the composition of amplicon sequence variants (ASVs) in samples was performed. Dots indicate the sample scores for the first two axes, while the vectors indicate the chemical attributes of the soil. H + Al = potential acidity; SB = sum of bases; and V = base saturation.

Table 3. The variability of alpha diversity (richness, Shannon index, and Gini–Simpson index) in soil samples after 60 days of application of cattle manure (CAM-S), poultry litter (POL-S), and control soil (CTL-S) which did not receive application of these residues.

	Mean	SD	*p* Value
	Richness		0.8752
CAM-S	893 ^{ns}	74	
CTL-S	876 ^{ns}	47	
POL-S	872 ^{ns}	103	
	Shannon		0.0608
CAM-S	6.18 ^{a}	0.15	
CTL-S	6.21 ^{a}	0.08	
POL-S	5.84 ^{b}	0.28	
	Gini–Simpson		0.0608
CAM-S	0.996 ^{a}	0.001	
CTL-S	0.996 ^{a}	0.001	
POL-S	0.988 ^{b}	0.006	

Differences between treatments were statistically evaluated using a Kruskal–Wallis test (p value ≤ 0.1) followed by Fisher's LSD post hoc test for multiple comparisons and grouping of means (p value ≤ 0.1). The difference between treatments is represented by different letters, while "ns" means there are no significant differences between treatments.

The CAM-F sample showed a higher abundance of Proteobacteria (49.91%), Actinobacteria (22.56%), and Bacteroidota (13.62%), and the POL-F sample had greater abundances of Firmicutes (50.13%) and Actinobacteria (43.40%). The prevalent phyla in CLT-S and CAM-S were Proteobacteria (25.24 and 27.13%, respectively) and Actinobacteriota (17.22 and 25.17%, respectively), while there was a prevalence of Proteobacteria (30.58%) and Verrucomicrobiota (18.89%) in POL-S (Figure 3).

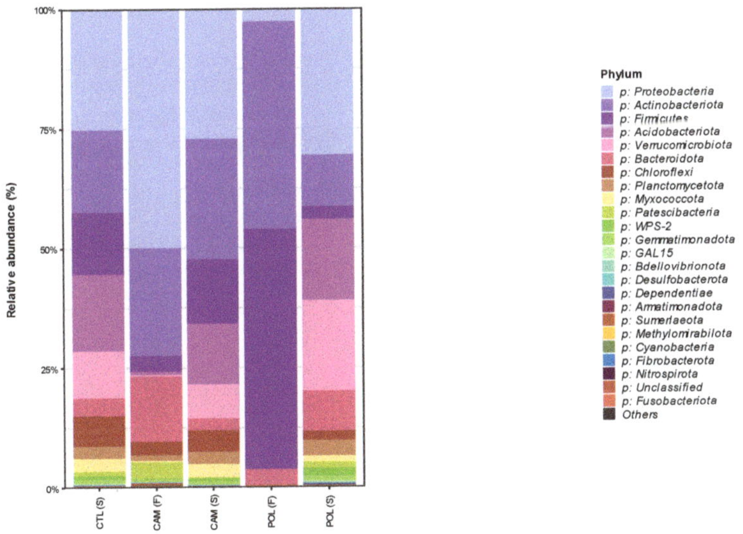

Figure 3. Relative abundance of bacterial groups at the phylum level, showing the 24 most abundant taxa in samples of cattle manure (CAM-F), poultry litter (POL-F), and soils 60 days post fertilization with cattle manure (CAM-S) and poultry litter (POL-S) along with a control without application of residue (CTL-S). The less abundant phyla are grouped under the category "Others".

The most prevalent family in CAM-F was Xanthomonadaceae (Proteobacteria) (14.73%), and POL-F presented the families Staphylococcaceae, Bacillaceae (both Firmicutes) (19.26 and 17.42%, respectively), and Brevibacteriaceae (Actinobacteriota) (15.17%). The three families found in POL-F corresponded to 52% of the ASVs. CTL-S and CAM-S showed a prevalence of the families Bacillaceae (Firmicutes) (10.31 and 11.21%, respectively), Xanthobacteraceae (Proteobacteria) (7.08 and 7.61%, respectively), and Chthoniobacteraceae (Verrucomicrobiota) (7.86 and 5.52%, respectively), while POL-S had the families Chthoniobacteraceae (Verrucomicrobiota) (15.59%), Sphingomonadaceae (Proteobacteria) (11.86%), and Chitinophagaceae (Bacteroidota) (7.02%) (Figure 4).

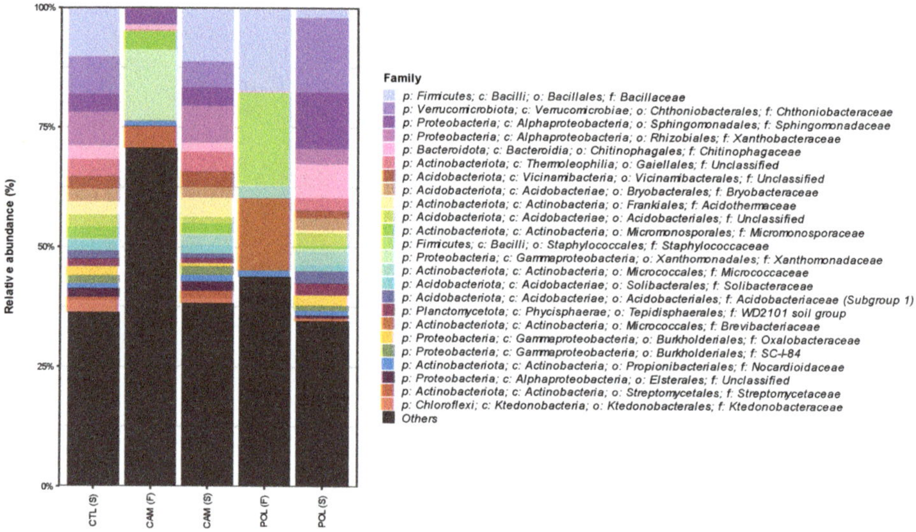

Figure 4. Relative abundance of bacterial groups at the family level, showing the 24 most abundant taxa in samples of cattle manure (CAM-F), poultry litter (POL-F), and soils 60 days post fertilization with cattle manure (CAM-S) and poultry litter (POL-S) as well as a control without application of residue (CTL-S). The less abundant families are grouped under the category "Others".

Among the identified ASVs, 704 were detected across all treatments, with 405 found in both CAM-S and CTL-S, 326 in POL-S and CTL-S, and 284 in POL-S and CAM-S (Figure 5). Additionally, only 66 ASVs were found exclusively in CTL-S, 110 were found in CAM-S, and 187 were in POL-S, 24 ASVs were found both in CAM-F and CAM-S, and nine were found between POL-F and POL-S.

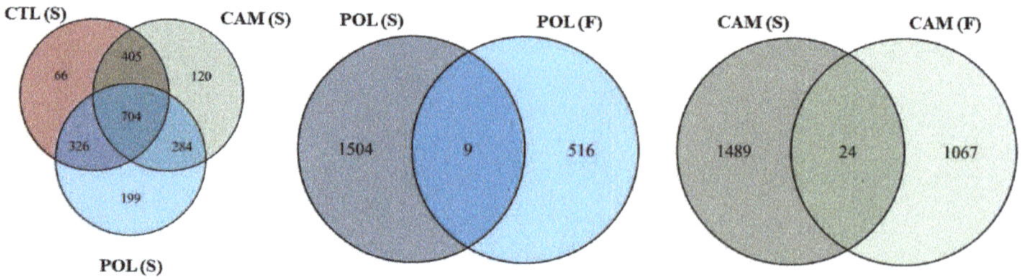

Figure 5. Venn diagram showing ASVs found in cattle manure (CAM-F), poultry litter (POL-F), and in soil samples 60 days after fertilization with cattle manure (CAM-S) or poultry litter (POL-S) as well as a control (CTL-S).

Among the 24 ASVs found in CAM-F and the nine ASVs detected in POL-F, which were also present in soils incorporated with the respective fertilizers, the representation of the 20 most abundant bacterial families was limited. Specifically, only three were classified as Planococcaceae, two as Sphingomonadaceae, and one as Micrococcaceae in both CAM-F and CAM-S. POL-F presented two ASVs classified as Bacillaceae, one as Planococcaceae, and one as Nocardioidaceae. These values represent a small portion of the 3673 ASVs found throughout the study.

A comparison between CAM-S and POL-S revealed 15 bacterial families with significantly different abundances between treatments (Figure 6). Among these, the families Chitinophagaceae and Sphingomonadaceae were more abundant in POL-S, while the families Streptomycetaceae, Planococcaceae, and Bacillaceae were more prevalent in CAM-S. A comparison between CAM-S and CTL-S showed significant differences only in the families Micrococcaceae and Ktedonobacteraceae, with Micrococcaceae being most abundant in CAM-S and Ktedonobacteraceae being most abundant in CTL-S. Among the 11 families that exhibited significant differences between POL-S and CTL-S, Micrococcaceae and Sphingomonadaceae were found to be more abundant in POL-S, while Planococcaceae, Bacillaceae, and Ktedonobacteraceae were more prevalent in CTL-S.

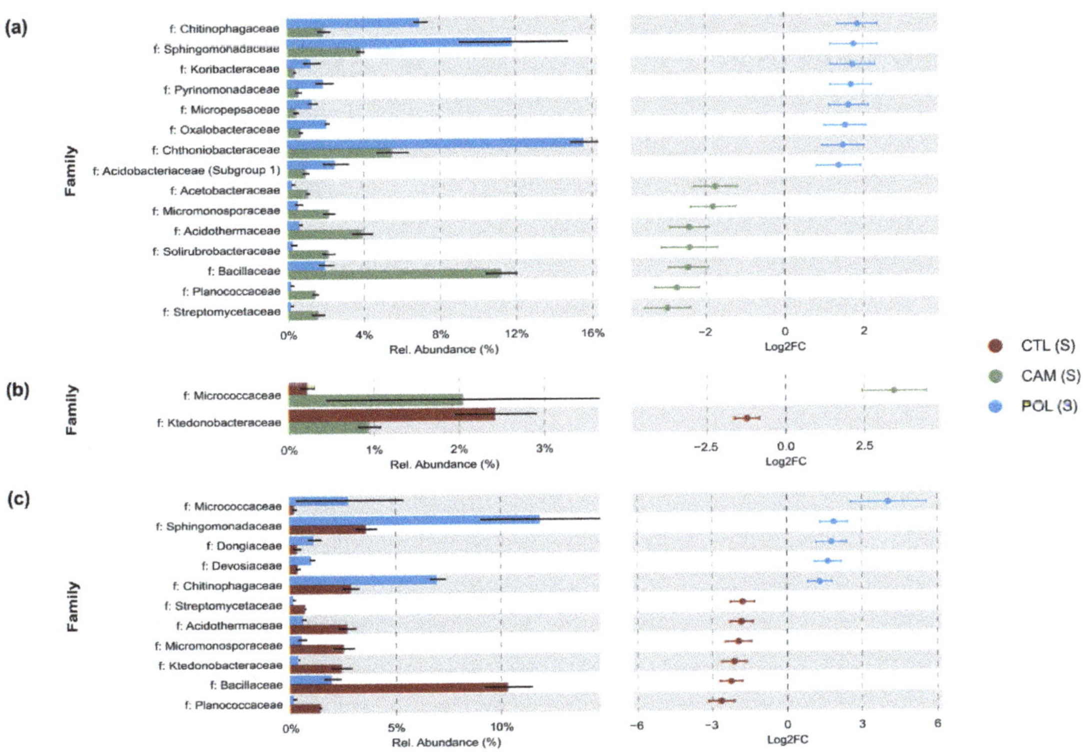

Figure 6. Differentially abundant taxa at family level in soil samples after 60 days of application of cattle manure (CAM-S) or poultry litter (POL-S) and a control soil (CTL-S) which did not receive application of these residues. (**a**) Comparison between CAM-S and POL-S, (**b**) comparison between CAM-S and CTL-S, (**c**) comparison between POL-S and CTL-S. Means were compared using a negative binomial model with a Wald test (adjusted p value < 0.05).

Among the 20 most abundant bacterial families observed in the analyzed soils, 17 exhibited correlation with the chemical characteristics, with 12 of them showing either positive or negative correlations with the SOC. Moreover, among the families correlated with chemical characteristics, 10 were significantly different between treatments (Figure 7). This means that 50% of the 20 families most abundant in the soil showed significant differences correlated with some chemical characteristic of the soil between treatments.

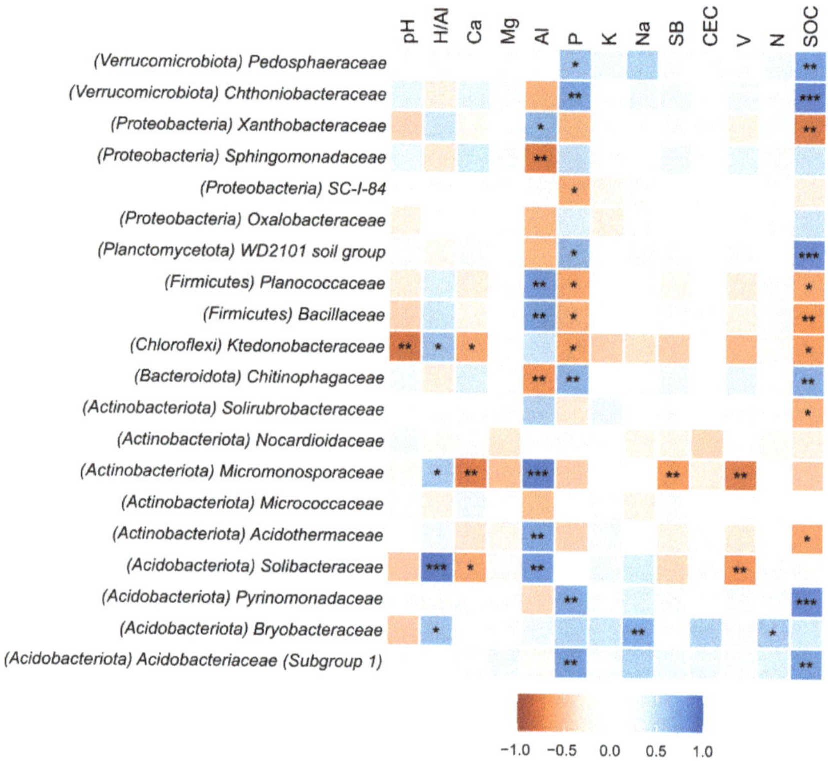

Figure 7. Correlation between chemical characteristics of the soil and the 20 most abundant bacterial families after 60 days of application of cattle manure (CAM-S) or poultry litter (POL-S) and in the control (CTL-S). Pearson's coefficient was used (*p* value: 0.05). Families that represented at least 1% of the sample abundance were considered. H + Al = potential acidity; SB = sum of bases; CEC = cation exchange capacity; and V = base saturation. Different numbers of asterisks (*) represent different levels of significance.

The families Chthoniobacteraceae, Sphingomonadaceae, Chitinophagaceae, Pyrinomonadaceae, and Acidobacteriaceae (subgroup 1) were most abundant in POL-F, showing a positive correlation with P and SOC and a negative correlation with Al. The families Planococcaceae, Bacillaceae, Micromonosporaceae, and Acidothermaceae were less abundant in POL-S, generally showing a positive correlation with Al and a negative correlation with P and SOC. The family Ktedonobacteraceae was more abundant in CTL-S than in the other treatments, presenting a positive correlation with H + Al and a negative correlation with pH, Ca, P, and SOC.

4. Discussion

This study suggests that the soil bacterial community 60 days after incorporation of POL-F, compared with soils with CAM-F and CTL-S, is likely influenced by factors such as the SOC, P, and Al contents and the pH level (Figure 2). However, POL-S and POL-F presented different bacterial compositions (Figures 2 and 3). The low number of ASVs found both in POL-F and POL-S indicates a low transfer of bacteria from the litter to the soil (Figure 5). These findings imply that POL-F changed the soil bacterial community by introducing new species and altering the characteristics of the soil.

In our study, we did not observe significant differences in the composition or alpha diversity (Table 3) of the soil community between samples without organic fertilizers (CTL) and those with cattle manure (CAM). Many bacteria present in manure originate from the intestinal microbiome of cattle and may be less competitive in the soil environment. This could be due to the fact that soil is more aerobic compared with the intestinal tract, leading to the disappearance of oxygen-sensitive genera [41]. Therefore, depending on the microbial composition of the manure, the survival of certain taxa may be impaired in the soil, which explains the similarity between the CAM and CTL soil samples.

This study aimed to determine whether the increase in microbial biomass and soil diversity after manure application is due to the survival of microbes introduced with the manure or activation of the soil microbiome. The data revealed that manure application increased the microbial biomass and diversity, in addition to activating many underrepresented taxa in the soil. However, manure microorganisms had low survival rates in the soil, indicating the soil's ability to act as an effective buffer against the entry of new microorganisms [42]. Thus, it is important to conduct more studies in this area, considering the differences inherent in waste production and soil conditions that can lead to varying compositions of the soil microbiota.

Regarding poultry litter (POL), previous studies have reported that long-term fertilization with poultry litter alters the edaphic microbial community through changes in soil chemical attributes [9,43]. Our study showed that changes in soil diversity already occurred 60 days after POL application (Table 3).

In this study, the disparity in the C:N ratio between CAM-F and POL-F likely contributed to the observed differences in soil bacterial compositions. Specifically, POL-F led to a relative increase in the abundance of the family Chthoniobacteraceae (phylum: Verrucomicrobiota), which has been positively correlated with SOC [44] and negatively correlated with the pH level [45]. Previous studies have also reported a positive correlation with soil N levels [45]. This family, which has only one cultivated representative (*Chthoniobacter flavus*), harbors denitrification genes and is involved in the decomposition of organic substrates [46]. N, normally associated with organic matter, requires mineralization to become available in soil, a process that largely depends on specific functional microorganism groups, including decomposers and ammonia oxidizers, whose abundance and activity depend on the C:N ratio of the soil [47]. Additionally, the genus RB41 represented 100% of the sequences in the family Pyrinomonadaceae (phylum: Acidobacteriota), known for its diverse metabolic processes, including the degradation of residues rich in lignocellulose and chitin and soil C cycling [48,49].

The family Sphingomonadaceae (phylum: Proteobacteria), which was also abundant in the soil after POL-F incorporation, was previously negatively correlated with Al [50]. These bacteria participate in the N cycle [51], degrade pollutants, and inhibit pathogens [52]. The family Chitinophagaceae (Bacteroidota), which was positively correlated with P and negatively correlated with Al in this study, tends to be relatively more prevalent in soils with higher pH levels [53]. The presence of these bacterial groups in soils incorporated with POL-F could be related to the common practice of using CaO, which is frequently used in POL-F incorporation and aims to reduce the presence of pathogenic bacteria such as *Salmonella* and *Clostridium* [54]. POL-F tends to increase soil's pH level due to its alkaline nature, also reducing the exchangeable acidity of the soil and, consequently, P availability [55].

Finally, the genus *Edaphobacter* was present in 70% of the sequences in the family Acidobacteriaceae (subgroup 1), which also belongs to the phylum Acidobacteriota. This family is associated with higher nutrient contents in the soil at pH levels close to neutral, being an indicator of soil fertility [56]. Therefore, in our study, the greater occurrence of these microorganisms in POL-S may indicate improved soil fertility with the use of POL-F.

Conversely, some bacterial families were relatively less abundant with POL-F incorporation. The family Planococcaceae (phylum: Firmicutes) may have been negatively affected by competition for resources with the family Sphingomonadaceae [57]. These bacteria are important for reducing nitrate and degrading organic pollutants in the soil [58,59]. The family Bacillaceae (phylum: Firmicutes) was also positively correlated with Al and negatively correlated with P, corroborating [60]. The genus *Bacillus*, which corresponded to 78% of the sequences in this family, is involved in N cycling [61] and P solubilization [62]. The family Micromonosporaceae (phylum: Actinobacteriota) showed a positive correlation with Al and H + Al and a negative correlation with Ca, SB, and V. These bacteria play important roles in soil ecology by decomposing complex organic substrates. Finally, the family Acidothermaceae (phylum: Actinobacteriota) showed a positive correlation with Al, being represented by thermophilic acidophilic bacteria that have thermostable enzymes capable of degrading cellulose. This family is currently represented only by the genus *Acidothermus* [63].

Therefore, despite the increased bacterial abundance indicating soil fertility improvement, it is crucial to acknowledge that families potentially important to the soil were negatively affected by the use of POL-F. Some of the functions of these bacteria include participating in specific N and P cycling processes, degrading complex carbohydrates, and bioremediation. The reduced abundance of these families can be justified by changes in the soil characteristics. POL-F reduced the microbial groups adapted to soils with lower nutrient contents and acidic pH levels. Therefore, POL-F increased nutrient incorporation and the soil's pH due to the increased presence of microorganisms adapted to these new soil conditions, decreasing bacterial groups adapted to less fertile soils.

The Ktedonobacteraceae family (phylum: Chloroflexi) includes aerobic filamentous bacteria, which were more abundant in CTL-S than in the other treatments, and has only the species *Ktedonobacter racemifer* described for it [64]. These bacteria were positively correlated with H + Al and negatively correlated with P, Ca, and the pH level. This outcome corroborates a previous study which reported increased relative abundance with lower soil pH levels [65].

Families such as Oxalobacteraceae and Micrococcaceae showed different relative abundance between treatments but without correlation with the chemical characteristics of the soil or a significant number of ASVs present in either the fertilizer or the incorporated soil. The increased or decreased relative abundance of these microorganisms in the soil with fertilizer incorporation may be associated with other factors, including the presence of specific organic compounds or heavy metals.

The results of this highlight the impact of animal waste incorporation, particularly poultry litter (POL-F), on soil bacteria. However, optimizing soil management for subsequent crops demands further studies into the correlation between the bacterial composition of organic fertilizers and their short- and long-term impacts on soil microbiota. It is essential to ensure that the functions of these microorganisms are sustained over time.

5. Conclusions

The introduction of poultry litter (POL-F) significantly changed the soil bacterial community, whereas the incorporation of cattle manure (CAM-F) demonstrated a negligible effect compared with the control (CTL-S). This shift does not appear to be primarily linked to the direct transfer of bacteria from the fertilizer to the soil. Rather, it seems to be driven by changes in the chemical characteristics of the soil subsequent to the incorporation of POL-F.

Author Contributions: Conceptualization, E.Z., I.d.S.C. and G.S.L.S.; methodology, G.S.L.S. and J.V.d.S.G.; formal analysis, G.S.L.S., L.A.L.d.C., M.S.R.d.A.d.S. and D.G.P.; writing—original draft preparation, G.S.L.S. and M.S.R.d.A.d.S.; writing—review and editing, L.A.L.d.C., M.S.R.d.A.d.S., J.V.d.S.G., D.G.P., E.Z. and I.d.S.C.; supervision, E.Z. and I.d.S.C. All authors have read and agreed to the published version of the manuscript.

Funding: This research was funded by the Conselho Nacional de Desenvolvimento Científico e Tecnológico (CNPq) (309002/2018-4), the Coordenaçao de Aperfeiçoamento de Pessoal de Nível Superior (CAPES) (finance code 001), and the Fundação Carlos Chagas Filho de Amparo à Pesquisa do Estado do Rio de Janeiro (FAPERJ) (Process SEI: E-26/205.892/2022).

Data Availability Statement: The original contributions presented in the study are included in the article, and further inquiries can be directed to the corresponding author.

Acknowledgments: We thank the Programa de Pós-graduação em Ciência, Tecnologia e Inovação em Agropecuária (PPGCTIA) of the UFRRJ for their support.

Conflicts of Interest: The authors declare no conflicts of interest.

References

1. USDA. *Livestock and Poultry: World Markets and Trade*; United States Department of Agriculture and Foreign Agricultural Service: Washington, DC, USA, 2024.
2. Biscarra-Bellio, J.C.; de Oliveira, G.B.; Marques, M.C.P.; Molento, C.F.M. Demand Changes Meat as Changing Meat Reshapes Demand: The Great Meat Revolution. *Meat Sci.* **2023**, *196*, 109040. [CrossRef] [PubMed]
3. Marchi, C.M.D.F.; Gonçalves, I. de O. Compostagem: A Importância Da Reutilização Dos Resíduos Orgânicos Para a Sustentabilidade de Uma Instituição de Ensino Superior. *Rev. Monogr. Ambient. Santa Maria* **2020**, *19*, e1. [CrossRef]
4. Sui, Q.; Zhang, J.; Chen, M.; Wang, R.; Wang, Y.; Wei, Y. Fate of Microbial Pollutants and Evolution of Antibiotic Resistance in Three Types of Soil Amended with Swine Slurry. *Environ. Pollut.* **2019**, *245*, 353–362. [CrossRef] [PubMed]
5. Xu, Y.; Li, J.; Zhang, X.; Wang, L.; Xu, X.; Xu, L.; Gong, H.; Xie, H.; Li, F. Data Integration Analysis: Heavy Metal Pollution in China's Large-Scale Cattle Rearing and Reduction Potential in Manure Utilization. *J. Clean. Prod.* **2019**, *232*, 308–317. [CrossRef]
6. Li, Z.; Zhang, X.; Xu, J.; Cao, K.; Wang, J.; Xu, C.; Cao, W. Green Manure Incorporation with Reductions in Chemical Fertilizer Inputs Improves Rice Yield and Soil Organic Matter Accumulation. *J. Soils Sediments* **2020**, *20*, 2784–2793. [CrossRef]
7. Jansson, J.K.; Hofmockel, K.S. The Soil Microbiome—From Metagenomics to Metaphenomics. *Curr. Opin. Microbiol.* **2018**, *43*, 162–168. [CrossRef] [PubMed]
8. Parente, C.E.T.; Brito, E.M.S.; Caretta, C.A.; Cervantes-Rodríguez, E.A.; Fábila-Canto, A.P.; Vollú, R.E.; Seldin, L.; Malm, O. Bacterial Diversity Changes in Agricultural Soils Influenced by Poultry Litter Fertilization. *Braz. J. Microbiol.* **2021**, *52*, 675–686. [CrossRef] [PubMed]
9. Ashworth, A.J.; DeBruyn, J.M.; Allen, F.L.; Radosevich, M.; Owens, P.R. Microbial Community Structure Is Affected by Cropping Sequences and Poultry Litter under Long-Term No-Tillage. *Soil Biol. Biochem.* **2017**, *114*, 210–219. [CrossRef]
10. Yang, Y.; Ashworth, A.J.; DeBruyn, J.M.; Willett, C.; Durso, L.M.; Cook, K.; Moore, P.A.; Owens, P.R. Soil Bacterial Biodiversity Is Driven by Long-Term Pasture Management, Poultry Litter, and Cattle Manure Inputs. *PeerJ* **2019**, *2019*, e7839. [CrossRef]
11. Jat, H.S.; Choudhary, M.; Datta, A.; Yadav, A.K.; Meena, M.D.; Devi, R.; Gathala, M.K.; Jat, M.L.; McDonald, A.; Sharma, P.C. Temporal Changes in Soil Microbial Properties and Nutrient Dynamics under Climate Smart Agriculture Practices. *Soil Tillage Res.* **2020**, *199*, 104595. [CrossRef]
12. Liu, W.R.; Zeng, D.; She, L.; Su, W.X.; He, D.C.; Wu, G.Y.; Ma, X.R.; Jiang, S.; Jiang, C.H.; Ying, G.G. Comparisons of Pollution Characteristics, Emission Situations, and Mass Loads for Heavy Metals in the Manures of Different Livestock and Poultry in China. *Sci. Total Environ.* **2020**, *734*, 139023. [CrossRef] [PubMed]
13. Li, J.; Chen, Q.; Li, H.; Li, S.; Liu, Y.; Yang, L.; Han, X. Impacts of Different Sources of Animal Manures on Dissemination of Human Pathogenic Bacteria in Agricultural Soils. *Environ. Pollut.* **2020**, *266*, 115399. [CrossRef] [PubMed]
14. Wu, J.; Wang, J.; Li, Z.; Guo, S.; Li, K.; Xu, P.; Ok, Y.S.; Jones, D.L.; Zou, J. Antibiotics and Antibiotic Resistance Genes in Agricultural Soils: A Systematic Analysis. *Crit. Rev. Environ. Sci. Technol.* **2023**, *53*, 847–864. [CrossRef]
15. Gurmessa, B.; Ashworth, A.J.; Yang, Y.; Savin, M.; Moore, P.A.; Ricke, S.C.; Corti, G.; Pedretti, E.F.; Cocco, S. Variations in Bacterial Community Structure and Antimicrobial Resistance Gene Abundance in Cattle Manure and Poultry Litter. *Environ. Res.* **2021**, *197*, 111011. [CrossRef]
16. Ye, C.; Huang, S.; Sha, C.; Wu, J.; Cui, C.; Su, J.; Ruan, J.; Tan, J.; Tang, H.; Xue, J. Changes of Bacterial Community in Arable Soil after Short-Term Application of Fresh Manures and Organic Fertilizer. *Environ. Technol.* **2022**, *43*, 824–834. [CrossRef]
17. U.S. EPA. Method 3050B: Acid Digestion of Sediments, Sludges, and Soils, Revision 2. Washington, DC. 1996. Available online: https://www.epa.gov/esam/epa-method-3050b-acid-digestion-sediments-sludges-and-soils (accessed on 18 June 2023).
18. Soil Survey Staff. *Keys to Soil Taxonomy*, 11th ed.; USDA-National Resources Conservation Service: Washington, DC, USA, 2014; Volume 12, ISBN 0926487221.
19. Dos Santos, H.G.; Jacomine, P.K.T.; dos Anjos, L.H.C.; de Oliveira, V.Á.; Lumbreras, J.F.; Coelho, M.R.; de Almeida, J.A.; de Araújo Filho, J.C.; de Oliveira, J.B.; Cunha, T.J.F. *Sistema Brasileiro de Classificação de Solos*, 5th ed.; Embrapa: Brasília, Brazil, 2018; ISBN 978-85-7035-198-2.
20. de Carvalho, D.F.; Silva, L.B.D.; Folegatti, M.V.; Costa, J.R.C.; da Cruz, F.A. Avaliação Da Evapotranspiração de Referência Na Região de Seropédica- RJ, Utilizando Lisímetro de Pesagem. *Rev. Bras. Agrometeorol.* **2006**, *14*, 1–9. [CrossRef]

21. Teixeira, P.C.; Donagemma, G.K.; Fontana, A.; Teixeira, W.G. *Manual de Métodos de Análise de Solo*, 3rd ed.; Embrapa Solos: Rio de Janeiro, Brazil, 2017; Volume 3, ISBN 9788570357717.
22. Adkins, N.L.; Hall, J.A.; Georgel, P.T. The Use of Quantitative Agarose Gel Electrophoresis for Rapid Analysis of the Integrity of Protein-DNA Complexes. *J. Biochem. Biophys. Methods* **2007**, *70*, 721–726. [CrossRef]
23. Andrews, S. FastQC: A Quality Control Tool for High Throughput Sequence Data. Available online: http://www.bioinformatics.babraham.ac.uk/projects/fastqc/ (accessed on 1 June 2023).
24. Edgar, R.C. Search and Clustering Orders of Magnitude Faster than BLAST. *Bioinformatics* **2010**, *26*, 2460–2461. [CrossRef]
25. Didion, J.P.; Martin, M.; Collins, F.S. Atropos: Specific, Sensitive, and Speedy Trimming of Sequencing Reads. *PeerJ* **2017**, *2017*, e3720. [CrossRef]
26. Chen, S.; Zhou, Y.; Chen, Y.; Gu, J. Fastp: An Ultra-Fast All-in-One FASTQ Preprocessor. *Bioinformatics* **2018**, *34*, i884–i890. [CrossRef]
27. Magoč, T.; Salzberg, S.L. FLASH: Fast Length Adjustment of Short Reads to Improve Genome Assemblies. *Bioinformatics* **2011**, *27*, 2957–2963. [CrossRef] [PubMed]
28. Callahan, B.J.; Mcmurdie, P.J.; Rosen, M.J.; Han, A.W.; Johnson, A.J.A.; Holmes, S.P. DADA2: High Resolution Sample Inference from Illumina Amplicon Data. *Nat. Methods* **2016**, *13*, 581–583. [CrossRef] [PubMed]
29. R Core Team. *R: A Language and Environment for Statistical Computing*; R Foundation for Statistical Computing: Vienna, Austria, 2021. Available online: https://www.R-project.org (accessed on 17 June 2024).
30. Quast, C.; Pruesse, E.; Yilmaz, P.; Gerken, J.; Schweer, T.; Yarza, P.; Peplies, J.; Glöckner, F.O. The SILVA Ribosomal RNA Gene Database Project: Improved Data Processing and Web-Based Tools. *Nucleic Acids Res.* **2013**, *41*, D590–D596. [CrossRef] [PubMed]
31. Cole, J.R.; Wang, Q.; Fish, J.A.; Chai, B.; McGarrell, D.M.; Sun, Y.; Brown, C.T.; Porras-Alfaro, A.; Kuske, C.R.; Tiedje, J.M. Ribosomal Database Project: Data and Tools for High Throughput RRNA Analysis. *Nucleic Acids Res.* **2014**, *42*, D633–D642. [CrossRef] [PubMed]
32. Parks, D.H.; Chuvochina, M.; Rinke, C.; Mussig, A.J.; Chaumeil, P.A.; Hugenholtz, P. GTDB: An Ongoing Census of Bacterial and Archaeal Diversity through a Phylogenetically Consistent, Rank Normalized and Complete Genome-Based Taxonomy. *Nucleic Acids Res.* **2022**, *50*, D785–D794. [CrossRef] [PubMed]
33. McMurdie, P.J.; Holmes, S. Phyloseq: An R Package for Reproducible Interactive Analysis and Graphics of Microbiome Census Data. *PLoS ONE* **2013**, *8*, e61217. [CrossRef] [PubMed]
34. Mikryukov, V.; Mahé, F. MetagMisc: Miscellaneous Functions for Metagenomic Analysis 2019. Available online: https://rdrr.io/github/vmikk/metagMisc/ (accessed on 20 June 2024).
35. Andersen, K.S.; Kirkegaard, R.H.; Karst, S.M.; Albertsen, M. Ampvis2: An R Package to Analyse and Visualise 16S RRNA Amplicon Data. *bioRxiv* **2018**. [CrossRef]
36. Lahti, L.; Shetty, S. Introduction to the Microbiome R Package. Bioconductor 2018. Available online: https://www.bioconductor.org/packages/devel/bioc/vignettes/microbiome/inst/doc/vignette.html (accessed on 10 June 2024).
37. de Mendiburu, F. Package "agricolae" Title Statistical Procedures for Agricultural Research. In *Statistical Procedures for Agricultural Research*; John Wiley & Sons: New York, NY, USA, 2020.
38. Love, M.I.; Huber, W.; Anders, S. Moderated Estimation of Fold Change and Dispersion for RNA-Seq Data with DESeq2. *Genome Biol.* **2014**, *15*, 550. [CrossRef]
39. Kassambara, A. Ggpubr: "ggplot2" Based Publication Ready Plots. R Package Version 0.2. 2020. Available online: https://CRAN.R-project.org/package=ggpubr (accessed on 19 June 2024).
40. Revelle, W. Package "Psych"—Procedures for Psychological, Psychometric and Personality Research. R Package 2024. Available online: https://cran.r-project.org/web/packages/psych/psych.pdf (accessed on 11 June 2024).
41. Semenov, A.V.; Van Overbeek, L.; Termorshuizen, A.J.; Van Bruggen, A.H.C. Influence of Aerobic and Anaerobic Conditions on Survival of Escherichia Coli O157:H7 and Salmonella Enterica Serovar Typhimurium in Luria-Bertani Broth, Farm-Yard Manure and Slurry. *J. Environ. Manag.* **2011**, *92*, 780–787. [CrossRef]
42. Semenov, M.V.; Krasnov, G.S.; Semenov, V.M.; Ksenofontova, N.; Zinyakova, N.B.; van Bruggen, A.H.C. Does Fresh Farmyard Manure Introduce Surviving Microbes into Soil or Activate Soil-Borne Microbiota? *J. Environ. Manag.* **2021**, *294*, 113018. [CrossRef]
43. Liu, P.; Jia, S.; He, X.; Zhang, X.; Ye, L. Different Impacts of Manure and Chemical Fertilizers on Bacterial Community Structure and Antibiotic Resistance Genes in Arable Soils. *Chemosphere* **2017**, *188*, 455–464. [CrossRef] [PubMed]
44. Wang, Z.; Zhang, Q.; Staley, C.; Gao, H.; Ishii, S.; Wei, X.; Liu, J.; Cheng, J.; Hao, M.; Sadowsky, M.J. Impact of Long-Term Grazing Exclusion on Soil Microbial Community Composition and Nutrient Availability. *Biol. Fertil. Soils* **2019**, *55*, 121–134. [CrossRef]
45. Tang, X.; Zhong, R.; Jiang, J.; He, L.; Huang, Z.; Shi, G.; Wu, H.; Liu, J.; Xiong, F.; Han, Z.; et al. Cassava/Peanut Intercropping Improves Soil Quality via Rhizospheric Microbes Increased Available Nitrogen Contents. *BMC Biotechnol.* **2020**, *20*, 13. [CrossRef] [PubMed]
46. Chen, Z.; Zhang, W.; Peng, A.; Shen, Y.; Jin, X.; Stedtfeld, R.D.; Boyd, S.A.; Teppen, B.J.; Tiedje, J.M.; Gu, C.; et al. Bacterial Community Assembly and Antibiotic Resistance Genes in Soils Exposed to Antibiotics at Environmentally Relevant Concentrations. *Environ. Microbiol.* **2023**, *25*, 1439–1450. [CrossRef] [PubMed]
47. Ye, J.; Perez, P.G.; Zhang, R.; Nielsen, S.; Huang, D.; Thomas, T. Effects of Different C/N Ratios on Bacterial Compositions and Processes in an Organically Managed Soil. *Biol. Fertil. Soils* **2018**, *54*, 137–147. [CrossRef]

48. Wang, M.; Lan, X.; Xu, X.; Fang, Y.; Singh, B.P.; Sardans, J.; Romero, E.; Peñuelas, J.; Wang, W. Steel Slag and Biochar Amendments Decreased CO2 Emissions by Altering Soil Chemical Properties and Bacterial Community Structure over Two-Year in a Subtropical Paddy Field. *Sci. Total Environ.* **2020**, *740*, 140403. [CrossRef] [PubMed]
49. Zheng, J.; Wang, Q.; Li, S.; Zhang, B.; Zhang, F.; Zhao, T.; Qiao, J.; Zhao, M. Mowing Effects on Soil Bacterial Community Assembly Processes in a Semiarid Grassland. *Plant Soil* **2023**, *493*, 309–324. [CrossRef]
50. Zhen, Z.; Luo, S.; Chen, Y.; Li, G.; Li, H.; Wei, T.; Huang, F.; Ren, L.; Liang, Y.Q.; Lin, Z.; et al. Performance and Mechanisms of Biochar-Assisted Vermicomposting in Accelerating Di-(2-Ethylhexyl) Phthalate Biodegradation in Farmland Soil. *J. Hazard. Mater.* **2023**, *443*, 130330. [CrossRef] [PubMed]
51. Shi, L.; Zhang, P.; He, Y.; Zeng, F.; Xu, J.; He, L. Enantioselective Effects of Cyflumetofen on Microbial Community and Related Nitrogen Cycle Gene Function in Acid-Soil. *Sci. Total Environ.* **2021**, *771*, 144831. [CrossRef]
52. Zhang, M.; Riaz, M.; Liu, B.; Xia, H.; El-desouki, Z.; Jiang, C. Two-Year Study of Biochar: Achieving Excellent Capability of Potassium Supply via Alter Clay Mineral Composition and Potassium-Dissolving Bacteria Activity. *Sci. Total Environ.* **2020**, *717*, 137286. [CrossRef]
53. Wang, Q.; Chu, C.; Zhao, Z.; Wu, S.; Zhou, D. Pre-Flooding Soil Used in Monocropping Increased Strawberry Biomass and Altered Bacterial Community Composition. *Soil Sci. Plant Nutr.* **2021**, *67*, 643–652. [CrossRef]
54. Dai Pra, M.A.; Corrêa, É.K.; Roll, V.F.; Eduardo, G.X.; Lopes, D.C.N.; Lourenço, F.F.; Zanusso, J.T.; Roll, A.P. Quicklime for Controlling Salmonella Spp. and Clostridium Spp in Litter from Floor Pens of Broilers. *Cienc. Rural* **2009**, *39*, 1189–1194. [CrossRef]
55. Masud, M.M.; Al Baquy, M.A.; Akhter, S.; Sen, R.; Barman, A.; Khatun, M.R. Liming Effects of Poultry Litter Derived Biochar on Soil Acidity Amelioration and Maize Growth. *Ecotoxicol. Environ. Saf.* **2020**, *202*, 110865. [CrossRef] [PubMed]
56. Zhang, Y.; Ding, K.; Yrjälä, K.; Liu, H.; Tong, Z.; Zhang, J. Introduction of Broadleaf Species into Monospecific Cunninghamia Lanceolata Plantations Changed the Soil Acidobacteria Subgroups Composition and Nitrogen-Cycling Gene Abundances. *Plant Soil* **2021**, *467*, 29–46. [CrossRef]
57. Silva, A.M.M.; Estrada-Bonilla, G.A.; Lopes, C.M.; Matteoli, F.P.; Cotta, S.R.; Feiler, H.P.; Rodrigues, Y.F.; Cardoso, E.J.B.N. Does Organomineral Fertilizer Combined with Phosphate-Solubilizing Bacteria in Sugarcane Modulate Soil Microbial Community and Functions? *Microb. Ecol.* **2021**, *84*, 539–555. [CrossRef] [PubMed]
58. Zhang, L.; Yi, M.; Lu, P. Effects of Pyrene on the Structure and Metabolic Function of Soil Microbial Communities. *Environ. Pollut.* **2022**, *305*, 119301. [CrossRef] [PubMed]
59. Luo, D.; Meng, X.; Zheng, N.; Li, Y.; Yao, H.; Chapman, S.J. The Anaerobic Oxidation of Methane in Paddy Soil by Ferric Iron and Nitrate, and the Microbial Communities Involved. *Sci. Total Environ.* **2021**, *788*, 147773. [CrossRef] [PubMed]
60. Barbosa Lima, A.; Cannavan, F.S.; Navarrete, A.A.; Teixeira, W.G.; Kuramae, E.E.; Tsai, S.M. Amazonian Dark Earth and Plant Species from the Amazon Region Contribute to Shape Rhizosphere Bacterial Communities. *Microb. Ecol.* **2015**, *69*, 855–866. [CrossRef] [PubMed]
61. Verbaendert, I. Denitrification in Gram-Positive Bacteria, with Focus on Members of the Bacillaceae. Ph.D. Thesis, Ghent University, Ghent, Belgium, 2014.
62. López-González, J.A.; Suárez-Estrella, F.; Vargas-García, M.C.; López, M.J.; Jurado, M.M.; Moreno, J. Dynamics of Bacterial Microbiota during Lignocellulosic Waste Composting: Studies upon Its Structure, Functionality and Biodiversity. *Bioresour. Technol.* **2015**, *175*, 406–416. [CrossRef]
63. Berry, A.M.; Barabote, R.D.; Normand, P. The Family Acidothermaceae. In *The Prokaryotes*; Springer: Berlin/Heidelberg, Germany, 2014; pp. 13–19. [CrossRef]
64. Cavaletti, L.; Monciardini, P.; Bamonte, R.; Schumann, P.; Ronde, M.; Sosio, M.; Donadio, S. New Lineage of Filamentous, Spore-Forming, Gram-Positive Bacteria from Soil. *Appl. Environ. Microbiol.* **2006**, *72*, 4360–4369. [CrossRef]
65. Kim, H.M.; Jung, J.Y.; Yergeau, E.; Hwang, C.Y.; Hinzman, L.; Nam, S.; Hong, S.G.; Kim, O.S.; Chun, J.; Lee, Y.K. Bacterial Community Structure and Soil Properties of a Subarctic Tundra Soil in Council, Alaska. *FEMS Microbiol. Ecol.* **2014**, *89*, 465–475. [CrossRef] [PubMed]

agronomy

MDPI

Article

Valorizing Combustible and Compostable Fractions of Municipal Solid Waste to Biochar and Compost as an Alternative to Chemical Fertilizer for Improving Soil Health and Sunflower Yield

Samreen Aslam and Aisha Nazir *

Environmental Biotechnology Laboratory, Institute of Botany, University of the Punjab, Lahore 54590, Pakistan
* Correspondence: aisha.botany@pu.edu.pk; Tel.: +92-300-4136566

Abstract: Reduced reliance on synthetic chemical fertilizers necessarily requires using renewable biomaterial-derived soil organic amendments (SOAs) in agriculture for sustained retention of nutrients through improvement in the soil organic matter (SOM). SOM replenishment through SOAs derived from wasted materials could help in its valorization by furthering the sustainability prospects of agronomic crop production systems. In the current study, compost (CP) and biochar (BC) were derived as SOAs from combustible and compostable fractions of municipal solid waste (MSW) for their potential valorization by adding SOAs as potential sustainable sources of nutrients as a replacement of chemical fertilizers (CF) for sunflower crops cultivated in potted soils. The experimental design included quadruplicated soil application of MSW-derived BC and CP in discrete and combined forms, each in three doses (% w:w), viz., low (L), medium (M), and high (H), i.e., BC-L, BC-M, BC-H; CP-L, CP-M, CP-H; and BC + CP-L, BC + CP-M, BC-CP-H. The results showed that, compared to the control (soil only), the sunflower growth and harvestable yield were significantly greater in BC + CF with a medium dose and were comparable to the growth and yield obtained in soils with CF. Sunflower growth in the discrete SOAs remained less than in the combined SOAs (BC + CP) and was attributed to the comprehensive soil health improvement rendered by the applied SOAs. The soil health improvement factors included SOM, CEC, and concentrations of total and available NPK. The dose-effect comparison of the SOAs showed highly variable trends, i.e., the sunflower growth did not correspond with the increase in dose of the SOAs. It is concluded that the combined application of BC + CP derived from MSW components at a medium dose could act as a potential alternative to CF. The developed approach resulted in MSW valorization, which improved soil health and yielded a better sunflower crop.

Keywords: combustible; biochar; compostable; compost; sunflower; municipal solid waste; pyrolysis; waste valorization

1. Introduction

Continuously increasing population pressure as well as climate change have a negative impact on soil fertility, so it leads to a decline in agricultural production in developing countries [1]. The decline in per capita land allotment and degrading soil quality have resulted in an escalation in the application rates of inorganic synthetic fertilizers for managing crop productivity. The literature generally agrees that most arid to semi-arid agricultural systems undergo more depletion of soil organic matter (SOM) due to a severe disparity between lower inputs than the long-term extractive agricultural approaches; hence, making agricultural systems unsustainable. Deriving organic amendments from waste and their application in soil is imperative to the effective management of SOM.

Overuse of chemical fertilizers in agricultural soils could lead to nutritional imbalances and depletion of soil fertility at the cost of global warming, which would proportionately lead to rapid loss of SOM and depletion of soil organic carbon (SOC) [2]. Chemical fertilizers contribute nearly 20% to the total global greenhouse gas emissions (GHGs) budget. It is crucial to explore cost-effective, environmentally friendly, and sustainable crop management techniques that enhance crop productivity, improve SOM, and lessen GHGs emissions from agricultural systems for potential mitigation of climate change. Such SOM management techniques could help in supporting soil preservation and conserving the long-term productivity potential of agroecosystems by boosting soils' ability to act as carbon sinks rather than carbon sources. A sustainable agricultural production system could improve per-unit area crop production to confirm food accessibility for the continuously increasing world population.

The role of organic amendments such as biochar (BC) and compost (CP) has been reported as feasible options for addressing the ever-intensifying challenges of sustaining soil health and productivity [3]. Soil organic matter enrichment stands as a vital factor for enhancing soil fertility as well as productivity due to its integrated improvement in soils' water holding capacity, nutrient availability, mineralization, and activity of soil microorganisms [4]. BC and CP have been some of the major sources of replenishing SOM for enhancing soil productivity potential and crop performance. Using discarded wastes as feedstock for deriving BC and CP could be an added benefit to their sustainable aspects as it could render waste valuable, which has been less reported in the literature in terms of its integration in agricultural crop production systems.

The valorization of wastes such as Municipal Solid Waste (MSW) could be described as its potential use by transforming it into a value-added product(s) coupled with a substantial reduction in its volume. The conversion of MSW into useful derivatives has been the subject of emerging research and development projects in recent years, with a focus on methods and procedures that lower emissions and incorporate life-cycle assessments of technologies and materials. Direct recycling of garbage and residues has been encouraged since the 1990s due to ever-intensifying socioeconomic concerns and the continuous depletion of non-renewable raw materials. Over time, there has been a continuous evolution of techniques that promote waste valorization by lowering GHGs emissions and minimizing the loss of resources as waste. Composting and pyrolysis of organic waste have been techniques for the conversion of waste components into CP and BC [5]. From the defined applications of the CP and BC derived from MSW for their valorization, application as SOAs has been a dominant approach for achieving comprehensive soil health improvement in agricultural crop production systems.

BC derived from MSW has exclusive physicochemical properties with great potential for improving soil health and enhancing plant crop yields [6]. CP derived from MSW through low-tech composting also has unique potential for comprehensive soil health improvement. The performance of BC and CP could become even better when both are applied as SOA by boosting soil microbial activity and refining soil physico-chemical properties such as bulk density, soil structure, water holding capacity, cation exchange capacity (CEC), pH, nutrient bioavailability, SOM, and SOC [7]. Application of BC and CP could also reduce nutrient loss from the soil profiles by minimizing nutrient leaching by storing them in pores with large surface areas and providing unique sites for nutrient sorption [8] and toxic element immobilization in the contaminated soils [9]. CP could make carbon and nutrients available to the crop plants through the integrated activities of its microbial and microbial consortia being stocked in the SOM, even for a short time lasting in the soil after application [10]. In recent times, research on potential synergic SOAs from BC and CP has been much demanded [11]. Hence, the application of BC and CP derived from no longer carried wastes as SOAs for substituting CF in plant crop production systems is worth exploration.

In the current study, BC and CP were derived as SOAs from combustible and compostable fractions of MSW for its potential valorization by adding SOAs as potential sustainable sources of nutrients, replacing chemical fertilizers (CF) for sunflower crops.

2. Materials and Methods

2.1. Collection of Materials

The combustible and compostable fractions of MSW were collected from the open dump location at the Mehmood Booti site in Lahore (Figure 1). The selected open dump site has been receiving fresh batches of unsorted MSW on a regular basis for several decades. The disposed MSW at the selected site mainly contained (≥65%) biodegradable components for potential composting and quite a fraction (≤4%) of combustibles. The composite samples of each of the combustible and compostable fractions of MSW were collected in a statistically representative way. The collected soil was mainly silty loam and was collected from the Botanical Garden, University of the Punjab Lahore-54590, Pakistan.

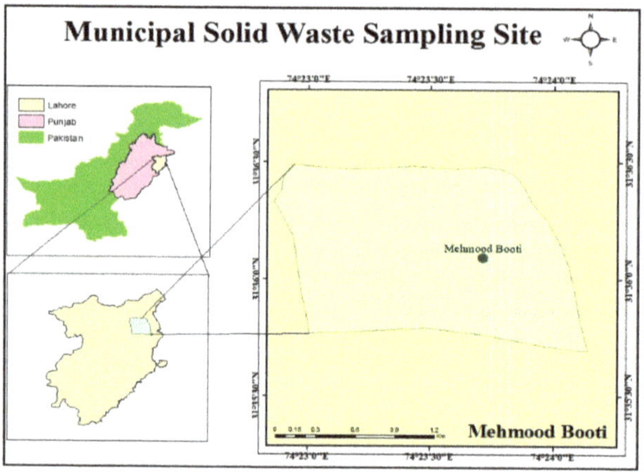

Figure 1. MSW component sampling site at Mehmood Booti open dump in Lahore, Pakistan.

2.2. Derivation of BC and CP from MSW Components

The combustible fraction of MSW included a mixture of paper, cardboard, textile waste, wood chips, dry leaves, and twigs. After washing and air drying for 24 h, the mixture of components was used as feedstock for deriving BC by loading it in a pyrolysis unit set at 550 °C under a limited supply of oxygen [12]. The process of BC derivation is given in Figure 2. Based on weight differences, the BC yield was determined before being grounded and saved in airtight bags for further analysis and a pot trial experiment.

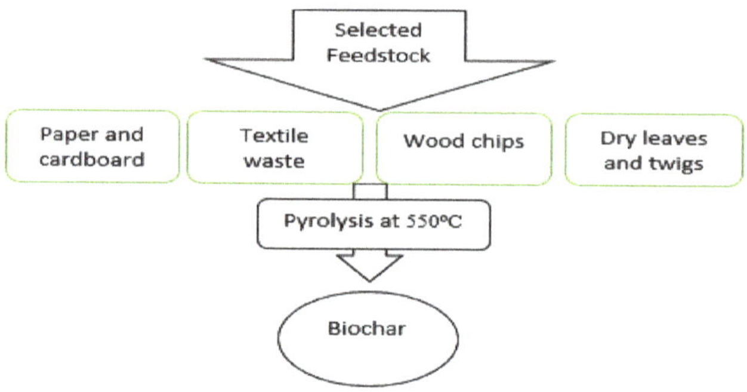

Figure 2. Combustible components of MSW used as feedstock for production of BC.

The compostable fraction of MSW included kitchen waste (fruits and vegetable leftovers) and dry leaves segregated from bulk of MSW sample collected from Mehmood Booti Open dump site Lahore (Figure 3). The compostable feedstock components were chopped (≈2 cm particle size), mixed in a proportion for optimizing cumulative moisture contents (≈60%), cumulative C:N ratio (C:N 25 ± 5:1) [13], and introduced in a pit (3 × 3 m) provided with an aeration assembly (interconnected perforated pipes). The composting pile in the pit was optimized for thermophilic to mesophilic composting with a total duration of 2.5–3 months.

Figure 3. (**A**) Unsorted bulk MSW sample collected for Mehmood Booti Open dump site of Lahore; (**B**) compostable fraction of MSW; (**C**) combustible fraction of MSW; (**D**) components of MSW left after derivation of combustible and compostables.

2.3. Characterization of BC and CP Derived from MSW Components

As the stability of BC in CP greatly depended on its surface area, mineral content, and pH, proximate (moisture, ash, and volatile content) and elemental analysis (carbon, hydrogen, nitrogen, and sulfur) were conducted by following the standard procedures [14]. The characterization parameters of BC and CP are given in Figure 4.

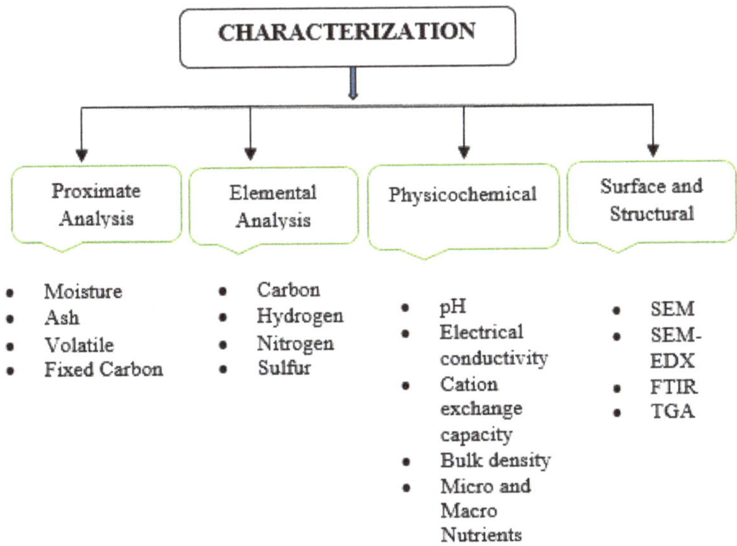

Figure 4. Characterization parameters of BC and CP.

2.3.1. Proximate and Elemental Analyses of BC and CP

The proximate analyses of BC and CP included the determination of moisture contents (MC), ash contents (AC), and volatile contents (VC), which were determined by following the standard methods [14]. The fixed carbon contents were determined by subtracting the values of MC, AC, and VC from a total of 100% [15]. The elemental analyses included the determination of C, H, N, and S contents in the feedstock components of MS, and derived BC and CP were determined by using an elemental analyzer [16].

2.3.2. Physicochemical Analyses of BC, CP and Soil

The physicochemical analyses included the determination of pH and electrical conductivity (EC) of the extracts of feedstock components of MSW and derived BC and CP. The extract was prepared by vigorously mixing 1 g of the feedstock components in 20 mL of distilled water [17]. The cation exchange capacity (CEC) of BC and CP was determined by determining the consumption of ammonium acetate solution, as given by Lu (2000) [18]. Bulk density (BD) was calculated by using the pre-weighed 100-mL measuring cylinder filled with the respective volumes of BC and CP and following the equation [19]:

$$Bulk\ Density(BD) = \frac{Mass\ of\ dried\ BC\ (or\ CP)}{Vol.\ of\ cylinder}$$

The loss-on-ignition (LOI) method was used to quantify the organic matter in soil [20]. The acid digestible (in $HCLO_4$:HNO_3 in a ratio of 4:1) samples (nearly 100 mL of digestate dilution with distilled H_2O) of BC, CP, harvested plants, and soil were used for the determination of micro- and macro-elements [21].

The soil used for the pot cultivation trial was analyzed for the determination of pH, CEC, ECe, BD, WHC, SOM, total nitrogen (TN), available phosphorous (P), and available potassium (K) by following the protocol used for BC and CP (as mentioned above). The Walkley [22] method was used to determine the SOM. For all the soil analyses, saturation paste was developed by mixing 100 g of soil with 100 mL of distilled water. The water-holding capacity of soil was determined by following the equation:

$$Water\ holding\ capacity\ (\%) = \frac{Retained\ water\ in\ saturation\ paste}{Total\ Volume\ of\ water} \times 100$$

The TN in the soil was determined through Kjeldahl's digestion protocol [23]. The Olsen's P [24] was determined by using a sodium bicarbonate solution and quantification at 700 nm on a spectrophotometer (Cintra 1010 GBC, Melbourne, Australia). The K was determined by running the soil digestate on a flame photometer (PFP7 and PFP7/C Jenway, UK) [25].

2.3.3. Surface and Structural Characterization of BC and CP

The surface and structural morphology of BC and CP were determined by scanning electron microscopy (SEM JEOL, 6480LV, Akishima, Japan) analyses [26]. The samples were put on non-two-edge carbon tape that stuck to an aluminum (Al) stub, and then the complete distribution of micro- and mesopores was determined by SEM. The composition of elements and functional groups associated with the surfaces of BC and CP were determined by energy dispersive X-ray spectroscopy (EDX) and Fourier transform infrared spectroscopy (FTIR), respectively. To detect the functional groups, the BC and CP samples were scanned at a specific resolution (4 cm^{-1}) with a wavelength range of 650–4000 cm^{-1}. The IR software was used to visualize the FTIR spectra. The thermal analyzer (model SDT Q600, TA-instruments, New Castle, DE, USA) was used for thermogravimetric analysis (TGA) of BC and CP conducted in an N-rich (100 mL/min) atmosphere up to 1000 °C ambient temperature with a 20 °C/min heating rate [27].

2.4. Crop Performance of Sunflower in Potted Soils Amended with BC and CP Derived from MSW

2.4.1. Sunflower Pot Cultivation Trial in Potted Soils Amended with BC and CP

A mesocosmic sunflower cultivation trial was conducted in pots filled with soil mixed with SOAs. The pot trial experiment was set in the Botanical Garden, University of the Punjab, Lahore-54590, Pakistan. The mesocosmic cultivation containers included PVC pots (height-diameter: 15 × 17 cm, respectively). The experimental layout included quadruplicated soil application of MSW-derived BC and CP in discrete and combined forms, each in three doses (% w:w), viz., low (L), medium (M), and high (H), i.e., BC-L, BC-M, BC-H; CP-L, CP-M, CP-H; and BC + CP-L, BC + CP-M, BC-CP-H (Table 1). The statistical design [28] consisted of a randomized complete block design (RCBD) developed for the experimental units based on the treatments and their replicates given in Table 1. Control included soil only and soil applied with the recommended dose of CF for sunflower, i.e., 150:70:50 kg/ha of NPK. The BC + CP consisted of the mixing of both SOAs in a ratio of 1:1 (v:v). The pot mixtures were left to equilibrate for 20 days. Certified sunflower seeds (Orisun oil-yielding variety) were sown in the middle of each pot with 100% germination in each pot. Watering in each pot was performed twice a week to maintain soil moisture according to the pot capacity.

Table 1. Treatments and their replicates for sunflower cultivation trial in potted soil amended with BC, CP, and their combinations.

Control								Dose	SOAs											
Soil only				Soil + CF					BC				CP				BC + CP			
Replicates									Replicates											
R_1	R_2	R_3	R_4	R_1	R_2	R_3	R_4		R_1	R_2	R_3	R_4	R_1	R_2	R_3	R_4	R_1	R_2	R_3	R_4
								L	BC-L-R_1	BC-L-R_2	BC-L-R_3	BC-L-R_4	CP-L-R_1	CP-L-R_2	CP-L-R_3	CP-L-R_4	BC + CP-L-R_1	BC + CP-L-R_2	BC + CP-L-R_3	BC + CP-L-R_4
S-R_1	S-R_2	S-R_3	S-R_4	S + CF R_1	S + CF R_2	S + CF R_3	S + CF R_{14}	M	BC-M-R_1	BC-M-R_2	BC-M-R_3	BC-M-R_4	CP-M-R_1	CP-M-R_2	CP-M-R_3	CP-M-R_4	BC + CP-M-R_1	BC + CP-M-R_2	BC + CP-M-R_3	BC + CP-M-R_4
								H	BC-H-R_1	BC-H-R_2	BC-H-R_3	BC-H-R_4	CP-H-R_1	CP-H-R_2	CP-H-R_3	CP-H-R_4	BC + CP-H-R_1	BC + CP-H-R_2	BC + CP-H-R_3	BC + CP-H-R_4

Here; BC: biochar; CP: compost; CF: chemical fertilizer; L: low; M: medium; H: high; SOAs: soil organic amendments; Rn: replicate no.

2.4.2. Sunflower Crop Performance in Pot Trial

After 75 days of cultivation trails, sunflower plants were harvested carefully from each pot after observing morphological and agronomic parameters [plant height (cm), disk diameter (cm), number of seeds, and seed weight per disk]. For post-harvest analyses, the soil samples from each pot were collected and examined by following the relevant methods as stated above.

2.4.3. Oil Contents in Sunflower Seeds Derived from Pot Cultivation Trial

For the determination of oil contents, 20 sunflower seeds were selected from each treatment on a random basis. The selected seeds were air dried and ground before oil extraction according to the method given by Abudlrazzaq et al. [29]. The Soxhlet apparatus was used for extracting oil from sunflower seeds by using a solvent of n-hexane at 60–65 °C. Through the distillation process, solvent was recovered from the extract. The mixture of oil and solvent was positioned in a water bath to let the solvent evaporate to obtain purified oil.

2.5. Statistical Analysis

For the allocation of positions to each of the experimental units of the experimental design, the random number approach was used, according to instructions [28] given for RCBD. The descriptive statistics on the quadruplicated observations for each of the variables were applied, mainly including the arithmetic mean and standard deviation. After ensuring normality of data for the selected treatments, the significance of the mean was determined by applying the F-significance test for single-factor experiments, i.e., one-way ANOVA (analysis of variance), after staying strictly compliant with the instructions given for the RCBD of a single-factor experiment [28]. After confirming F-significance ($p = 0.05$), the mean grouping was executed by computing the least significance difference ($LSD_{0.05}$) values for each of the treatment combinations considered for comparisons. For all the descriptive and analytical statistics, the Data Analysis package of MS Excel (Microsoft Office 2016) was used. The same software was used for developing data graphs.

3. Results and Discussion

3.1. Obtained Yield of BC and CP

The BC derived from selected components of MSW showed a 33% yield under the chosen pyrolysis conditions (Figure 5). The reduction in BC yield was mainly due to the thermal degradation of the components of MSW selected as feedstock, which contained components such as lignin, cellulose, and hemicellulose. The reduction in BC yield was also due to the hydroxyl groups present in the selected components of MSW considered as feedstock [16]. An average 70–80% reduction in feedstock contents has been reported for BC derived from components of MSW [29]. An average CP yield was 38% (Figure 5), i.e., the obtained CP was of medium to high quality based on yield. Mostly, yield is inversely proportional to the quality of CP, i.e., the greater the yield of CP, the lower its quality, and vice versa. The low yield and high quality of CP are generally attributed to the high losses during the rigorous maturation and curing phases of the composting process, which have not happened in the case of CP derived over a short span of time.

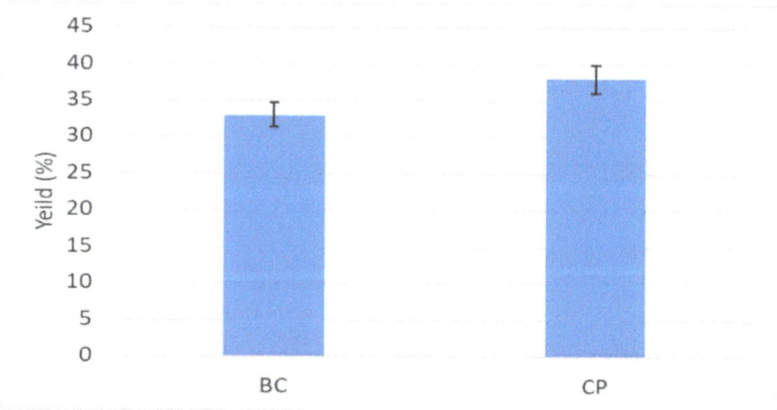

Figure 5. Biochar (BC) and compost (CP) yields (%) derived from MSW components.

3.2. Characterization of BC and CP

3.2.1. Proximate and Ultimate Analyses of BC and CP

Proximate analyses of combustible and compostable components of MSW used as feedstock for deriving BC and CP are given in Figure 6. Both BC and CP had low VC (%) contents as compared to their respective feedstocks. The AC (%) contents in the BC and CP were greater than their respective feedstocks. Perhaps it was the high mineral contents that resulted in enrichment after a reduction in the mass of feedstock up to 1/3rd rendered by the pyrolysis and composting processes, respectively. Such an increase in AC (%) contents due to enrichment has been well documented [30,31]. The VC (%) contents in BC could also be lower because of the high temperature provided during the pyrolysis process [32], which induced a great amount of porosity in BC [33]. The AC (%) in the CP had increased due to the increase in organic acid contents rendered by the microbial activity on the biodegradable components of the MSW selected as feedstock. Hence, AC (%) in CP increased due to the accumulation of minerals and organic residues due to the production and accumulation of organic acids [34], especially during the thermophilic phase of the composting process, when temperature had sustained around 65–70 °C for 6–8 weeks.

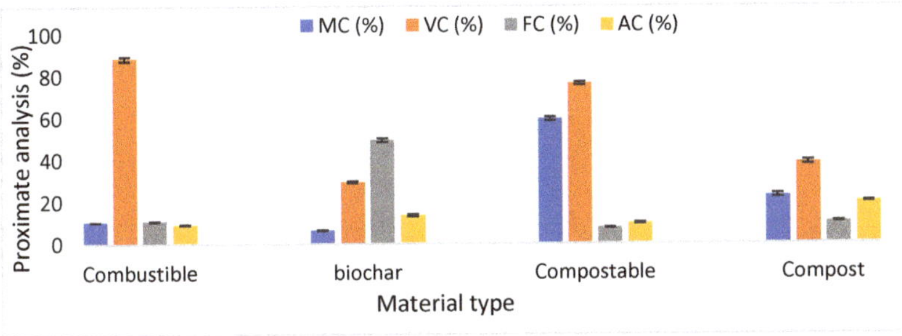

Figure 6. Proximate analysis of compostable and combustible fractions of MSW used as feedstock and derived BC and CP.

The ultimate analyses of the BC and CP are given in Table 2. The C, H, and S contents (Wt. %) were significantly greater in BC than CP. However, N contents were significantly greater in CP than BC. The CHS contents in BC were potentially due to mineral enrichment in the pyrolyzed derivatives left after the loss of most of the volatiles from the feedstock selected from MSW. The N contents in the CP were perhaps because of the mineralization rendered by the microbial activity on biodegradable feedstock derived from MSW.

Table 2. Ultimate analyses of BC and CP derived from MSW components. Mean ± S.D., followed by difference letter, is considered significantly different according to LSD 0.05.

Elemental Constituents	Units	SOAs	
		BC	CP
C	Wt.%	42.86 ± 1.16 [a]	17.44 ± 1.07 [b]
H		2.85 ± 0.44 [a]	1.43 ± 0.18 [b]
S		0.55 ± 0.02 [a]	0.36 ± 0.05 [b]
N		0.90 ± 0.10 [b]	1.92 ± 0.01 [a]

Here: SOAs: soil organic amendments; BC: biochar; CP: compost.

3.2.2. Physico-Chemical and Elemental Analyses of Soil, BC, and CP before Cultivation of Sunflower

The physicochemical and elemental results of BC and CP are given in Table 3. The BC derived from MSW was alkaline in nature with a pH of 8.73, whereas the CP had an alkaline neutral pH of 7.67. From recent studies, it has been clear that BC mostly has an alkaline pH, which, however, depends upon the temperature and feedstock type [35]. The high pH is generally due to the accumulation of alkaline salts from carbon-based waste [36], while the acidic pH is due to the volatilization of ammonia during the composting process [37].

The high values of ECe and CEC in BC and CP were due to the presence of functional groups (O-bonded) on the surface of SOAs derived from MSW feedstock [38]. The WHC of BC was greater than that of CP, being 33.5 and 24.41%, respectively. It was due to the high porosity of BC, as revealed by its surface properties as determined by SEM results. The porous structure of BC has been responsible for better WHC. The BD of CP was low because of the presence of mineralization of nutrients under aerobic conditions, with a loss of over 60% of the mass of feedstock derived from MSW. The carbon contents in BC were greater than those in CP due to the high cellulosic contents of the selected feedstocks, viz., paper and cardboard. BC has a remarkable amount of mineral nutrients [36]. The K contents were lower in CP than BC because the biodegradable components of MSW selected as composting feedstock had an inherent low concentration of K as well as N and P. During composting, microbial activity typically mineralizes small portions of nutrients, increasing the end product's nutrient content. So, composting and pyrolysis yield CP and BC as potential sources of nutrient recycling, such as when applied to the soil as SOAs [37,38]. Hence, BC and CP derived during the current study had potential for application as SOAs in the soil for the cultivation of plant crops.

Table 3. Physicochemical and elemental analysis of soil, BC, and CP.

Anlyses Type	Parameter	Units	Soil	BC	CP
Physico-chemical analysis	pH $_{(1:20)}$	-	7.39 ± 0.11	8.73 ± 0.11	7.67 ± 0.17
	$ECe_{1:20}$	mS cm^{-1}	364.3 ± 2.1	422.3 ± 1.45	553.66 ± 2.33
	WHC	(wt.%)	19.91 ± 0.62	33.5 ± 0.97	24.41 ± 0.22
	SOM		0.55 ± 0.07	7.14 ± 0.78	15.78 ± 1.66
	BD	(g cm^{-3})	1.25 ± 0.02	1.09 ± 0.11	0.60 ± 0.02
	CEC	($cmol_c$ kg^{-1})	22.29 ± 1.58	46.69 ± 0.94	43.06 ± 0.74
Elemental analyses	N	(wt.%)	0.17 ± 0.02	0.90 ± 0.10	1.92 ± 0.01
	P		0.06 ± 0.01	1.5 ± 0.01	1.69 ± 0.14
	K		0.03 ± 0.006	1.90 ± 0.49	0.89 ± 0.11
	C		ND	42.86 ± 1.16	17.44 ± 1.07
	H		ND	2.85 ± 0.44	1.43 ± 0.18
	S		ND	0.55 ± 0.02	0.36 ± 0.05
	Ca	ppm	1.003 ± 0.06	1.17 ± 0.04	2.22 ± 0.55
	Mg		1.36 ± 0.07	3.03 ± 0.18	1.52 ± 0.67
	Zn		0.23 ± 0.02	0.49 ± 0.01	0.56 ± 0.01
	Cu		0.02 ± 0.01	0.11 ± 0.01	0.14 ± 0.005

Here; BC: biochar; CP: compost; ND: not detected.

3.2.3. Surface and Structural Characterization of BC and CP

BC derived from MSW components was stable based on its structure, surface functional groups, elemental, and ultimate analyses [39]. The characterization techniques of BC have mostly been based on the determination of its structure, surface functional groups, elemental analysis, and ultimate analysis [40,41] by using different characterization methods, including SEM, EDX, FTIR, XRD, TGA, proximate analysis, and ultimate analysis [16]. The stability of BC under consideration is mainly indicated by its pore structure, pH, minerals, sorption process, particle size, and surface area [22].

FTIR spectra were analyzed to investigate functional groups present on the surfaces of BC and CP, as shown in Figure 7. Several peaks were observed at different wavelengths (3500–1000 cm^{-1}). Medium peaks were detected at 3412.4 cm^{-1}, with the OH functional group being the main component of cellulose (Figure 7A). The hydroxyl (OH) present on the surface of biochar is one of the important functional groups responsible for its adsorption capability [42]. Feedstock type and temperature of pyrolysis have been reported as main factors for the determination of functional groups on the surface of BC [43]. The O-H stretching was associated with the peaks observed at 3748 and 3252 cm^{-1}, due to the dehydration of hemicellulose and cellulose compounds. Thus, the peak at 3412 cm^{-1} could be allocated to O-H stretching. The aromatic C=O, C=C, and functional groups were associated with 1636 cm^{-1}. Li et al. [22] identified aromatic C=O and C=C peaks at 1636 cm^{-1} in *Eichornia crassipes* derived biochar, which might be glossed at a higher temperature. Between 700 and 873 cm^{-1} C-H stretching peaks were observed. In the current study, the bonds were observed at around 776.2 cm^{-1} for BC.

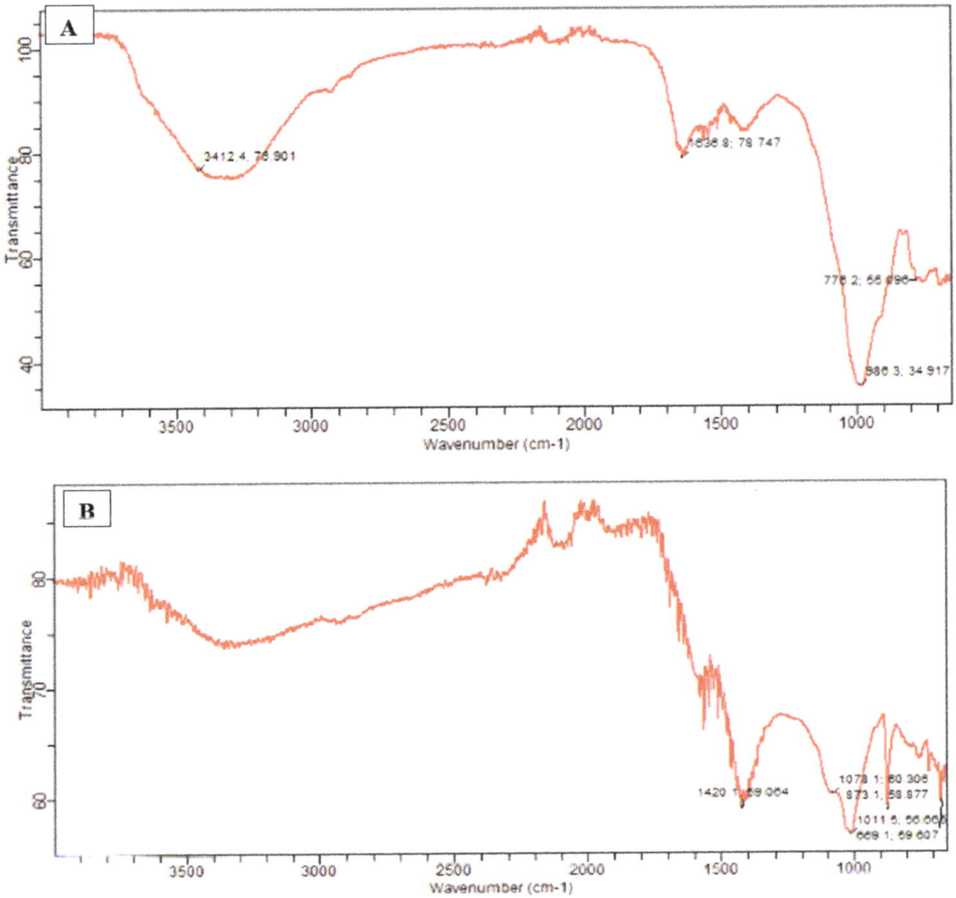

Figure 7. FTIR spectra obtained for BC (**A**) and CP (**B**) derived from MSW components demonstrate potential functional groups present on their surfaces.

As shown in Figure 7B, the FTIR spectra obtained for CP were in the 4000–500 cm^{-1} range. The peak at 1420 cm^{-1} characterized the aliphatic C-H stretching. A change in the functional groups was the result of metabolite exchange in MSW due to the biodegradation process. The peaks declined at 1633.74 and 1078.1 cm^{-1} and were associated with C=C, C=O, and S=O stretching. In the composting process, microbes used aliphatic substances as well as polysaccharides as energy sources, which ultimately enhanced the process of biodegradation and mineralization of nutrients. Similar interpretations about the functional groups of composts have been described in the literature [31].

The surface structures of BC and CP characterized by SEM are given in Figure 8. Various micro- and mesopores were present on the surface of both BC and CP. The results revealed the presence of smooth structures on the surface of CP at lower magnification, but pours and tubular structures were observed in BC at high magnification [44]. SEM of BC depicted that the pores and fibrous structures were present on its surface, as can be seen in Figure 8A. Micro- and macro-pores were present on the surface of BC, which could be responsible for enhancing WHC [45]. Usually, biochar with a greater surface area and high porosity will have a greater adsorption capacity. Thus, pore structure is produced as the result of the dehydration phenomenon during pyrolysis [46]. SEM images of BC showed that various processes and temperatures led to substantial changes in the surface morphology of the original particles while still largely maintaining their visible shape [47]. The enlargement of pores in BC with temperature rises might have greatly enhanced the pore properties of BC. Moreover, it was likely that the rise in the pyrolysis temperature increased the crystallinity of mineral sections, giving rise to highly favorable aromatic structures in BC [32]. SEM images of CP at different magnifications showed a spongy and deposited structure on its surface, as shown in Figure 8B. In the previous study [48], the MSW-derived CP contained fluffy and spongy-type structures, usually due to the microorganisms involved in the degradation process.

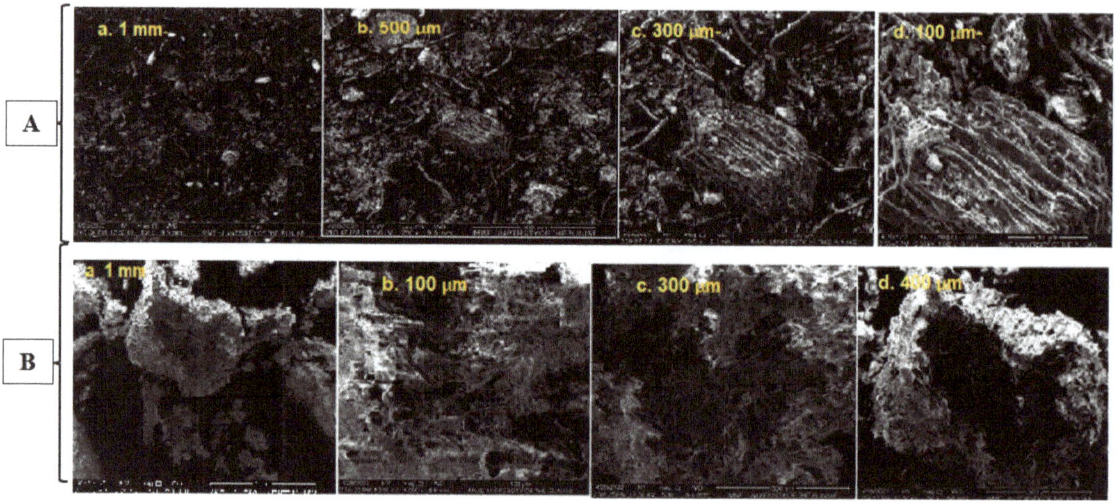

Figure 8. Scanning electron micrographs (SEM): (**A**) SEM view of combustible fraction of MSW derived biochar (**B**) SEM view of compostable fraction of MSW derived compost.

SEM-EDX spectral assays of the BC and CP were made to determine the structural configuration of both SOAs (Table 4 and Figure 9), which revealed that carbon contents in BC were significantly higher than CP and O and Si contents were significantly greater in CP than BC. The significant variation in SEM-EDX assays of BC and CP was mainly because of feedstock type and the temperature of processing [49]. The 550 °C processing temperature applied to the combustible fraction of MSW taken as feedstock rendered carbon enrichment per unit mass of the derived BC, resulting in a 2/3rd loss of actual feedstock mass. On the other hand, composting of the compostable fraction of MSW considered as feedstock at 60–65 °C (only during the thermophilic phase) over 2–2.5 months rendered low carbon enrichment per unit mass of the derived CP. The SEM-EDX spectra showed that there was less nutrient residual percentage on the surface of BC, mainly including minerals of Mg, Na, K, Ca, O, B, and Si, and the highest percentage of carbon (Table 4 and Figure 9).

The relationship between the carbon content and the yield of BC is negative, suggesting that carbonization sped up with the rise in temperature in the feedstock chamber of the pyrolysis unit. The decrease in hydrogen and oxygen content at higher pyrolysis temperatures was believed to be due to the breaking of weak bonds in the structure of BC [50]. The SEM-EDX spectra (Figure 9B) revealed elemental composition on the surface of CP, mainly including O, Al, Na, Si, C, and K, being greater than P, Mg, Fe, and Na. Phosphates and dissolved nitrates may be adsorbed into the pores of the fresh BC. However, the mineral deposition on the surface of BC may vary depending on the feedstock type and pyrolysis conditions. For example, K, Na, and Ca minerals were in higher percentages on the surface of BC derived from *Wodyetia bifurcata* feedstock [51].

Table 4. SEM-EDX data achieved from the elemental investigation of biochar and compost derived from their respective municipal solid waste fractions.

	Biochar		Compost	
Elements	**Weight (%)**	**Atomic (%)**	**Weight (%)**	**Atomic (%)**
C	77.73 ± 0.01	87.3 ± 0.006	10.6 ± 0.41	15.31 ± 0.61
O	9.8 ± 0.10	8.36 ± 0008	46.01 ± 2.48	53.59 ± 6.19
Na	0.36 ± 0.005	0.24 ± 0.01	1.75 ± 0.13	1.29 ± 0.07
Mg	0.39 ± 0.005	0.26 ± 0.01	2.33 ± 0.49	2.01 ± 0.54
Si	1.48 ± 0.003	0.76 ± 0.008	22.28 ± 2.53	16.04 ± 2.68
K	2.91 ± 0.003	1.1 ± 0.003	3.31 ± 0.49	2.11 ± 0.54
Ca	12.41 ± 006	4.52 ± 0.01	1.88 ± 0.42	1.12 ± 0.17
Al	0.63 ± 0.01	0.33 ± 0.01	9.50 ± 1.25	6.92 ± 1.22
Fe	-	-	2.27 ± 0.55	0.54 ± 0.03
P	-	-	0.25 ± 0.01	0.14 ± 0.01

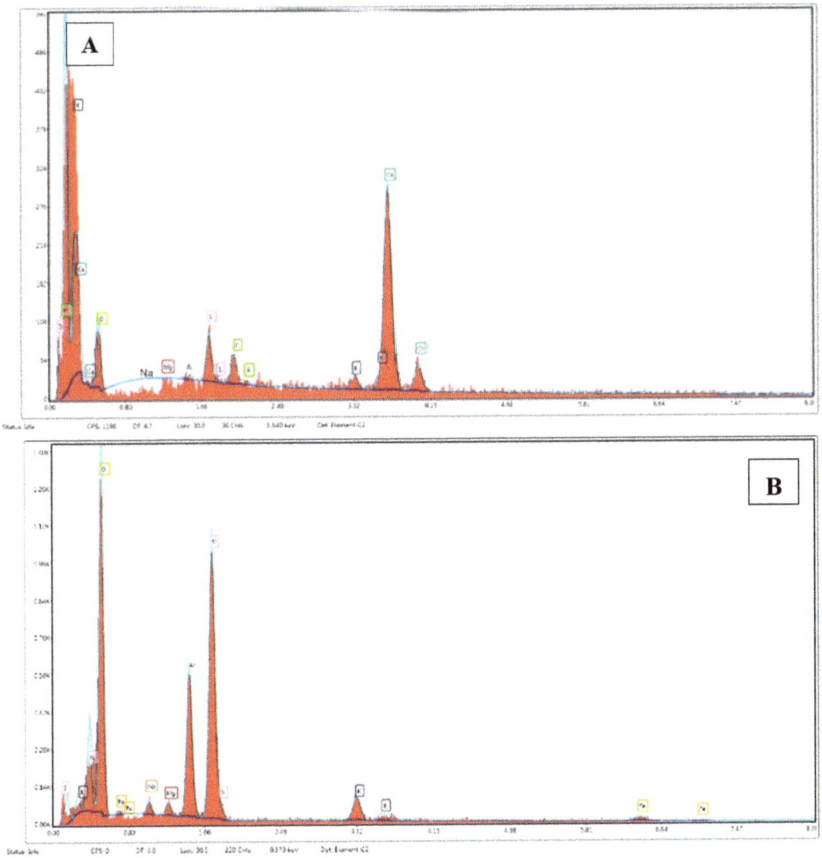

Figure 9. Spectra from Energy dispersive X-ray spectroscopy of (**A**) MSW-derived biochar (**B**) MSW-derived compost.

The thermal stability as well as heat response of the BC surface are shown in the TGA spectra (Figure 10A), which demonstrated that by increasing the pyrolysis temperature, combustion of feedstock took place, which indicated carbon oxidization and the production of VC (%) at high temperatures [52]. In the case of CP, the TGA spectra showed that there was significantly less carbon oxidation due to the very low feedstock processing temperature compared to BC. There have been reports that composting temperature and air circulation through the feedstock pile directly affect its thermal decomposition behavior and other associated kinetic parameters [53]. Results of the TGA-BC showed that the weight loss was about 27.3% at 1000 °C and remained stable thereafter. TGA-CP (Figure 10B) showed that the weight loss percentage was about 60% at 1000 °C and was stable from 600 to 1000 °C. Results revealed that between 33 and 180 °C, the first weight loss happened with the loss of the main percentage of moisture contents in the sample. About 55–60% weight loss was observed at 600 °C, which was perhaps because of the decomposition of lignin and cellulosic contents in the sample. In the case of CP, there were three phases of degradation of the sample, viz., the 1st between 25 and 120 °C, the 2nd between 200 and 350 °C, and the 3rd between 350 and 700 °C. The TGA spectra have been one of the most commonly used parameters for observing thermal degradation behaviors and associated kinetics of feedstocks during the processing of plant-based feedstock [54]. At the beginning of thermal decomposition, dehydration of the sample occurred, which transited to the thermal degradation of organic acids and volatile contents such as cellulose and hemicellulose, i.e., the breakdown of aliphatic compounds. Weight loss of the sample was proportionate to the loss of moisture contents driven by a continuous increase in temperature, i.e., dehydration of the sample over temperature processing time [11]. Hence, the stability of the derived BC and CP was higher at high temperatures.

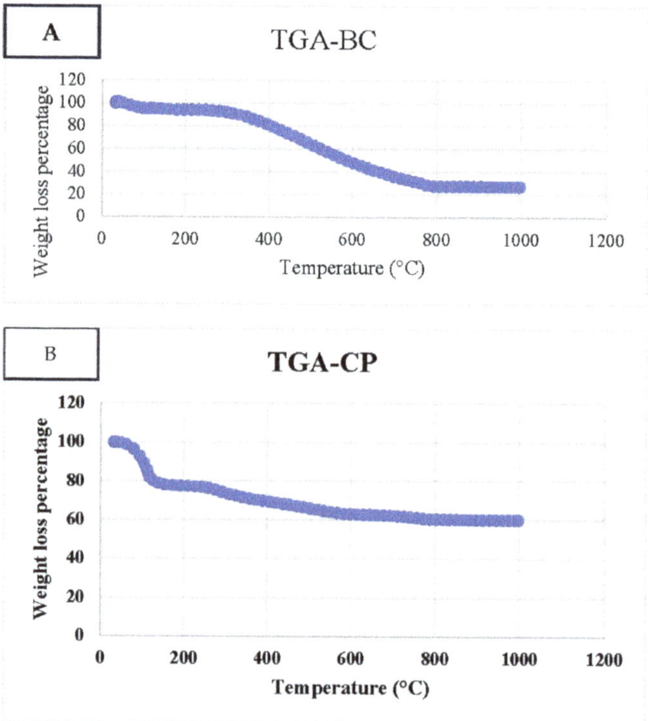

Figure 10. Showing TGA-BC curve of combustible fraction-derived biochar (**A**), TGA-CP curve of compostable fraction-derived compost (**B**).

3.3. Soil Health Improvement Induced by BC, CP, and Its Combinations

The application of SOAs in the soil, i.e., BC, CP, and their combinations, significantly ($p < 0.05$) affected its pH, CEC, WHC, and SOM (Table 5). The BD of soil got reduced with the increasing dose of SOAs, while soil pH kept significantly increasing with the increasing dose of SOAs. The combined application of BC and CP had resulted in better results than individual applications of either BC or CP in terms of inducing cumulative soil health, which aligned with previously reported findings about the combined effects of BC and CP on soil health improvement [55]. However, an increase in the soil pH was perhaps because of a soluble ionic salt bank deposited on the surface of BC, which got drained into the soil over time, leading to the gradual increase in the pH value over time, as reported by Khan et al. [56]. The actual pH values of BC and CP were 8.73 and 7.67, respectively. On combined application of BC and CP in the basic neutral soil (pH 7.39), the mutual interactions of the soluble salty cations and anions led to an initial increase and then a decrease in the soil pH. The high surface area rendered by the high micro- and microporosity of the BC could have a significant contribution to soil pH fluctuations due to its own surface, electrostatic charge, and the emergence and propagation of microbial communities on the BC surface over time [57]. Hence, the application of BC, CP, and their combination in the soil induced changes in the redox dynamism-related parameters of cumulative soil health, which were more turbulent at the early phase of experimentation (on the addition of freshly prepared BC and CP) than later in the study.

Table 5. Physicochemical characterization of post-harvested soil amended with selected treatments after growth period of sunflower.

Treatments		pH	EC (mS cm^{-1})	BD (g cm^{-3})	CEC (Cmolc kg^{-1})	SOM	WHC	N	P
						Wt. (%)			
Control	Soil only	7.49 ± 0.03	415.6 ± 1.97	1.22 ± 0.01	20.22 ± 0.52	1.23 ± 0.04	53.72 ± 0.02	0.175 ± 0.04	0.21 ± 0.03
	Soil + CF	8.56 ± 0.09	538.5 ± 0.52	1.12 ± 0.002	28.63 ± 0.66	5.40 ± 0.11	84.55 ± 0.17	1.61 ± 0.02	1.52 ± 0.10
BC-L		8.13 ± 0.05	23.42 ± 0.47	1.16 ± 0.01	23.42 ± 0.47	5.12 ± 0.05	70.41 ± 0.02	0.675 ± 0.02	1.21 ± 0.16
BC-M		8.37 ± 0.09	519.25 ± 0.84	1.14 ± 0.001	25.75 ± 0.50	4.98 ± 0.14	66.57 ± 0.13	0.59 ± 0.03	1.38 ± 0.11
BC-H		8.06 ± 0.01	510.52 ± 0.14	1.11 ± 0.007	27.27 ± 0.45	5.42 ± 0.05	66.41 ± 0.02	0.74 ± 0.01	1.59 ± 0.09
CP-L		8.59 ± 0.02	557 ± 1.44	1.18 ± 0.002	27.16 ± 0.8	8.33 ± 0.12	63.79 ± 1.00	1.18 ± 0.06	1.47 ± 0.17
CP-M		8.05 ± 0.01	563.7 ± 0.09	1.11 ± 0.002	29.44 ± 0.55	8.44 ± 0.11	59.75 ± 0.01	1.66 ± 0.01	1.63 ± 0.1
CP-H		8.65 ± 0.01	548.075 ± 2.5	1.10 ± 0.01	30.38 ± 0.65	7.81 ± 0.40	57.61 ± 0.95	1.73 ± 0.02	1.66 ± 0.09
BC + CP-L		8.63 ± 0.03	536.15 ± 1.88	1.06 ± 0.01	29.32 ± 0.33	8.57 ± 0.16	68.86 ± 0.01	1.78 ± 0.01	1.85 ± 0.01
BC + CP-M		8.24 ± 0.29	583.37 ± 0.13	1.05 ± 0.004	31.90 ± 0.86	9.26 ± 0.23	62.73 ± 0.57	1.82 ± 0.01	1.83 ± 1.0.03
BC + CP-H		86.26 ± 0.01	574 ± 0.075	1.008 ± 0.15	32.31 ± 1.51	13.21 ± 0.43	86.26 ± 0.01	1.85 ± 0.01	1.91 ± 0.02

Here: BC: biochar; CP: compost; CF: chemical fertilizer; L: low; M: medium; H: high; SOM: soil organic matter; EC: electrical conductivity; CEC: cation exchange capacity; WHC: water holding capacity; BD: bulk density of soil.

The soil EC of the soil amended with the combined application of BC and CP was significantly higher (583.3 μS cm^{-1}) than the control soil only (415.6 μS cm^{-1}). However, control soil with CF had greater EC than soil alone, which was perhaps because of the mineral salts rendered by CF [58]. The WHC and CEC of the soil improved (86.26% and 32.31 cmol kg^{-1}, respectively) with the combined application of BC and CP. This was perhaps because of the high porosity, mineral, and ash contents in the BC and CP, as observed in EDX analyses. However, the dose effect of both BC and CP had no corresponding effects on the variation in indicators of cumulative soil health, i.e., the increase in dose of BC and CP from low to medium to high did not result in an increase in the values of the soil health parameters; rather, they fluctuated in a variable manner. It was perhaps because of complex interactions between the soluble redox dynamics of the SOAs that an induced shift in the kinetic equilibrium of their physical, chemical, and biological properties was less compatible with the apparent increase in dose of BC, CP, and their combinations.

Crop performance is generally significantly influenced by the variation in WHC, which is influenced by the shift in BD rendered by the types and doses of SOAs applied to the soil. Adding BC and CP to the soil improved its ability to retain water, which improved the soluble mineral contents of the soil [59]. The combined application of BC and CP proved more effective in improving SOM (13.12%) as compared to the control. At high doses of BC + CP, the SOM was the highest. Compost enhances SOM in soil over time if continuous replenishments are made on a recurrent basis. Soils with low SOM require more recurrent and constant replenishment of CP, as it degrades more rapidly than BC. Soils in Punjab (Pakistan) contain very low SOM due to climatic conditions and unhealthy soil management practices [60]. Regular replenishments of SOAs, as developed in the current study, would be helpful in restoring and improving SOM over time. Applying SOAs like CP made from MSW-based feedstock helped improve SOM in the field soil over time [61]. It has also been observed that the long-term application of CF alone in soil without the addition of SOAs could cause a gradual reduction in SOM and a cumulative drop in the crop productivity potential of the soil.

The results of the current study showed that the available N increased in soils applied with BC, CP, and their combinations (Table 5). The highest available N was observed in the soil treated with the combined application of BC and CP, i.e., compared to the control treatment (0.175%), there was up to 1.85%, 1.73%, and 0.74% increase in soil with the application of CP, BC, and its combination, respectively. The BC potentially acted as a slow-release N source [62]. Generally, applying BC along with other fertilizers has more favorable impacts on counteracting the potentially inaccessible N from applied BC [63]. This is why there was more soluble N in soils applied with a combined dose of BC and CP, as observed in the current study. The combined application of BC and CP led to the highest increase in the level of available P in the soil (Table 5). The available K contents at the pre-sowing stage were below the adequate level in soils applied with either BC or CP. However, soil applied with both BC and CP had greater available K contents (1.91%). Overall, application of BC, CP, and their combinations at any time has improved cumulative soil health by stabilizing soil pH, increasing SOM, CEC, and the level of available NPK, which could be attributed to the complex microbial consortium added through CP.

3.4. Crop Performance of Sunflower in the Presence of BC, CP and Its Combinations

Sunflower plant growth and yield performance are influenced by the presence of BC, CP, and their combinations, as shown in Figure 11. Compared to the control, the sunflower dry biomass increased dramatically under the individual and combined effects of BC and CP at different doses. The sunflower yield obtained from soil applied with combined BC + CP was almost equal to the yield derived from soil applied with CF (control). From the dose of SOAs applied, it was observed that BC + CP applied at a medium dose performed noticeably better. This could be due to the provision of NPK by the SOAs to the sunflower crop in a range that is optimal and equal to the recommended NPK in the form of CF. This could also be because of the slow solubilization of NPK in the soil due to the microbial activity of the consortia added to the soil through CP. In soil treated with combined BC + CP at a medium dose, sunflower showed greater disc diameter (13.54 cm) and seed yield (29.64 g), as given in Figure 11c,d. This may be because different organic volatiles deposited on the surface of the biochar caused plants to undergo hormonal changes, leading to early inflorescence and a notable ($p < 0.05$) increase in seed yield. The activity of microbes rendered by CP in combination with BC has been well documented [64]. However, the soil with CF (control) yielded the highest sunflower dry biomass (32.38 g and 31.42 g). During the current study, at high doses of the SOAs, the sunflower dry biomass decreased, along with a small drop in the seed yield. The high doses of the SOAs might have caused nutrient sorption on the electrostatically active surfaces of the SOAs, which might have compromised sunflower growth. Another potential reason could be an apparently hidden abiotic stress in the soil solution rendered by the salty minerals released by the freshly prepared BC combined with CP. When BC and CP were applied individually, the soil showed better quality; nevertheless, the sunflower seed yield got reduced in comparison to soil with CF (control) but remained better than soil alone (control). Hence, it could be inferred that the combined application of BC and CP at a medium dose could be an effective

SOA for improving soil health and yielding sunflower growth and yield comparable to the CF, i.e., the BC and CP derived from the combustible and compostable fractions of MSW have potentially acted as an alternative to the CF for sunflower crop performance.

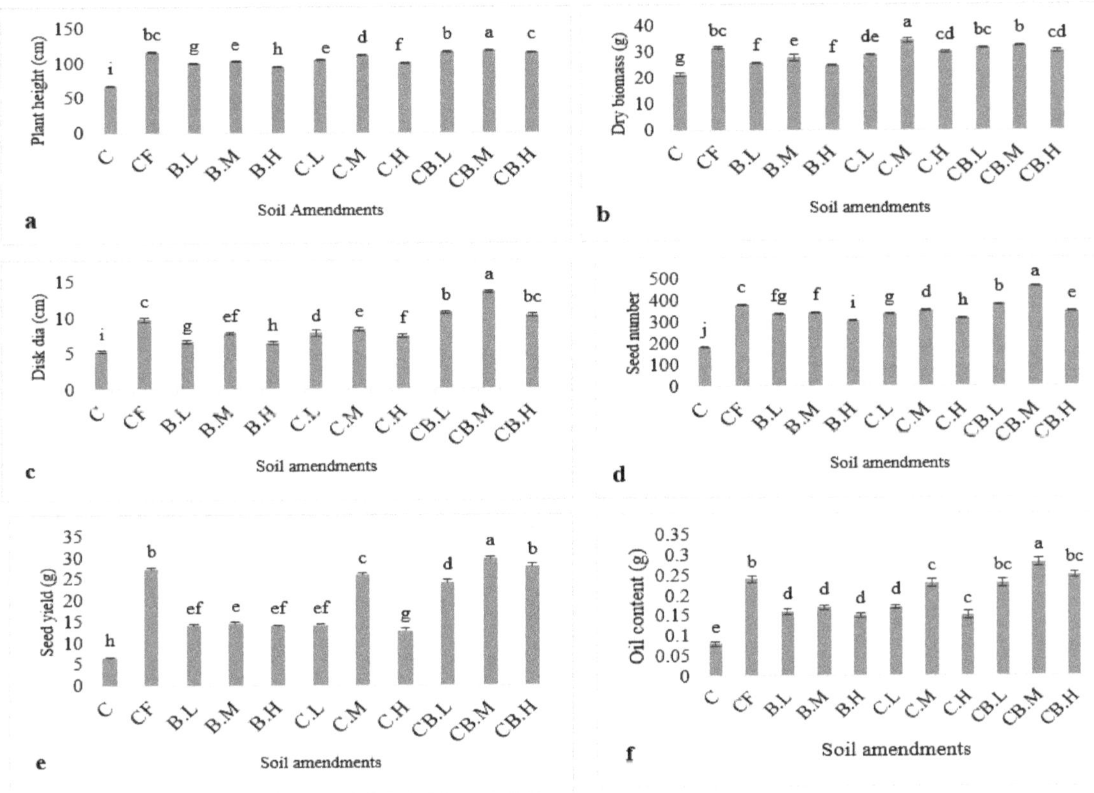

Figure 11. Showing the growth and yield performance of sunflower; (**a**) plant height (cm); (**b**) dry biomass (g); (**c**) disk diameter (cm); (**d**) seed number; (**e**) seed yield; (**f**) oil content. Different letters above bars show the Duncan's multiple range test (DMRT) to determine for comparisons among means.

4. Conclusions

The study concluded that the combined application of BC and CP at medium doses resulted in soil improvement (SOM + 90.7% and CEC + 37.4% compared to the control). The soil health improvement effects of the SOAs were mainly attributed to the significant increases in SOM, CEC, total, and available NPK. The BC + CP at medium dose yielded the highest sunflower dry biomass and seed yield, being 30.8 and 68.9% greater than the control, respectively. The biomass yield in combined SOAs was comparable to the biomass yield in CF, viz., the former being just 4% less than the latter. In the case of the dry biomass of sunflower, the SOAs derived from combustible and compostable fractions of MSW acted as an alternative to the CF. However, seed yield in BC + CP at a medium dose was 28.5% less than in CF. The crop growth performance of sunflower in the presence of SOAs derived from combustible and compostable fractions of MSW being comparable to the CF could create new avenues for developing BC + CF composites in the future. Recycling components from MSW open dumps in developing countries could create avenues for producing SOAs with the benefit of reducing waste stream loads.

Author Contributions: S.A.: Methodology, Data curation, data analysis, writing the original draft. A.N.: Supervision, conceptualization, methodology, data analysis, Editing and reviewing the draft. All authors have read and agreed to the published version of the manuscript.

Funding: This research received no external funding.

Data Availability Statement: Raw data will be available upon request.

Conflicts of Interest: The authors declare no competing interest.

References

1. Vijay, V.; Shreedhar, S.; Adlak, K.; Payyanad, S.; Sreedharan, V.; Gopi, G.; Aravind, P.V. Review of large-scale biochar field-trials for soil amendment and the observed influences on crop yield variations. *Front. Energy. Res.* **2021**, *9*, 710766. [CrossRef]
2. Agegnehu, G.; Bass, A.M.; Nelson, P.N.; Bird, M.I. Benefits of Biochar, Compost and Biochar-Compost for Soil Quality, maize Yield and Greenhouse Gas Emissions in a Tropical Agricultural Soil. *Sci. Total Environ.* **2016**, *543*, 295–306. [CrossRef] [PubMed]
3. Nair, V.D.; Nair, P.K.R.; Dari, B.; Freitas, A.M.; Chatterjee, N.; Pinheiro, F.M. Biochar in the Agroecosystem-Climate-Change-Sustainability Nexus. *Front. Plant Sci.* **2017**, *8*, 2051. [CrossRef] [PubMed]
4. Bizuhoraho, T.; Kayiranga, A.; Manirakiza, N.; Mourad, K.A. The effect of land use systems on soil properties; A case study from Rwanda. *Sustain. Agric. Res.* **2018**, *7*, 30–40. [CrossRef]
5. Gumisiriza, R.; Hawumba, J.F.; Okure, M.; Hensel, O. Biomass waste-to-energy valorisation technologies: A review case for banana processing in Uganda. *Biotechnol. Biofuels.* **2017**, *10*, 1–29. [CrossRef] [PubMed]
6. Jeyasubramanian, K.; Thangagiri, B.; Sakthivel, A.; Dhaveethu Raja, J.; Seenivasan, S.; Vallinayagam, P. A Complete Review on Biochar: Production, Property, Multifaceted Applications, Interaction Mechanism and Computational Approach. *Fuel* **2021**, *292*, 120243. [CrossRef]
7. Godlewska, P.; Ok, Y.S.; Oleszczuk, P. The Dark Side of Black Gold: Ecotoxicological Aspects of Biochar and Biochar-Amended Soils. *J. Hazard. Mater.* **2021**, *403*, 123833. [CrossRef] [PubMed]
8. Xiao, X.; Chen, B.; Chen, Z.; Zhu, L.; Schnoor, J.L. Insight into Multiple and Multilevel Structures of Biochars and Their Potential Environmental Applications: A Critical Review. *Environ. Sci. Technol.* **2018**, *52*, 5027–5047. [CrossRef] [PubMed]
9. Chen, P.; Liu, Y.; Mo, C.; Jiang, Z.; Yang, J.; Lin, J. Microbial Mechanism of Biochar Addition on Nitrogen Leaching and Retention in tea Soils from Different Plantation Ages. *Sci. Total Environ.* **2021**, *757*, 143817. [CrossRef]
10. Khan, N.; Bolan, N.; Jospeh, S.; Anh, M.T.L.; Meier, S.; Kookana, R.; Qiu, R. Complementing compost with biochar for agriculture, soil remediation and climate mitigation. *Adv. Agron.* **2023**, *179*, 1–90.
11. Agegnehu, G.; Srivastava, A.K.; Bird, M.I. The role of biochar and biochar-compost in improving soil quality and crop performance: A review. *Appl. Soil Ecol.* **2017**, *119*, 156–170. [CrossRef]
12. Jayawardhana, Y.; Mayakaduwa, S.S.; Kumarathilaka, P.; Gamage, S.; Vithanage, M. Municipal solid waste-derived biochar for the removal of benzene from landfill leachate. *Environ. Geochem. Health* **2019**, *41*, 1739–1753. [CrossRef] [PubMed]
13. Rynk, R.; Van de Kamp, M.; Willson, G.B.; Singley, M.E.; Richard, T.L.; Kolega, J.J.; Gouin, F.R.; Laliberty, L.; Kay, D.; Murphy, D.; et al. *On-Farm Composting Handbook (NRAES 54)*; Northeast Regional Agricultural Engineering Service: Ithaca, NY, USA, 1992.
14. ASTM International. *Standard Practice for Proximate Analysis of Coal and Coke*; ASTM International: West Conshohocken, PA, USA, 2013.
15. ASTM. Standard test method for chemical analysis of wood charcoal-D1762-84. *Am. Soc. Test Mater.* **2001**, *84*, 1–2.
16. Nazir, A.; Laila, U.E.; Bareen, F.E.; Hameed, E.; Shafiq, M. Sustainable management of peanut shell through biochar and its application as soil ameliorant. *Sustainability* **2021**, *13*, 13796. [CrossRef]
17. Bordoloi, N.; Narzari, R.; Chutia, R.S.; Bhaskar, T.; Ataki, R. Pyrolysis of *Mesua ferrea* and *Pongamia glabra* seed cover: Characterization of bio-oil and its sub-fractions. *Bioresour. Technol.* **2015**, *178*, 83–89. [CrossRef]
18. Lu, R.K. *Assay on Agro-Chemical Properties of Soil*; China Agricultural Science and Technology Press: Beijing, China, 2000.
19. Brewer, C.E.; Levine, J. Weight or Volume for Handling Biochar and Biomss? In the Biochar Journal, 2015, Arbaz, Switzerland. Available online: http://www.biochar-journal.org/en/Ct/71 (accessed on 15 June 2022).
20. Schulte, E.E.; Hopkins, B.G. Estimation of soil organic matter by weight loss-on-ignition. *Soil Org. Matter Anal. Interpret.* **1996**, *46*, 21–31.
21. Baird, R.B.; Eaton, A.D. *Standard Methods for the Examination of Water and Wastewater*; Rice, E.W., Ed.; American Public Health Association: Washington, DC, USA, 2017; Volume 23, pp. 5–41.
22. Li, F.; Shen, K.; Long, X.; Wen, J.; Xie, X.; Zeng, X.; Liang, Y.; Wei, Y.; Lin, Z.; Huang, W. Preparation and characterization of biochars from *Eichornia crassipes* for cadmium removal in aqueous solutions. *PLoS ONE* **2016**, *11*, 0148132. [CrossRef] [PubMed]
23. Bremner, J.M.; Tabatabai, M.A. Use of an ammonia electrode for determination of ammonium in Kjeldahl analysis of soils. Commun. *Soil. Sci. Plant. Anal.* **1972**, *3*, 159–165. [CrossRef]
24. Watanabe, F.S.; Olsen, S.R. Test of an ascorbic acid method for determining phosphorus in water and $NaHCO_3$ extracts from soil. *Soil. Sci. Soc. Am. J.* **1965**, *29*, 677–678. [CrossRef]
25. Lu, R.K. *Analytical Methods of Soil Agrochemistry*; Chinese Agricultural Science and Technology Publishing House: Beijing, China, 2000; pp. 18–99.

26. Waqas, M.; Aburiazaiza, A.S.; Miandad, R.; Rehan, M.; Barakat, M.A.; Nizami, A.S. Development of biochar as fuel and catalyst in energy recovery technologies. *J. Clean. Prod.* **2018**, *188*, 477–488. [CrossRef]
27. Zornoza, R.; Moreno-Barriga, F.; Acosta, J.A.; Muñoz, M.A.; Faz, A. Stability, nutrient availability and hydrophobicity of biochars derived from manure, crop residues, and municipal solid waste for their use as soil amendments. *Chemosphere* **2016**, *144*, 122–130. [CrossRef] [PubMed]
28. Gomez, K.A.; Gomez, A.A. Chapter 2 Single-Factor Experiments. In *Statistical Procedures for Agricultural Research*; A Wiley-Interscience Publication; John Wiley & Sons: New York, NY, USA; Chichester, UK; Brisbane, QC, Australia; Toronto, ON, Canada; Singapore, 1984.
29. Abdulrazzaq, H.; Jol, H.; Husni, A.; Abu-Bakr, R. Characterization and stabilisation of biochars obtained from empty fruit bunch, wood, and rice husk. *BioResources* **2014**, *9*, 2888–2898. [CrossRef]
30. Chen, T.; Zhang, Y.; Wang, H.; Lu, W.; Zhou, Z.; Zhang, Y.; Ren, L. Influence of pyrolysis temperature on characteristics and heavy metal adsorptive performance of biochar derived from municipal sewage sludge. *Bioresour. Technol.* **2014**, *164*, 47–54. [CrossRef] [PubMed]
31. Li, J.; Shen, F.; Yang, G.; Zhang, Y.; Deng, S.; Zhang, J.; Zeng, Y.; Luo, T.; Mei, Z. Valorizing Rice Straw and Its Anaerobically Digested Residues for Biochar to Remove Pb(II) from Aqueous Solution. *Int. J. Polym. Sci.* **2018**, *2018*, 2684962. [CrossRef]
32. Zhang, X.; Zhang, P.; Yuan, X.; Li, Y.; Han, L. Effect of pyrolysis temperature and correlation analysis on the yield and physicochemical properties of crop residue biochar. *Bioresour. Technol.* **2020**, *296*, 122318. [CrossRef] [PubMed]
33. Leng, L.; Xiong, Q.; Yang, L.; Li, H.; Zhou, Y.; Zhang, W.; Jiang, S.; Li, H.; Huang, H. An overview on engineering the surface area and porosity of biochar. *Sci. Total Environ.* **2021**, *763*, 144204. [CrossRef] [PubMed]
34. Qureshi, T.; Shoaib, H.M.; Ali, U.; Siddiqi, M.H.; Hussain, M.A.; Lateef, H.U. Comparison of Physio-Chemical Characteristics of Different Compost Samples. *Int. J. Thermal. Environ. Eng.* **2021**, *18*, 09–17.
35. Agrafioti, E.; Kalderis, D.; Diamadopoulos, E. Arsenic and chromium removal from water using biochars derived from rice husk, organic solid wastes and sewage sludge. *J. Environ. Manag.* **2014**, *133*, 309–314. [CrossRef] [PubMed]
36. Yuan, J.H.; Xu, R.K. The amelioration effects of low temperature biochar generated from nine crop residues on an acidic Ultisol. *Soil Use Manag.* **2011**, *27*, 110–115. [CrossRef]
37. Wei, Y.; Liang, Z.; Zhang, Y. Evolution of physicochemical proper ties and bacterial community in aerobic composting of swine manure based on a patent compost tray. *Bioresour. Technol.* **2022**, *343*, 126136. [CrossRef]
38. Alkurdi, S.S.; Al-Juboori, R.A.; Bundschuh, J.; Bowtel, L.; McKnight, S. Effect of pyrolysis conditions on bone char characterization and its ability for arsenic and fluoride removal. *Environ. Pollut.* **2020**, *262*, 114–221. [CrossRef] [PubMed]
39. Singh, A.; Chandel, M.K. Physicochemical and FTIR spectroscopic analysis of fine fraction from a municipal solid waste dumpsite for potential reclamation of materials. *Waste Manag. Res.* **2021**, *39*, 374–385. [CrossRef] [PubMed]
40. Han, L.; Sun, K.; Yang, Y.; Xia, X.; Li, F.; Yang, Z.; Xing, B. Biochar's stability and effect on the content, composition and turnover of soil organic carbon. *Geoderma* **2020**, *364*, 114184. [CrossRef]
41. Chen, S.; Qin, C.; Wang, T.; Chen, F.; Li, X.; Hou, H.; Zhou, M. Study on the adsorption of dyestuffs with different properties by sludge-rice husk biochar: Adsorption capacity, isotherm, kinetic, thermodynamics and mechanism. *J. Mol. Liq.* **2019**, *285*, 62–74. [CrossRef]
42. Yaashikaa, P.R.; Kumar, P.S.; Varjani, S.; Saravanan, A. A critical review on the biochar production techniques, characterization, stability and applications for circular bioeconomy. *Biotechnol. Rep.* **2020**, *28*, e00570. [CrossRef] [PubMed]
43. Rashid, J.; Tehreem, F.; Rehman, A.; Kumar, R. Synthesis using natural functionalization of activated carbon from pumpkin peels for decolourization of aqueous methylene blue. *Sci. Total Environ.* **2019**, *671*, 369–376. [CrossRef]
44. Hoslett, J.; Ghazal, H.; Mohamad, N.; Jouhara, H. Removal of methylene blue from aqueous solutions by biochar prepared from the pyrolysis of mixed municipal discarded material. *Sci. Total Environ.* **2020**, *714*, 136–832. [CrossRef]
45. Sigmund, G.; Hüffer, T.; Hofmann, T.; Kah, M. Biochar total surface area and total pore volume determined by N2 and CO2 physisorption are strongly influenced by degassing temperature. *Sci. Total Environ.* **2017**, *580*, 770–775. [CrossRef]
46. Papurello, D.; Gandiglio, M.; Kafashan, J.; Lanzini, A. Biogas purification: A comparison of adsorption wood-derived char using isotherm equations. *Processes* **2019**, *7*, 774. [CrossRef]
47. Mujtaba, G.; Hayat, R.; Hussain, Q.; Ahmed, M. Physico-Chemical Characterization of Biochar, Compost and Co-Composted Biochar Derived from Green Waste. *Sustainability* **2021**, *13*, 4628. [CrossRef]
48. Sun, Y.; Xiong, X.; He, M.; Xu, Z.; Hou, D.; Zhang, W.; Tsang, D.C. Roles of biochar-derived dissolved organic matter in soil amendment and environmental remediation: A critical review. *J. Chem. Eng.* **2021**, *424*, 130387. [CrossRef]
49. Mohammad, M.I.; Al-Wabel, A.; Al-Omran, A.M.; El-Naggar, M.; Nadeem, A.R.A. Usman. Pyrolysis temperature induced changes in characteristics and chemical composition of biochar produced from conocarpus wastes. *Bioresour. Technol.* **2014**, *131*, 374–379.
50. Mary, G.S.; Sugumaran, P.; Niveditha, S.; Ramalakshmi, B.; Ravichandran, P.; Seshadri, S. Production, characterization and evaluation of biochar from pod (*Pisum sativum*), leaf (*Brassica oleracea*) and peel (*Citrus sinensis*) wastes. *Int. J. Recycl. Org. Waste Agric.* **2016**, *5*, 43–53. [CrossRef]
51. Jean, K.; Emanuella, G.; Souza, D.; Mychel, Í.; Leite, G.; Sá, D.; Meili, L. Bioresource Technology Wodyetia bifurcata biochar for methylene blue removal from aqueous matrix. *Bioresour. Technol.* **2019**, *293*, 122093.

52. Saffari, N.; Hajabbasi, M.A.; Shirani, H.; Mosaddeghi, M.R.; Mamedov, A.I. Biochar type and pyrolysis temperature effects on soil quality indicators and structural stability. *J. Environ. Manag.* **2020**, *261*, 110190. [CrossRef] [PubMed]
53. Dhyani, V.; Awasthi, M.K.; Wang, Q.; Kumar, J.; Ren, X.; Zhao, J.; Chen, H.; Wang, M.; Bhaskar, T.; Zhang, Z. Effect of composting on the thermal decomposition behavior and kinetic parameters of pig manure-derived solid waste. *Bioresour. Technol.* **2018**, *252*, 59–65. [CrossRef] [PubMed]
54. Ma, Z.; Wang, J.; Yang, Y.; Zhang, Y.; Zhao, C.; Yu, Y.; Wang, S. Comparison of the thermal degradation behaviors and kinetics of palm oil waste under nitrogen and air atmosphere in TGA-FTIR with a complementary use of model-free and model-fitting approaches. *J. Anal. Appl. Pyrol.* **2018**, *134*, 12–24. [CrossRef]
55. Martinsen, V.; Alling, V.; Nurida, N.L.; Mulder, J.; Hale, S.E.; Ritz, C.; Cornelissen, G. pH effects of the addition of three biochars to acidic Indonesian mineral soils. *Soil. Sci. Plant. Nutr.* **2015**, *61*, 821–834. [CrossRef]
56. Khan, N.; Clark, I.; Sanchez-Monedero, M.A.; Shea, S.; Meier, S.; Qi, F.J.; Kookana, R.S.; Bolan, N. Physical and chemical properties of biochars co-composted with biowastes and incubated with a chicken litter compost. *Chemosphere* **2016**, *142*, 14–23. [CrossRef]
57. Ozlu, E.; Kumar, S. Response of soil organic carbon, ph, electrical conductivity, and water stable aggregates to long-term annual manure and inorganic fertilizer. *Soil Sci. Soc. Am. J.* **2018**, *82*, 1243–1251. [CrossRef]
58. Williams, A.; Hunter, M.C.; Kammerer, M.; Kane, D.A.; Jordan, N.R.; Mortensen, D.A. Soil water holding capacity mitigates downside risk and volatility in US rainfed maize: Time to invest in soil organic matter. *PLoS ONE* **2016**, *11*, e0160974. [CrossRef] [PubMed]
59. Manna, M.C.; Rahman, M.M.; Naidu, R.; Sahu, A.; Bhattacharjya, S.; Wanjari, R.H.; Khanna, S.S. Bio-waste management in subtropical soils of India: Future challenges and opportunities in agriculture. *Adv. Agron.* **2018**, *152*, 87–148.
60. Yousra, M.; Hussain, Q.; Khan, K.S.; Ansar, M.; Sarwar, S.; Khan, M.Z. Soil Organic Carbon Pools in Benchmark Soils of Punjab Pakistan. *Commun. Soil Sci. Plant Anal.* **2023**, *54*, 571–585. [CrossRef]
61. Chan, K.Y.; Xu, Z. Biochar: Nutrient properties and their enhancement. In *Biochar for Environmental Management*; Lehmann, J. Joseph, S., Eds.; Earthscan: London, UK, 2009; pp. 67–84.
62. Doan, T.T.; Tureaux, T.H.; Rumpel, C.; Janeau, J.L.; Jouquet, P. Impact of compost, vermicompost andbiochar on soil fertility maize yield and soil erosion in Northern Vietnam: A three-year mesocosm experiment. *Sci. Total Environ.* **2015**, *514*, 147–154. [CrossRef] [PubMed]
63. Akdeniz, N. A systematic review of biochar use in animal waste composting. *Waste Manag.* **2019**, *88*, 291–300. [CrossRef]
64. Farid, I.M.; Siam, H.S.; Abbas, M.H.; Mohamed, I.; Mahmoud, S.A.; Tolba, M.; Shaheen, S.M. Co-composted biochar derived from rice straw and sugarcane bagasse improved soil properties, carbon balance, and zucchini growth in a sandy soil: A trial for enhancing the health of low fertile arid soils. *Chemosphere* **2022**, *292*, 133389. [CrossRef]

agronomy

MDPI

Article

Valorisation of Deinking Paper Sludge for Fertiliser Purposes: New Perspective in Sustainable Agriculture

Halyna Kominko *, Katarzyna Gorazda, Katarzyna Łoś and Zbigniew Wzorek

Department of Chemical Technology and Environmental Analytics, Faculty of Chemical Engineering and Technology, Cracow University of Technology, Warszawska 24, 31-155 Cracow, Poland; katarzyna.goradza@pk.edu.pl (K.G.); loskasia27@gmail.com (K.Ł.); zbigniew.wzorek@pk.edu.pl (Z.W.)
* Correspondence: halyna.kominko@pk.edu.pl

Abstract: The growth of the global population, coupled with concomitant economic development, has resulted in the generation of a substantial quantity of waste. The transition of the European Union's economy towards a closed-loop model is prompting a comprehensive search for waste management concepts across a range of industrial sectors. The objective of this study is to valorise deinking paper sludge, which has a high potential for soil formation due to its high organic matter content. To produce organic–mineral fertiliser, the deinking sludge was subjected to acid hydrolysis, then neutralised with KOH solution and enriched with poultry litter ash. The final products were characterised in terms of their nutrient and heavy metal content. The bioavailability of phosphorus, along with the forms in which it occurs in fertilisers, was determined through the implementation of a five-step fractionation procedure. Furthermore, an eight-week incubation period was conducted to assess the fertilisers' performance in soil. Soil samples were tested on a weekly basis for pH, water-soluble and bioavailable phosphorus content using the spectroscopic method after previous extraction in water and Bray's solution, and catalase activity using the titrimetric method. The resulting fertilisers were found to meet the requirements for organo-mineral fertilisers and were categorised as PK-type fertilisers with a total nutrient content of 24.6–39.3%. Fractionation studies demonstrated that the fertilisers contained 20–30% of the total potentially bioavailable phosphorus. Furthermore, the long-term release of phosphorus from the fertilisers was confirmed through incubation studies. Additionally, the fertilisers were observed to contribute to an increase in catalase activity in the soil.

Keywords: fertilisers; organic matter recycling; deinking paper sludge; phosphorus fractionation; waste management

1. Introduction

The global agricultural sector is confronted with a multitude of challenges that impinge upon its capacity to ensure food security for an expanding population [1]. On the one hand, there is the adaptation of crops to changing vegetation zones and extreme weather conditions [2], disrupted supply chains due to armed conflicts or depleted raw material resources [3]. On the other hand, the degradation of productive functions and ecosystems due to the loss of biodiversity [4], the impact of fertilised crops on environmental degradation and the adaptation of agriculture to the requirements of the European Union (EU) strategy [5]. For example, the EU agricultural strategy has the objective of reducing nutrient losses by a minimum of 50% by 2030. This is of particular significance in the context of soil degradation and the eutrophication of water bodies. The strategy encourages the utilisation of organic fertilisers, precise fertilisation techniques and crop rotation. Moreover, the EU strategy advocates the recovery and reuse of nutrients from organic waste streams, thereby facilitating the closure of nutrient cycles. A significant focus is placed on the enhancement of organic matter content within the soil, which serves as a reservoir for nutrients [6].

In addition to crop rotation, mulching, conservation tillage [7], and maintaining plant cover, the use of organic and organo-mineral fertilisers represents a further key practice [8].

These fertilisers enrich the soil with organic matter, thereby improving its structure, water retention capacity and aeration. They also support the development of soil microorganisms and provide the necessary nutrients in a timely manner [5].

The implementation of the principles of the circular economy has led to an increase in the use of biomass as a source of organic matter in fertilisers [9,10]. This biomass is produced from municipal and industrial waste materials, including plant residues, manure, lignocellulosic biomass, paper waste as well as compost and digestate [11]. The mineral component of these fertilisers is ash produced after incineration of biomass like sewage sludge, chicken manure, or wood [12,13]. In light of the EU strategies and the considerable quantity of lignocellulosic biomass produced (exceeding 18 billion tons per year) as well as waste generated from the paper industry and recycling (11 million tons in the EU, of which 7.7 million are from recycling), it can be seen that its intrinsic properties, including the capacity to enhance soil structure, mitigate erosion, and stimulate microbial activity, render it a promising renewable organic raw material with significant fertilising potential following appropriate processing [14–17]. These processes include composting [18,19], fermentation [15], and enzymatic or chemical hydrolysis [20–22].

The utilisation of waste materials derived from the paper industry as components of substrates or soil improvers has been the subject of extensive analysis [23,24]. The waste in question is characterised by a high cellulose content and a high carbon-to-nitrogen ratio, in addition to a high biological oxygen demand BOD, which causes short-term nitrogen sequestration in the soil [24]. Following the processes of composting or hydrolysis, this waste has the potential to serve as an organic component of fertilisers [25]. The findings of recent research [14,19,26] indicate its use as a means of supplementing compost with organic matter and minerals, thereby representing a potential component of gardening substrates. Zawadzińska et al. [14] corroborated the assertion that a compost of superior quality, formulated from pulp and paper mill sludge, fruit and vegetable waste, processed mushroom substrate and rye straw, has a beneficial impact on the yield and quality of cherry tomato fruit (*Lycopersicon esculentum* Mill). The impact of paper waste and peat mixtures on the yield of marigold (*Calendula officinalis* L.), petunia (*Petunia* × *hybrid* L.) and Matthiola (*Matthiola incana* L.) plants was investigated by [26]. It was demonstrated that up to 30% of paper waste can be substituted for peat in the cultivation of marigold and petunia plants, resulting in an increased number of buds but a concomitant reduction in chlorophyll content and a decrease in both height and diameter of the plants.

A further problematic by-product of the recycling of paper is the residue resulting from the deinking process. The deinking stage, which occurs subsequent to repulping, serves to augment the strength and visual appeal of the final product. Flotation is the most common method employed for this process [23]. These wastes are distinguished by a high concentration of carbon and nutrients, which renders them suitable for utilisation in organic plant fertilisation and mulching [27–29].

A review of the literature reveals a dearth of information concerning the utilisation of this specific type of waste in the production of organo-mineral fertilisers. The objective of the presented research is to investigate the potential use of deinking paper sludge, following hydrolysis, as a source of organic matter in the composition of organo-mineral fertilisers based on other waste materials, specifically ashes from the incineration of poultry litter. The resulting fertilisers were evaluated in accordance with the relevant regulatory framework and their efficacy was assessed through fractionation tests and a six-week incubation period in soil. The potential for utilising valuable components from waste materials to support both organic and mineral fertilisation represents a significant advancement in the implementation of circular economy principles.

2. Materials and Methods

2.1. Materials

2.1.1. Waste Paper Pulp

Deinking paper sludge was obtained from a local company (Małopolska region, Poland) specialising in the recycling of paper products. The material is generated as a by-product of the flotation stage. The characterisation of the material is presented in Table 1.

Table 1. Characterisation of deinking paper sludge (values expressed on dry weight basis, except moisture content).

	pH	8.75 ± 0.05
	Moisture content (%)	65.1 ± 0.5
	Organic matter content (%)	68.2 ± 0.005
Main nutrients	K_2O (%)	0.15 ± 0.01
	P_2O_5 (%)	0.01 ± 0.002
Secondary nutrients	Ca (%)	1.22 ± 0.02
	Mg (%)	0.25 ± 0.01
Micronutrients	Zn (mg/kg)	226 ± 7
	Fe (mg/kg)	3724 ± 26
	Cu (mg/kg)	178 ± 10
	Mn (mg/kg)	338 ± 4
	Co (mg/kg)	22 ± 1
Heavy metals	Cd (mg/kg)	5.0 ± 0.4
	Pb (mg/kg)	<18
	Cr (mg/kg)	42 ± 1
	Ni (mg/kg)	56 ± 8

The high moisture content of deinking paper sludge makes impossible its thermal utilisation. We analysed waste rich in organic matter (68.2%), which is beneficial for soil. In addition, it contains secondary nutrients and micronutrients. The concentration of heavy metals in deinking paper sludge was found to be higher than that reported in the literature [27,30], which is primarily attributable to the source of the paper. The agricultural utilisation of such sludge can pose some environmental hazards. Heavy metals are characterised by their persistence in the environment and their capacity to be absorbed by soil particles and taken up by plants, which can result in a reduction in crop yield and the contamination of food items. Furthermore, heavy metals have the ability to cause a decline in microbial diversity, penetrate into aquatic systems and subsequently lead to adverse effects on ecosystems and human health [31,32].

2.1.2. Poultry Litter Ash

The poultry litter ash used in the study was derived from an industrial incinerator. The ash was found to contain 24.7% K_2O and 18.7% P_2O_5 of potassium and phosphorus, respectively, which makes it a promising alternative raw material for the production of fertilisers. Additionally, poultry litter ash is a rich source of secondary nutrients and micronutrients, which are essential for the optimal growth and development of plants. Given the alkaline pH and relatively high content of Cr and Ni, the direct use of ash as a fertiliser is not recommended (Table 2). The availability of heavy metals from poultry litter ash is contingent upon the prevailing soil conditions, particularly the pH value. In acidic conditions, the heaviest metals become more soluble, thereby increasing their potential for leaching. An elevated concentration of H^+ ions reduces the absorption of metal ions in the soil, thereby releasing them into the soil solution. The potential leaching of heavy metals from the soil is also significantly influenced by organic matter and water content, as well as soil structure [33,34].

Table 2. Characterisation of poultry litter ash (values are expressed on dry mass).

pH		12.0 ± 0.2
Main nutrietns	K_2O (%)	24.7 ± 0.8
	P_2O_5 total (%)	18.7 ± 0.3
	P_2O_5 water soluble (%)	0.07 ± 0.01
Secondary nutrients	Ca (%)	9.68 ± 0.47
	Mg (%)	6.27 ± 0.03
Micronutrients	Zn (mg/kg)	2288 ± 29
	Fe (mg/kg)	782 ± 1
	Cu (mg/kg)	694 ± 19
	Mn (mg/kg)	28 ± 1
	Co (mg/kg)	76 ± 5
Heavy metals	Cd (mg/kg)	5.3 ± 0.1
	Pb (mg/kg)	19.1 ± 0.4
	Cr (mg/kg)	175 ± 3
	Ni (mg/kg)	110 ± 8

2.1.3. Soil

The fertilisers were incubated using soil that had been previously subjected to a drying process at 105 °C and subsequently ground. The detailed characteristics of the soil are presented in Table 3.

Table 3. Characterisation of soil.

pH	pH_{H_2O}	5.26 ± 0.02
	pH_{KCl}	4.40 ± 0.05
Organic matter content (%)		4.94 ± 0.01
Main nutrients	K_2O (%)	0.014 ± 0.002
	P_2O_5 total (%)	0.118 ± 0.005
	P_2O_5 water soluble (%)	0.010 ± 0.001
Secondary nutrients	Ca (mg/kg)	405 ± 20
	Mg (mg/kg)	392 ± 14
Microelements	Zn (mg/kg)	77 ± 3
	Fe (mg/kg)	9203 ± 41
	Cu (mg/kg)	10 ± 1
Heavy metals	Cd (mg/kg)	2.1 ± 0.3
	Pb (mg/kg)	<12
	Cr (mg/kg)	39 ± 3
	Ni (mg/kg)	13 ± 1
Bulk density (g/cm^3)		0.943 ± 0.032

The soil was distinguished by a relatively low concentration of essential nutrients. It exhibited notable quantities of iron (9203 mg/kg). Based on the determined pH, it can be concluded that the analysed soil belongs to the category of acidic soils. The soil pH value has a considerable influence on numerous factors that determine its physicochemical properties, including the availability of nutrients. The concentration of heavy metals in the soil was found to be at a relatively low level.

2.2. Methods

2.2.1. Fertiliser Manufacture Process

Acid hydrolysis was carried out to partially decompose deinking paper sludge and obtain a consistency that would allow the production of a homogeneous fertiliser product. Table 4 shows the conditions for acid hydrolysis of waste. Concentrated sulphuric acid, 75% sulphuric acid solution, and a mixture of concentrated sulphuric acid and phosphoric acid in a volume ratio of 1:1 were tested as hydrolysis agents. In addition, different ratios of liquid (LF) to solid phase (SF) were tested.

Table 4. Deinking paper sludge hydrolysis conditions.

Lp	LF:SF	Hydrolysis Agent	Neutralisation Agent		pH Value	Organic Matter Content (%)
			KOH Solution	Poultry Litter Ash		
1	1:1	H_2SO_4 (98%)	–	+	1–2	6.64
2	1:1		+	+	6–7	5.80
3	2:1		–	+	1–2	12.1
4	2:1		+	+	5–6	9.80
5	3:1		–	+	1–2	16.5
6	3:1		+	+	6–7	16.7
7	1:1	H_2SO_4 (75%)	+	+	6–7	7.43
8	2:1		+	+	6–7	14.0
9	3:1		+	+	6–7	21.1
10	1:1	H_2SO_4 (98%) + H_3PO_4 (75%)	+	+	6–7	7.30
11	2:1		+	+	6–7	14.7
12	3:1		+	+	6–7	21.7

When deinking paper sludge was hydrolysed with concentrated sulphuric acid, an attempt was made to neutralise the resulting pulp using only alkaline poultry litter ash. However, the amount of ash added was too low to neutralise an acidity of the pulp to the required level due to its thick consistency, so a 40% potassium hydroxide solution had to be used for this purpose. The addition of the KOH solution increased the content of the main component of the fertiliser, potassium, and also made it possible to obtain a pH of the mixes at the level of 6–7, which is beneficial for the soil.

The selection of fertiliser mixtures to the next stages of research was based on their organic matter content and pH. According to the requirements [35], the minimum organic matter content in organo-mineral fertilisers should be 20%. In the case of mixtures 9 and 12, the organic matter content was higher than the required value, allowing them to be included in this group. These mixtures also met the pH requirements, which allowed them to be selected as potential fertiliser products.

In the next stage, fertiliser mixtures 9 and 12 were produced on a larger scale (500 g). The diagram of the production process is shown in Figure 1. Deinking paper sludge was hydrolysed with 75% sulphuric acid or a mixture of concentrated sulphuric and phosphoric acid (1:1) in the ratio SF:LF 3:1, neutralised with 40% potassium hydroxide solution and enriched with poultry litter ash until the mixture reached pH 6–7. The resulting fertiliser mixture was then pelletised and dried at 105 °C.

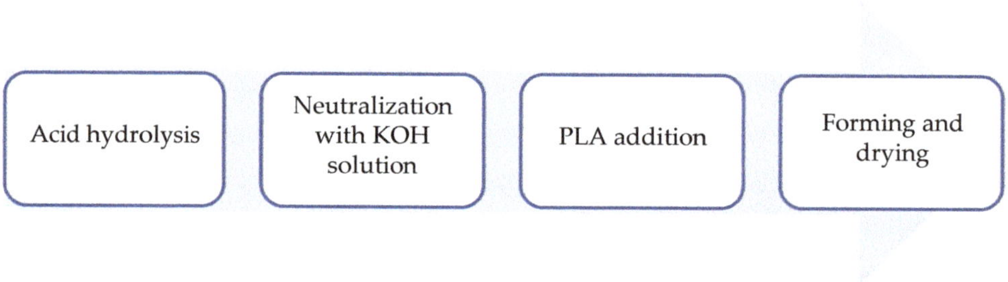

Figure 1. Diagram of fertiliser production process.

Figure 2 shows the composition of the fertilisers produced. In the case of both fertilisers, the mass of deinking paper sludge was about 30% of the total mass of the fertiliser. The addition of PLA, which constituted 11–13% of the total mass of the fertiliser, made the content of alternative raw materials in the fertilisers 41–42%, depending on the fertiliser.

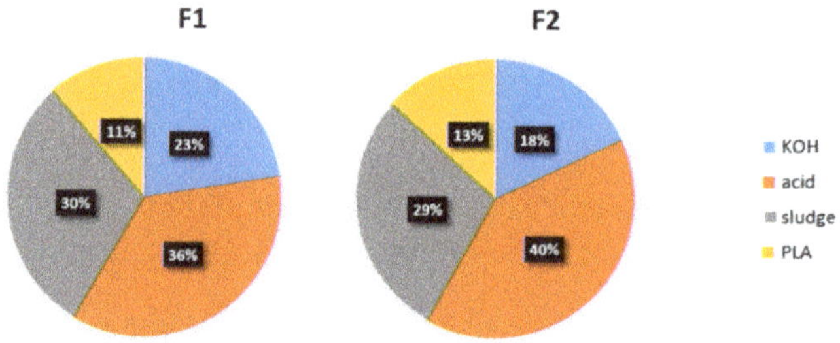

Figure 2. Composition of fertilisers.

2.2.2. Elemental Analysis

The raw materials (deinking paper sludge, poultry litter ash), fertilisers and soil were subjected to analysis with the objective of determining the concentration of selected elements. The atomic absorption spectroscopy (AAS) (AAnalyst 300, PerkinElmer, Waltham, MA, USA) was employed to determine the content of calcium (Ca), magnesium (Mg), potassium (K), iron (Fe), zinc (Zn), copper (Cu), manganese (Mn), cobalt (Co), nickel (Ni), chromium (Cr), lead (Pb), and cadmium (Cd) after digestion in aqua regia [36].

2.2.3. Phosphorus Determination

Two colorimetric methods were used to determine the phosphorus content in the tested samples. The molybdenum blue method was employed for the analysis of water-soluble and available phosphorus content in soil samples after previous extraction in water and Bray's solution the answers [37]. The vanadate-molybdate method according to [38] was used for the determination of phosphorus content in poultry litter ash, soil, and fertilisers.

2.2.4. Organic Matter Determination

The organic matter content of deinking paper sludge and fertilisers was determined by subjecting the samples to a three-hour roast at 550 °C in a muffle furnace, after which the mass loss on ignition was calculated based on the mass difference [39].

2.2.5. pH Determination

In order to determine the pH of the soil, both a solution in water and a 1 M KCl solution were employed [40]. To determine the pH of the fertilisers and soil during incubation, 10% aqueous solutions were prepared. Following a period of two hours, the pH value of the supernatant solution was determined using a CX-701 metre (Elmetron, Zabrze, Poland).

2.2.6. Phosphorus Fractionation

Phosphorus fractionation in fertilisers and poultry litter ash was conducted by adding 50 mL of the extraction solution to the analysed material (weighing approximately 2 g) and agitating for a specified time period, as outlined in Table 5. Subsequently, the samples were subjected to centrifugation for 20 min at 6000 rpm. After separation, the liquid phase was transferred to a separate container and allowed to stand for spectrophotometric molybdenum blue analysis. The sediment was then rinsed with distilled water and covered with a further extraction solution. The sediments remaining after fraction IV were subjected to mineralisation in aqua regia in order to determine the residual phosphorus content [41].

Table 5. Scheme of phosphorus fractionation.

Fractions	Extraction Solution	Time	Phosphorus Occurence Forms
I	1 M NH_4Cl	1 h	Potentially available p
II	0.11 M $Na_2S_2O_4/NaHCO_3$	1 h	Fe-bounded
III	1 M NaOH	18 h	Al-bounded
IV	0.5 M HCl	18 h	Ca-bounded
V	Aqua regia		Residual

2.3. Soil Incubation

In order to ascertain the impact of fertilisers on soil properties, as well as on the content of nutrients and the forms in which they occur, an eight-week incubation period was conducted, during which fertilisers were introduced into the soil. For this purpose, 50 g of previously dried and ground soil (with a particle size of less than 0.2 mm) was weighed into 100 mL of plastic containers. Subsequently, 1 g of fertiliser was weighed and an amount of water equivalent to the maximum sorption capacity of the soil at 70% was added. Additionally, control samples of soil devoid of fertiliser were prepared. The containers were hermetically sealed and placed in an incubator maintained at 25 °C. Subsequently, after each of the eight weeks, the following parameters were subjected to analysis:

The following parameters were analysed at regular intervals:

- pH
- water-soluble phosphorus content
- bioavailable phosphorus content (Bray test)
- organic carbon content
- catalase activity

2.4. Catalase Activity

In order to ascertain the activity of catalase in soil, the manganometric method was employed [42]. The method entails titrating the residual hydrogen peroxide present in the sample, which has not been decomposed by the catalase present in the sample. The titrant employed is potassium permanganate in an acidic environment. In order to perform the determination, the soil sample was incubated for 20 min with 5 mL of a 0.3% hydrogen peroxide solution. Subsequently, 5 mL of 1.5 M sulfuric acid was added to the sample and filtered into volumetric flasks. The determination of catalase activity entails titrating the filtrates with 0.01 M potassium permanganate until a pale pink colour is observed.

2.5. Organic Carbon Content

The organic carbon content was determined by employing the modified Tiurin method [43]. The process entails the oxidation of organic carbon to carbon dioxide in a strongly acidic environment, utilising $K_2Cr_2O_7$ as the oxidant. A solution of 0.1 M Mohr's salt is employed as the reducing agent for the excess oxidant. To perform the determination, approximately 2 g of previously dried and ground soil sample was weighed into an Erlenmeyer flask. Subsequently, 20 mL of a 0.067 M potassium dichromate solution and 15 mL of concentrated sulfuric acid were added. The contents of the flask were heated until boiling point and maintained at this temperature for a period of five minutes. Subsequently, the solution was cooled and transferred to a 200 mL flask, where it was filtered. The organic carbon content was determined spectrophotometrically at a wavelength of 590 nm.

3. Results

3.1. Characterisation of Fertilisers

Table 6 shows the characterisation of the fertilisers obtained. Based on the content of the main nutrients, the fertilisers can be classified as PK-type fertilisers. Phosphorus fertilisation of plants is extremely important, especially at the beginning of their vegetation. It is responsible for proper rooting, plant tillering, increases resistance to disease and also limits the accumulation of harmful nitrates. Phosphorus deficiency results in poor root development, making it difficult for plants to absorb water and other nutrients [44]. The F2 fertiliser was characterised by a phosphorus content five times higher than the F1 fertiliser, due to the use of phosphoric acid in the F2 fertiliser production process. Potassium is a nutrient that influences most of the biochemical and physiological processes that affect the proper growth and metabolism of plants [45]. The potassium content in both fertilisers was similar. The source of potassium in the fertilisers was poultry litter ash and potassium hydroxide solution. For the produced organo-mineral fertilisers F1 and F2, the sum of the main nutrient contents (P and K) was 24.6% and 39.3%, respectively, which is characteristic for mineral fertilisers. The fertilisers met the requirements for organic matter content in accordance with Polish regulations.

Table 6. Characterisation of fertilisers (values are expressed on dry mass).

		F1	F2	Requirements [35]
	Organic matter content (%)	21.0 ± 0.1	23.1 ± 0.1	20
Main nutrients	N (%)	0.22 ± 0.05	0.27 ± 0.1	1
	P_2O_5 (%)	3.12 ± 0.05	15.3 ± 0.02	0.5
	K_2O (%)	21.5 ± 0.6	23.9 ± 0.3	1
Secondary nutrients	Mg (%)	0.69 ± 0.02	0.68 ± 0.02	
	Ca (%)	8.4 ± 0.4	6.1 ± 0.2	
Microelements	Zn (mg/kg)	157 ± 7	164 ± 2	
	Fe (mg/kg)	945 ± 11	1128 ± 10	
	Co (mg/kg)	7 ± 0.2	8 ± 0.3	
	Cu (mg/kg)	252 ± 8	275 ± 6	
	Mn (mg/kg)	627 ± 8	687 ± 12	
Heavy metals	Pb (mg/kg)	<31	<31	140
	Cr (mg/kg)	16 ± 0.2	17 ± 0.4	100
	Ni (mg/kg)	17 ± 1	16 ± 1	60
	Cd (mg/kg)	4 ± 0.4	3 ± 0.2	5

Fertiliser products were characterised by a high content of secondary nutrients such as calcium and magnesium. The presence of magnesium affects the photosynthetic process, nitrogen management in plants and also activates enzymes. Calcium, in turn, supports the functioning of the root system and is also a building block of cell walls [46]. The F1 and F2 fertilisers produced are enriched with microelements (Fe, Zn, Cu, Mn). Based on the characteristics of fertilisers, it can be expected that their addition to soils poor in nutrients has a beneficial effect on the development and proper functioning of plants. The content of heavy metals in fertilisers did not exceed the maximum levels allowed by Polish legislation.

3.2. Phosphorus Fractionation Results

One of the most crucial elements in the context of plant utilisation of nutrients is the form in which they are presented in fertilisers. The fractionation analysis, results of which is presented in Figure 3, enabled the isolation of five fractions of phosphorus and an evaluation of its potential bioavailability for plants.

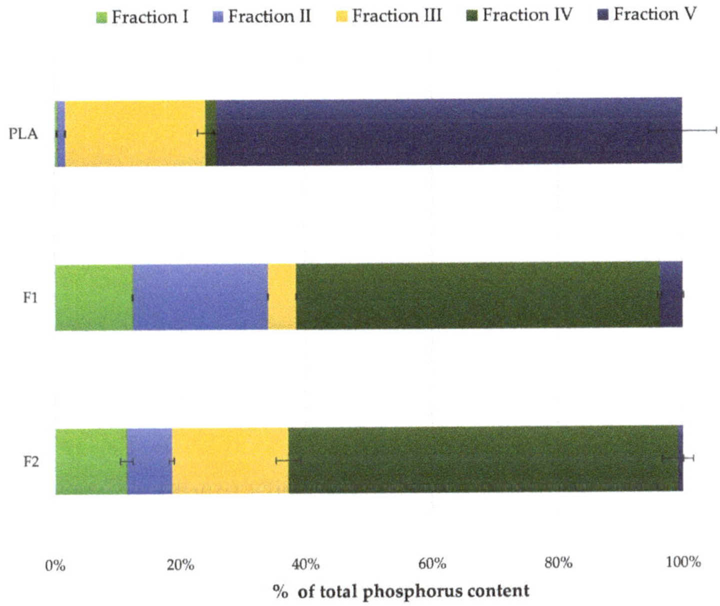

Figure 3. Phosphorus fractionation results.

In the case of poultry litter ash, the phosphorus content of fraction V was found to be 74%. This represents permanently immobilised phosphorus, which is unavailable to plants. It is phosphorus in the form of amorphous calcium phosphate and aluminosilicates associated with potassium, calcium, manganese and iron [47,48]. The ash was found to contain only 0.4% readily available phosphorus, with phosphorus bound to aluminium, iron or calcium representing 26% of the total phosphorus content. The results of the analysis demonstrate that chemical reactions occurring during the production of fertilisers result in a change in the forms of occurrence of the phosphorus present in the ash. The proportion of readily available phosphorus in fertilisers F1 and F2 was found to be 12% and 11% of the total phosphorus content, respectively. The proportion of phosphorus bound to aluminium, iron, and calcium is approximately 85% of the total phosphorus content. It is hypothesised that this form of phosphorus may become available to plants over time, contingent on soil conditions such as moisture content, pH or temperature [41]. The phosphorus fraction designated as V, or residual fraction, constituted 4% of the total phosphorus content in fertiliser F1 and 1% in fertiliser F2.

The outcomes of phosphorus fractionation permit an initial evaluation of the accessibility of phosphorus from fertilisers for plants. However, the behaviour of phosphorus in soil is a highly complex issue, shaped by a number of factors including chemical reactions, interactions with soil components, and the environmental conditions of the soil. The soluble form of phosphorus can be absorbed onto soil particles over time, which consequently reduces its immediate availability to plants. Moreover, phosphorus can undergo transformation into stable mineral forms, including aluminium and iron phosphates in acidic conditions and calcium phosphates in alkaline conditions. The unused phosphorus may leach out, which could have implications for the environment [49].

3.3. Soil Incubation Results

3.3.1. pH Value

The optimal soil pH has a considerable influence on the absorption of phosphorus by plants. Phosphorus is predominantly absorbed in the form of dihydrogen phosphate ions ($H_2PO_4^-$), which are present in the soil solution within a pH range of 5.5 to 7.0 [50]. In both cases, when the pH level is below or above the optimum range for phosphorus uptake, phosphorus can be absorbed on soil particles or precipitate as iron or aluminium phosphates (pH < 5.5) or calcium phosphates (pH > 7.5). As a result, it becomes unavailable to plants. Figure 4 illustrates the fluctuations in soil pH throughout the eight-week incubation study. A similar trend was observed for all samples, with an increase in pH value up to six weeks of incubation and then a slight decrease. In the case of control soil pH value increased from 6.31 to 6.92 and after eight weeks it was 6.49. Due to its slightly acidic pH, the addition of F1 fertiliser lowered the soil pH to 6.18. The maximum pH value of the soil (6.71) was reached in the sixth week of incubation studies.

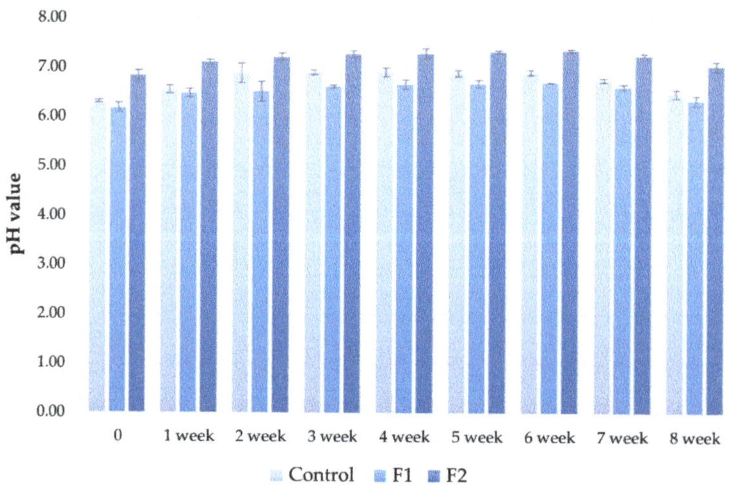

Figure 4. Changes in soil pH values during incubation studies.

In soils with an acidic pH, this fertiliser may contribute to an increase in soil pH, which in turn may lead to an increase in the solubility of phosphates associated with iron or aluminium. The effect of F1 fertiliser on the soil is diametrically opposed. The addition of the fertiliser to the soil resulted in a slight decrease in the pH value throughout the course of the experiment, which increased the solubility of calcium phosphates.

3.3.2. Organic Carbon Content

The presence of organic matter in soil can markedly enhance the accessibility of nutrients for plants, thereby influencing their storage within the soil. It serves as the primary reservoir for nitrogen and sulphur, in addition to iron and copper. Furthermore, it has the capacity to strongly bind toxic heavy metals, rendering them unavailable to plant roots [51]. Figure 5 presents a comparison of the organic carbon content in soil treated with F1 and F2 fertiliser and in the control sample.

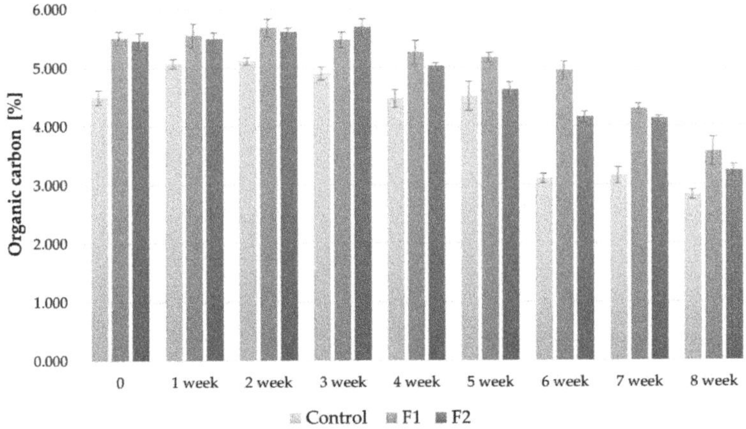

Figure 5. Changes in organic carbon content during incubation studies.

The addition of fertilisers has been demonstrated to result in a notable increase in the organic carbon content of soil, with a mean increase of approximately 20%. The organic carbon content remained at a comparable level for the samples treated with fertilisers throughout the initial three-week period of the experiment, exhibiting a slight increase in the case of the control sample. A decline in organic matter content was observed from week three onwards, which can be attributed to the mineralisation processes occurring in the soil, leading to the decomposition of organic matter. Soil microorganisms play a pivotal role in these processes. They are involved in the decomposition of a wide range of organic compounds, including cellulose, starch, and fats [52]. Their activity is closely correlated with the soil pH. The greatest microbial activity in soil is observed at a pH of 6 to 7. From week four to week eight, a distinct decline in organic carbon content is evident across all soil samples. In the case of the control sample, there was a 37% reduction in organic carbon content. In comparison, the fertilised soil samples exhibited a decrease of 32% and 36% for fertilisers F1 and F2, respectively.

3.3.3. Phosphorus Availability

Water-soluble phosphorus is characterised by immediate availability to plants. Figure 6 illustrates the alterations in water-soluble phosphorus concentration within the soil over the eight-week incubation period. The graph illustrates that the addition of fertilisers to the soil significantly affects the content of water-soluble phosphorus. The addition of fertilisers to soil results in a markedly elevated phosphorus content in comparison to the control samples. Furthermore, soil fertilised with F2 fertiliser contains 5–8 times more water-soluble phosphorus, which is a consequence of the higher phosphorus content in the fertiliser. It was observed that for soil fertilised with F1 fertiliser, the phosphorus content exhibited a slight increase up to week three, after which it remained at a similar level. From week seven, however, a decrease in the content of water-soluble phosphorus was noted. In the case of soil fertilised with F2, the content of this parameter remains constant up to week six, after which a slight decrease is observed. A comparable trend is evident in the phosphorus content of unfertilised soil.

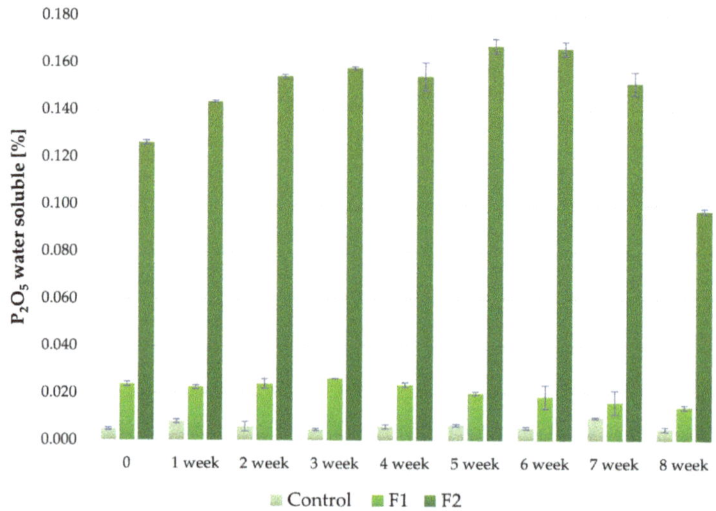

Figure 6. Changes in water-soluble phosphorus content in soil during incubation studies.

Phosphorus is distinguished by its low mobility in soil, which is a consequence of its strong binding to particles in the soil solution. A reduction in moisture content and temperature levels diminishes the rate of phosphorus diffusion towards the roots. The availability of phosphorus is also influenced by soil microbiota, which can compete with plants for phosphorus by producing mycorrhizae in order to enhance the efficiency of phosphorus acquisition [53]. Figure 7 illustrates the temporal dependence of bioavailable phosphorus content. The content of available phosphorus throughout the duration of the experiment remains at a consistent level for both the soil with the addition of fertilisers and the control sample.

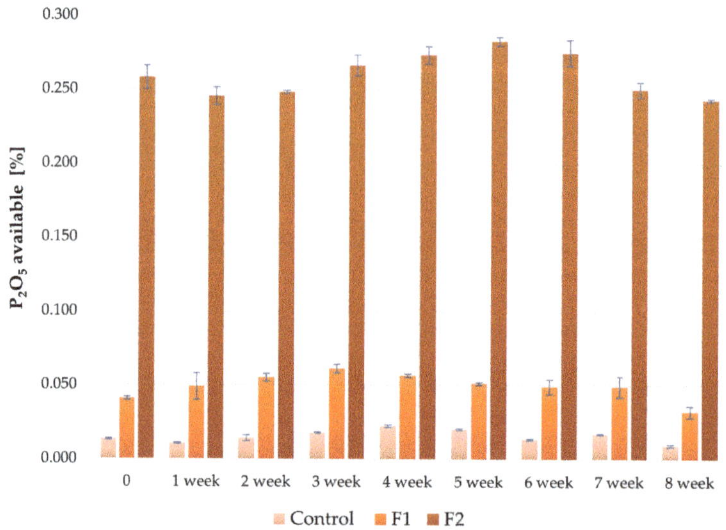

Figure 7. Changes in soil available phosphorus content during incubation studies.

This suggests that the soil underwent gradual, prolonged changes throughout the experiment, resulting in a transformation in the form of phosphorus occurrence. From the fifth week of the experiment, a reduction in the concentration of available phosphorus was observed across all samples. This phenomenon may be attributed to the increasing pH, which has the effect of reversing the chemical equilibrium of phosphorus.

3.3.4. Catalase Activity

Figure 8 illustrates the alterations in catalase activity observed in soil samples treated with fertilisers F1, F2 and the control sample.

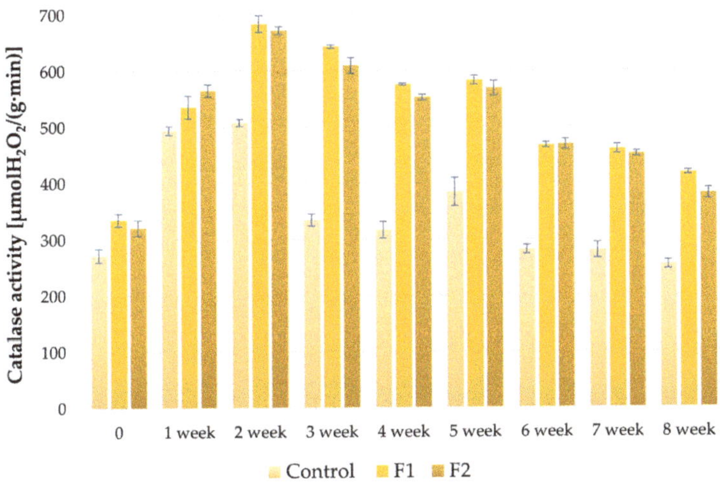

Figure 8. Changes in catalase activity during incubation studies.

During the initial two-week period of the experiment, a notable elevation in catalase activity was discerned, reaching 87% in the control sample, 104% and 109% in the samples fertilised with F1 and F2 fertilisers, respectively. The favourable incubation conditions (in terms of humidity and temperature) and the addition of organic matter in the case of soil with fertilisers stimulated the microbiological population and its activity. A significant correlation was observed between soil catalase activity and the content of organic carbon in the soil, with the former proportional to the latter. As organic carbon is utilised, there is a corresponding decrease in microbial activity, as evidenced by the decline in catalase activity observed in all samples from week two.

4. Discussion

The high organic matter content of deinking paper sludge makes it a potentially valuable resource for agricultural applications. This approach could contribute to the management of this waste and improvement of soil fertility. There are known studies confirming the positive impact of composts containing deinking paper sludge on the physicochemical properties of soil and plant growth (Table 7). Vannucchi et al. [54] demonstrated that a substrate comprising pelletised deinking paper sludge, compost and tephra (pumice 30% + lapillus 60%) is suitable for use in green roofs. The addition of waste reduced the fertility of the substrate, facilitating the coexistence of native stress-tolerant species with sedums and enhancing plant diversity. In additional studies, a substrate composed of mature compost and pellets derived from deinking paper sludge has been demonstrated to be an effective means of promoting the growth of trees in pots within nursery settings, while simultaneously enhancing their resilience to transplanting-related stressors [55].

Table 7. Summary of research results on deinking paper sludge agricultural use.

Type of Waste	Type of Trials, Rate	Results	Country
Compost (deinking paper sludge + manure)	2-years field study, 0, 14, 28 and 42 Mg/ha	• increase in snap bean yield in 17–24% depending on dose of compost • increase in the percentage of P recovered by snap bean in 4.2–15% depending on dose of compost • increase in the percentage of K recovered by snap bean in 39.1–57.5% depending on dose of compost • the concentration of heavy metals in bean tissues was not affected by compost treatment	Canada [56]
Deinking paper sludge	4.5-month greenhouse pot experiment, clay substitution 0, 20, 40, 60, 80% with deinking paper sludge, sand substitution 0, 20, 30% with deinking paper sludge	• decrease in alder and aspen growth with clay substitution more than 20% • increase in shoot dry weight, height and stem diameter od alder with the increase in sand substitution with deinking sludge	Canada [57]
Deinking paper sludge	2-years field study, 0, 30, 60 Mg/ha	• increase in soil pH with the increase in sludge dose • increase in soil organic matter content in 36–73% in comparison with control • increase in content of available P with the increase in sludge dose • increase in microbial biomass carbon up top 500% at dose of 60 Mg/ha • increase in biomass nitrogen in 16–83% depending on dose	Tunisia [58]

The fertilisers obtained provide not only valuable organic matter but also the main nutrients (P and K) and microelements required by plants, thereby increasing their market value. The results of the incubation study showed that fertiliser application resulted in an increase in soil pH. The highest increase in pH of the fertilised soil was observed in week six. It was 8.6% for the soil treated with F1 fertiliser and 7.4% for the soil treated with two fertilisers. This can be attributed to the fact that the fertilisers were rich in calcium and Ca^{2+} ions could replace the H^{+} ions on the clay–humic complex and consequently decrease the soil acidity [58]. Another reason for the increase in pH, which also occurred in the control soil samples, could be the decarboxylation of organic anions during the decomposition of organic matter [59]. However, during the sixth and seventh weeks, the soil pH slightly decreases, which may be related to the buffering properties of the soil.

The use of organo-mineral fertilisers contributed to an increase in soil organic carbon of around 20%. The source of organic carbon in the fertiliser was deinking sludge, which consists mainly of cellulose, hemicellulose, and lignin. These components serve as a carbon source for many different microorganisms and their degradation is an essential part of the global carbon cycle [60]. Their degradation takes place under the influence of enzymes secreted by specific aerobic–anaerobic bacteria and fungi. Temperature, aeration, moisture content and soil pH have a major influence on the decomposition process [61,62]. In general, the degradation process of biopolymers is rather slow. However, the soil samples were incubated under favourable conditions, which positively influences the degradation of organic components in fertilisers by microorganisms, which was observed in the third week of the study.

Microorganisms are considered the mainly source of enzymes of soils, and enzyme activities are strongly associated with the soil microbiological properties [63]. Catalase is an enzyme that is commonly found in animal cells, fungi, aerobic bacteria, and photosynthetic plants. The function of catalase is to protect organisms from the deleterious effects of oxidative stress [64]. Soil organic matter is the primary source of enzyme substrate [65]. Therefore, the introduction of organo-mineral fertilisers into the soil contributed to a significant increase in catalase activity (92% and 82% for F1 and F2 fertiliser compared with control, respectively), which was most visible in the third week of incubation.

The availability of phosphorus is crucial for its availability to plants. Water-soluble phosphorus is considered to be the fraction of phosphorus that is most susceptible to loss. Studies show a positive correlation between the content of soluble phosphorus in water and the content of phosphorus in leachate during leaching study [66]. Mineral fertilisers are usually characterised by a high content of water-soluble phosphorus, only part of which is used by plants and the remainder can be lost, contributing to eutrophication, loss of biodiversity, and environmental degradation in aquatic systems [67,68]. The content of water-soluble phosphorus in the soil during incubation was very low and remained at a similar level from week one to six for the soil treated with F1 and F2 fertilisers, which is beneficial from an environmental point of view. From the sixth week of incubation, the content of water-soluble phosphorus decreased by 36% and 42% for soils with F1 and F2 fertilisers, respectively, compared to the maximum value. This decrease can be attributed to the with various physicochemical processes occurring in the soil. Water-soluble phosphorus can be absorbed on the surface of Fe, Al oxides, kalionite, clay materials, etc. or precipitated [69]. These processes can occur simultaneously and are dependent on pH, temperature, concentration and ratio of individual ions in the soil solution [49]. Increasing pH promotes the precipitation of calcium phosphates, which probably occurred from week six of the soil incubation studies. The same trend was observed for the potential bioavailable phosphorus content, which was extracted with Bray's solution. This solution is designed to extract adsorbed forms of phosphate and is intended for use in soils with a pH < 7.5. From week five of the study, the phosphorus content in the soil began to decrease, reaching a decrease of 48% for the soil with F1 fertiliser and 14% for the soil with F2 fertiliser, with respect to the maximum content in week eight.

The next step in fertiliser research is to determine the effect of fertilisers on plants through application tests. Phytotoxicity tests will allow the effect of the resulting fertiliser products on the early stages of plant growth to be assessed. Pot tests provide information on the long-term effects of fertilisers on plants, but also on the environment.

5. Conclusions

Deinking sludge is produced in large quantities during the paper recycling process and is often difficult to handle. Since it is characterised by a high content of organic matter, mainly consisting of cellulose, hemicellulose, and lignin, it can be used in agriculture as a soil improver. A promising solution is the use of deinking paper sludge for the production of organic–mineral fertilisers, which has the advantage of providing not only organic matter, but also nutrients and microelements. The fertilisers produced in the study were qualified as PK-type fertilisers, containing 24.6% and 39.3% of the total content of the main nutrients, respectively. The content of readily available phosphorus in fertilisers was in the range of 11–12%, and the content of medium-available phosphorus, which are associated with iron, aluminium and calcium, was approximately 85%. The potassium in the fertilisers was in a readily available form, which indicates that the fertilisers are highly efficacious. However, the release dynamics of nutrients are contingent upon the prevailing soil conditions. Fertilisers were enriched with secondary nutrients (Ca, Mg) and microelements (Fe, Zn, Cu) and met the standard requirements for heavy metal content according to Polish legislation. Incubation studies have shown that the addition of fertilisers to the soil contributes to the increase in pH, organic matter content and the content of soil-soluble and potentially bioavailable phosphorus. In addition, fertilisers have a positive effect on the microbiological activity of the soil. To the authors' knowledge, no previous studies have been conducted on the possibility of producing fertilisers based on deinking paper sludge. Accordingly, further studies are required to ascertain the long-term impact of these materials on soil properties and plant growth and development.

Author Contributions: Conceptualization, H.K.; methodology, H.K.; formal analysis, H.K., K.G. and K.Ł.; investigation, H.K. and K.Ł.; resources, Z.W.; data curation, H.K. and K.Ł.; writing—original draft preparation, H.K., K.G. and K.Ł.; writing—review and editing, H.K. and K.G.; visualisation, H.K. and K.Ł.; supervision, Z.W. All authors have read and agreed to the published version of the manuscript.

Funding: This research received no external funding.

Data Availability Statement: The data presented in this study are available within the article.

Conflicts of Interest: The authors declare no conflicts of interest.

References

1. FAO. *The State of Food Security and Nutrition in the World*; FAO: Rome, Italy, 2018; ISBN 9789251305713.
2. Rezaei, E.E.; Webber, H.; Asseng, S.; Boote, K.; Durand, J.L.; Ewert, F.; Martre, P.; MacCarthy, D.S. Climate Change Impacts on Crop Yields. *Nat. Rev. Earth Environ.* **2023**, *4*, 831–846. [CrossRef]
3. El Bilali, H.; Hassen, T. Ben Disrupted Harvests: How Ukraine–Russia War Influences Global Food Systems–A Systematic Review. *Policy Stud.* **2024**, *45*, 310–335. [CrossRef]
4. Telo da Gama, J. The Role of Soils in Sustainability, Climate Change, and Ecosystem Services: Challenges and Opportunities. *Ecologies* **2023**, *4*, 552–567. [CrossRef]
5. Mcdonald, H.; Frelih-Larsen, A.; Lóránt, A.; Duin, L.; Andersen, S.P.; Costa, G.; Bradley, H. *Carbon Farming Making Agriculture Fit for 2030 Policy Department for Economic, Scientific and Quality of Life Policies Directorate-General for Internal Policies*; Coherent Digital: Ghent, Belgium, 2021.
6. Eeropean Commission. *Farm to Fork Strategy. For a Fair, Healthy and Environmentally-Friendly Food System*; Eeropean Commission: Brussels, Belgium, 2020.
7. Mattila, T.J.; Hagelberg, E.; Söderlund, S.; Joona, J. How Farmers Approach Soil Carbon Sequestration? Lessons Learned from 105 Carbon-Farming Plans. *Soil Tillage Res.* **2022**, *215*, 105204. [CrossRef]
8. Li, Z.; Zhang, Q.; Li, Z.; Qiao, Y.; Du, K.; Yue, Z.; Tian, C.; Leng, P.; Cheng, H.; Chen, G.; et al. Responses of Soil Greenhouse Gas Emissions to No-Tillage: A Global Meta-Analysis. *Sustain. Prod. Consum.* **2023**, *36*, 479–492. [CrossRef]
9. Khan, A.; Wichern, F.; Uporova, M.; Kuzyakov, Y. Mineralization and Temperature Sensitivity of Soil Organic Matter Pools of Contrasting Lability. *Eur. J. Soil Sci.* **2024**, *75*, e13451. [CrossRef]
10. Chojnacka, K.; Moustakas, K.; Witek-Krowiak, A. Bio-Based Fertilizers: A Practical Approach towards Circular Economy. *Bioresour. Technol.* **2020**, *295*, 122223. [CrossRef]
11. Kowalczewska, K.; Aardenne, J.; Iversen, P.; Petersen, J. *The European Biomass Puzzle. Challenges, Opportunities and Trade-Offs Around Biomass Production and Use in the EU*; Publications Office of the European Union: Luxemburg, 2023.
12. Kominko, H.; Gorazda, K.; Wzorek, Z. Potentiality of Sewage Sludge-Based Organo-Mineral Fertilizer Production in Poland Considering Nutrient Value, Heavy Metal Content and Phytotoxicity for Rapeseed Crops. *J. Environ. Manag.* **2019**, *248*, 109283. [CrossRef]
13. Kominko, H.; Gorazda, K.; Wzorek, Z. Formulation and Evaluation of Organo-Mineral Fertilizers Based on Sewage Sludge Optimized for Maize and Sunflower Crops. *Waste Manag.* **2021**, *136*, 57–66. [CrossRef]
14. Zawadzińska, A.; Salachna, P.; Nowak, J.S.; Kowalczyk, W.; Piechocki, R.; Łopusiewicz, Ł.; Pietrak, A. Compost Based on Pulp and Paper Mill Sludge, Fruit-Vegetable Waste, Mushroom Spent Substrate and Rye Straw Improves Yield and Nutritional Value of Tomato. *Agronomy* **2022**, *12*, 13. [CrossRef]
15. Izydorczyk, G.; Skrzypczak, D.; Mironiuk, M.; Mikula, K.; Samoraj, M.; Gil, F.; Taf, R.; Moustakas, K.; Chojnacka, K. Lignocellulosic Biomass Fertilizers: Production, Characterization, and Agri-Applications. *Sci. Total Environ.* **2024**, *923*, 171343. [CrossRef] [PubMed]
16. Mujtaba, M.; Fernandes Fraceto, L.; Fazeli, M.; Mukherjee, S.; Savassa, S.M.; Araujo de Medeiros, G.; do Espírito Santo Pereira, A.; Mancini, S.D.; Lipponen, J.; Vilaplana, F. Lignocellulosic Biomass from Agricultural Waste to the Circular Economy: A Review with Focus on Biofuels, Biocomposites and Bioplastics. *J. Clean. Prod.* **2023**, *402*, 136815. [CrossRef]
17. Monte, M.C.; Fuente, E.; Blanco, A.; Negro, C. Waste Management from Pulp and Paper Production in the European Union. *Waste Manag.* **2009**, *29*, 293–308. [CrossRef] [PubMed]
18. Atiwesh, G.; Parrish, C.C.; Banoub, J.; Le, T.A.T. Lignin Degradation by Microorganisms: A Review. *Biotechnol. Prog.* **2022**, *38*, e3226. [CrossRef] [PubMed]
19. Hazarika, J.; Khwairakpam, M. Evaluation of Biodegradation Feasibility through Rotary Drum Composting Recalcitrant Primary Paper Mill Sludge. *Waste Manag.* **2018**, *76*, 275–283. [CrossRef]
20. Liu, Y.; Tang, Y.; Gao, H.; Zhang, W.; Jiang, Y.; Xin, F.; Jiang, M. Challenges and Future Perspectives of Promising Biotechnologies for Lignocellulosic Biorefinery. *Molecules* **2021**, *26*, 5411. [CrossRef]

21. Srivastava, R.K.; Nedungadi, S.V.; Akhtar, N.; Sarangi, P.K.; Subudhi, S.; Shadangi, K.P.; Govarthanan, M. Effective Hydrolysis for Waste Plant Biomass Impacts Sustainable Fuel and Reduced Air Pollution Generation: A Comprehensive Review. *Sci. Total Environ.* **2023**, *859*, 160260. [CrossRef]
22. Zahoor; Wang, W.; Tan, X.; Guo, Y.; Zhang, B.; Chen, X.; Yu, Q.; Zhuang, X.; Yuan, Z. Mild Urea/KOH Pretreatment to Enhance Enzymatic Hydrolysis of Corn Stover with Liquid Waste Recovery for Plant Growth. *J. Clean. Prod.* **2021**, *284*, 125392. [CrossRef]
23. Abushammala, H.; Masood, M.A.; Ghulam, S.T.; Mao, J. On the Conversion of Paper Waste and Rejects into High-Value Materials and Energy. *Sustainability* **2023**, *15*, 6915. [CrossRef]
24. Cazier, E.A.; Pham, T.N.; Cossus, L.; Abla, M.; Ilc, T.; Lawrence, P. Exploring Industrial Lignocellulosic Waste: Sources, Types, and Potential as High-Value Molecules. *Waste Manag.* **2024**, *188*, 11–38. [CrossRef]
25. Wahyuningsih, S. Application of Pulp and Paper Sludge Compost on Anthocephalus Cadamba Seedlings in Ultisol and Peat Media. *IOP Conf. Ser. Earth Environ. Sci.* **2020**, *415*, 012021. [CrossRef]
26. Chrysargyris, A.; Stavrinides, M.; Moustakas, K.; Tzortzakis, N. Utilization of Paper Waste as Growing Media for Potted Ornamental Plants. *Clean Technol. Environ. Policy* **2019**, *21*, 1937–1948. [CrossRef]
27. Bretzel, F.; Tassi, E.L.; Rosellini, I.; Marouani, E.; Khouaja, A.; Koubaa, A. Evaluating the Properties of Deinking Paper Sludge from the Mediterranean Area for Recycling in Local Areas as a Soil Amendment and to Enhance Growth Substrates. *Euro-Mediterr. J. Environ. Integr.* **2024**, 1–12. [CrossRef]
28. Beauchamp, C.J.; Charest, M.H.; Gosselin, A. Examination of Environmental Quality of Raw and Composting De-Inking Paper Sludge. *Chemosphere* **2002**, *46*, 887–895. [CrossRef] [PubMed]
29. Baziramakenga, R.; Simard, R.R.; Lalande, R. Effect of De-Inking Paper Sludge Compost Application on Soil Chemical and Biological Properties. *Can. J. Soil Sci.* **2001**, *81*, 561–575. [CrossRef]
30. Paz-Ferreiro, J.; Plasencia, P.; Gascó, G.; Méndez, A. Biochar from Pyrolysis of Deinking Paper Sludge and Its Use in the Remediation of Zn-Polluted Soils. *L. Degrad. Dev.* **2017**, *28*, 355–360. [CrossRef]
31. Briffa, J.; Sinagra, E.; Blundell, R. Heavy Metal Pollution in the Environment and Their Toxicological Effects on Humans. *Heliyon* **2020**, *6*, e04691. [CrossRef]
32. Mitra, S.; Chakraborty, A.J.; Tareq, A.M.; Emran, T.B.; Nainu, F.; Khusro, A.; Idris, A.M.; Khandaker, M.U.; Osman, H.; Alhumaydhi, F.A.; et al. Impact of Heavy Metals on the Environment and Human Health: Novel Therapeutic Insights to Counter the Toxicity. *J. King Saud Univ.-Sci.* **2022**, *34*, 101865. [CrossRef]
33. Zhong, X.; Chen, Z.; Li, Y.; Ding, K.; Liu, W.; Liu, Y.; Yuan, Y.; Zhang, M.; Baker, A.J.M.; Yang, W.; et al. Factors Influencing Heavy Metal Availability and Risk Assessment of Soils at Typical Metal Mines in Eastern China. *J. Hazard. Mater.* **2020**, *400*, 123289. [CrossRef]
34. Wang, F.; Li, W.; Wang, H.; Hu, Y.; Cheng, H. The Leaching Behavior of Heavy Metal from Contaminated Mining Soil: The Effect of Rainfall Conditions and the Impact on Surrounding Agricultural Lands. *Sci. Total Environ.* **2024**, *914*, 169877. [CrossRef]
35. Fertilizer Regulation. *Regulation of the Minister of Agriculture and Rural Development of 9 August 2024 on the Implementation of Certain Provisions of the Act on Fertilizers and Fertilization*; Ministerstwo Rolnictwa i Rozwoju Wsi: Warszawa, Poland, 2024; pp. 1–9. (In Polish)
36. *CEN EN 13346:2000*; Characterization of Sludge—Determination of Trace Elements and Phosphorus—Aqua Regia Extraction Methods. European Committee for Standardization: Brussels, Belgium, 2000; pp. 1–28.
37. FAO. *Standard Operating Procedure for Soil Available Phosphorus*; FAO: Rome, Italy, 2021; pp. 1–17.
38. *PN-88/C-87015*; Synthetic Fertilizers—Methods of Testing the Content of Phosphates. PKN (Polish Committee for Standardization): Warsaw, Poland, 1988.
39. CEN. *Chemical Analyses—Determination of Loss on Ignition in Sediment, Sludge, Soil, and Waste*; European Committee for Standardization: Brussels, Belgium, 2003; pp. 1–10.
40. FAO. *Standard Operating Procedure for Soil PH Determination*; FAO: Rome, Italy, 2021; pp. 1–23.
41. Lee, M.; Kim, D.J. Identification of Phosphorus Forms in Sewage Sludge Ash during Acid Pre-Treatment for Phosphorus Recovery by Chemical Fractionation and Spectroscopy. *J. Ind. Eng. Chem.* **2017**, *51*, 64–70. [CrossRef]
42. Brzezińska, M.; Włodarczyk, T. Intracellular Redox Enzymes (Oxidoreductases). *Acta Agrophysica, Rozpr. Monogr.* **2005**, *2005*, 11–26. (In Polish)
43. FAO. Standard Operating Procedure for Soil Organic Carbon. *Glob. Soil Lab. Netw.* **2019**, *1*, 1–25.
44. FAO. *Efficiency of Soil and Fertilizer Phosphorus Use*; FAO: Rome, Italy, 2008; ISBN 9789251059296.
45. Wang, M.; Zheng, Q.; Shen, Q.; Guo, S. The Critical Role of Potassium in Plant Stress Response. *Int. J. Mol. Sci.* **2013**, *14*, 7370–7390. [CrossRef] [PubMed]
46. Sigel, H.; Sigel, A. *Metal Ions in Biological Systems: Volume 26: Compendium on Magnesium and Its Role in Biology, Nutrition, and Physiology*; CRC Press: Boca Raton, FL, USA, 1990.
47. Fahimi, A.; Bilo, F.; Assi, A.; Dalipi, R.; Federici, S.; Guedes, A.; Valentim, B.; Olgun, H.; Ye, G.; Bialecka, B.; et al. Poultry Litter Ash Characterisation and Recovery. *Waste Manag.* **2020**, *111*, 10–21. [CrossRef] [PubMed]
48. Pandey, D.S.; Yazhenskikh, E.; Müller, M.; Ziegner, M.; Trubetskaya, A.; Leahy, J.J.; Kwapinska, M. Transformation of Inorganic Matter in Poultry Litter during Fluidised Bed Gasification. *Fuel Process. Technol.* **2021**, *221*, 106918. [CrossRef]
49. Penn, C.J.; Camberato, J.J. A Critical Review on Soil Chemical Processes That Control How Soil Ph Affects Phosphorus Availability to Plants. *Agriculture* **2019**, *9*, 120. [CrossRef]

50. Jančaitienė, K.; Šlinkšienė, R.; Zvirdauskienė, R. Properties of Potassium Dihydrogen Phosphate and Its Effects on Plants and Soil. *Open Agric.* **2023**, *8*, 20220167. [CrossRef]
51. Gerke, J. The Central Role of Soil Organic Matter in Soil Fertility and Carbon Storage. *Soil Syst.* **2022**, *6*, 33. [CrossRef]
52. Raza, T.; Qadir, M.F.; Khan, K.S.; Eash, N.S.; Yousuf, M.; Chatterjee, S.; Manzoor, R.; ur Rehman, S.; Oetting, J.N. Unraveling the Potential of Microbes in Decomposition of Organic Matter and Release of Carbon in the Ecosystem. *J. Environ. Manag.* **2023**, *344*, 118529. [CrossRef]
53. Khan, F.; Siddique, A.B.; Shabala, S.; Zhou, M.; Zhao, C. Phosphorus Plays Key Roles in Regulating Plants' Physiological Responses to Abiotic Stresses. *Plants* **2023**, *12*, 2861. [CrossRef] [PubMed]
54. Vannucchi, F.; Pini, R.; Scatena, M.; Benelli, G.; Canale, A.; Bretzel, F. Deinking Sludge in the Substrate Reduces the Fertility and Enhances the Plant Species Richness of Extensive Green Roofs. *Ecol. Eng.* **2018**, *116*, 87–96. [CrossRef]
55. Vannucchi, F.; Scartazza, A.; Scatena, M.; Rosellini, I.; Tassi, E.; Cinelli, F.; Bretzel, F. De-Inked Paper Sludge and Mature Compost as High-Value Components of Soilless Substrate to Support Tree Growth. *J. Clean. Prod.* **2021**, *290*, 125176. [CrossRef]
56. Baziramakenga, R.; Simard, R.R. Effect on Deinking Paper Sludge Compost on Nutrient Uptake and Yields of Snap Bean and Potatoes Grown in Rotation. *Compost Sci. Util.* **2001**, *9*, 115–126. [CrossRef]
57. Filiatrault, P.; Camiré, C.; Norrie, J.P.; Beauchamp, C.J. Effects of De-Inking Paper Sludge on Growth and Nutritional Status of Alder and Aspen. *Resour. Conserv. Recycl.* **2006**, *48*, 209–226. [CrossRef]
58. Marouani, E.; Benzina, N.K.; Ziadi, N.; Bouslimi, B.; Abida, K.; Tlijani, H.; Koubba, A. CO_2 Emission and Change in the Fertility Parameters of a Calcareous Soil Following Annual Applications of Deinking Paper Sludge (The Case of Tunisia). *Agronomy* **2020**, *10*, 956. [CrossRef]
59. Yan, F.; Schubert, S.; Mengel, K. Soil PH Increase Due to Biological Decarboxylation of Organic Anions. *Soil Biol. Biochem.* **1996**, *28*, 617–624. [CrossRef]
60. Tan, B.; Yin, R.; Yang, W.; Zhang, J.; Xu, Z.; Liu, Y.; He, S.; Zhou, W.; Zhang, L.; Li, H.; et al. Soil Fauna Show Different Degradation Patterns of Lignin and Cellulose along an Elevational Gradient. *Appl. Soil Ecol.* **2020**, *155*, 103673. [CrossRef]
61. Datta, R. Enzymatic Degradation of Cellulose in Soil: A Review. *Heliyon* **2024**, *10*, e24022. [CrossRef]
62. Sharker, B.; Islam, M.A.; Hossain, M.A.A.; Ahmad, I.; Al Mamun, A.; Ghosh, S.; Rahman, A.; Hossain, M.S.; Ashik, M.A.; Hoque, M.R.; et al. Characterization of Lignin and Hemicellulose Degrading Bacteria Isolated from Cow Rumen and Forest Soil: Unveiling a Novel Enzymatic Model for Rice Straw Deconstruction. *Sci. Total Environ.* **2023**, *904*, 166704. [CrossRef]
63. Tian, S.; Xia, Y.; Yu, Z.; Zhou, H.; Wu, S.; Zhang, N.; Yue, X.; Deng, Y.; Xia, Y. Improvement and the Relationship between Chemical Properties and Microbial Communities in Secondary Salinization of Soils Induced by Rotating Vegetables. *Sci. Total Environ.* **2024**, *921*, 171019. [CrossRef] [PubMed]
64. McSweeney, P.L.H. Enzymes Exogenous to Milk in Dairy Technology | Catalase, Glucose Oxidase, Glucose Isomerase and Hexose Oxidase. In *Encyclopedia of Dairy Sciences*, 2nd ed.; Academic Press: Cambridge, MA, USA, 2011; pp. 301–303. ISBN 9780123744029.
65. Zhang, Y.; Cui, D.; Yang, H.; Kasim, N. Differences of Soil Enzyme Activities and Its Influencing Factors under Different Flooding Conditions in Ili Valley, Xinjiang. *PeerJ* **2020**, *8*, e8531. [CrossRef] [PubMed]
66. Herrera, D.; Mylavarapu, R.S.; Harris, W.G.; Colee, J. Soil Phosphorus Sources and Their Relative Water Solubility and Extractability. *Commun. Soil Sci. Plant Anal.* **2022**, *53*, 1445–1455. [CrossRef]
67. Zhang, G.; Xu, T.; Song, J.; Li, Q.; Li, T.; He, B. Characterization of Runoff Phosphorus Loss from a Combination of Long-Term Fertilizer Application and Cultivation on Sloping Croplands. *J. Hydrol. Reg. Stud.* **2024**, *54*, 101907. [CrossRef]
68. Hata, S.; Kobae, Y.; Banba, M. Interactions Between Plants and Arbuscular Mycorrhizal Fungi. In *International Review of Cell and Molecular Biology*; Academic Press: Cambridge, MA, USA, 2010; Volume 281, pp. 1–48.
69. Simonsson, M.; Östlund, A.; Renfjäll, L.; Sigtryggsson, C.; Börjesson, G.; Kätterer, T. Pools and Solubility of Soil Phosphorus as Affected by Liming in Long-Term Agricultural Field Experiments. *Geoderma* **2018**, *315*, 208–219. [CrossRef]

agronomy

MDPI

Article

Geospatial Modelling Predicts Agricultural Microplastic Hotspots from Biosolid Application Risks

Kieran McNulty [1] and Kamrun Nahar [2,*]

[1] School of Engineering, University of Southern Queensland, Toowoomba, QLD 4350, Australia; kieranmcnulty.km@gmail.com
[2] School of Agriculture and Environmental Science, University of Southern Queensland, Toowoomba, QLD 4350, Australia
* Correspondence: kamrun.nahar@unisq.edu.au; Tel.: +61-7-4631-1733

Abstract: Microplastics are emerging as widespread modern pollutants, posing a variety of health and environmental risks. Microplastics are found in agriculture; they are often introduced via biosolids from wastewater treatment plants and are sold as alternatives to inorganic fertilizers. In Australia, there has been limited research on the agricultural concentrations of microplastics, and there has been no predictive modelling to identify which geographies are most at risk for pollution. Based on global emerging trends, this study uses geospatial modelling to map potential high-risk areas for agricultural microplastics within an area of the Murray-Darling Basin in New South Wales, Australia. In doing so, this study demonstrates the use of a geospatial methodology that may be used in future risk assessments, both within Australia and globally. Risk index mapping was conducted for three different pollutant transport pathways: rainfall-runoff of microplastics, in-soil retention of microplastics, and groundwater infiltration of microplastics. Particular areas of risk were identified for each transport pathway, providing visualised mapping results that represent the value of the study and its methodology.

Keywords: GIS; plastic; agriculture; sludge; risk assessment; Australia; pollutant; mapping

1. Introduction

Globally, two megatons (Mt) of plastic were produced in 1950, and by 2015, this figure was closer to 400 Mt, with an annual compound growth rate of roughly 8% between 1950 and 2015 [1]. By the year 2050, it is estimated that 26,000 Mt of plastic waste will have been generated, and that 45% of that amount will be disposed of in landfills or the environment [1]. The majority of plastics are not fully biodegradable; however, they can break down into smaller fragmented microplastics. In general, microplastics are defined as "any solid or polymeric matrix of a regular or irregular shape, ranging from 1 mm to 5 mm in size, of primary or secondary manufacturing origin, that is water insoluble" [2].

In today's environment, microplastics are found in significant concentrations almost everywhere, e.g., in the air, atmosphere, water, the bodies of edible fish, blood, and even the snow on top of Mount Everest [3–7]. Upon exposure via contact, inhalation, or ingestion, microplastics have been shown to cause numerous negative health effects in humans and other animals, including intestinal damage and digestion disruption [8], decreased capacity for immune response systems [9], neurotoxicity (resulting in reduced learning and memory functions) [10], hormonal and endocrine system damage [11], decreased fertility, adverse effects on offspring [12], and oxidative stress in the lungs and airways (resulting in respiratory dysfunction) [13].

In agricultural contexts, microplastics have been shown to reduce soil fertility and soil microbe content [14], reduce plant growth and crop yield [15], increase soil–water evaporation [16], and decrease livestock fertility and transport to human food chains [17]. These agricultural microplastics also exhibit the potential to be transported to downstream aquatic environments [18], or infiltrate into groundwater [19], carrying adsorbed pollutants and heavy metals [20]. Microplastics are mostly introduced into agricultural settings as byproducts of wastewater treatment plant (WWTP) processes called biosolids. The use of biosolids in agriculture is based on their organic content as well as their potential to provide an environmentally friendly alternative to traditional fertilizers, promoting circular reuse as well as reducing greenhouse gas emissions associated with fertilizer production [21].

In Australia, total biosolid production is estimated to be around 349,000 tonnes of dry solids per year [22], with two-thirds of this material being applied to the soil as fertilisers or other soil-applied products [23]. Whilst there are state and federal regulations around biosolid application in Australia, current regulations neither recognise the risk of biosolids transporting microplastics to the agricultural environment nor provide any controls or measures to prevent this pollution.

Studies have shown that approximately 0.5–3% of biosolids contain microplastics [24], with concentrations of up to 287 μg/g of microplastics found in biosolid sludge [25]. According to a recent Australian study, approximately 4700 metric tons of plastics are released annually into the environment via biosolids, 3700 metric tons of which are released directly into agricultural settings [26].

In agricultural fields treated with biosolids in the long term, total microplastic concentrations have been found to be as high as 18,760 microplastics kg^{-1} in Chinese soils; 5190 microplastics kg^{-1} in Spanish soils; and 10,400 microplastics kg^{-1} in Chile [27–29]. Slightly lower concentrations were found elsewhere in Spain (~300 microplastics kg^{-1}) and in England (~900 microplastics kg^{-1}) [30,31]. There are few studies on microplastic field concentrations in Australia, but a recent study on biosolid-amended fields near Toowoomba, Queensland, found an average concentration of 1137 microplastics kg^{-1}, compared to 36 microplastics kg^{-1} at a non-amended reference site [32]. Substantial groundwater concentrations of microplastics were found in a Victorian groundwater bore, attributed to its proximity to agricultural land use, which may have been biosolid-applied land [33].

Utilising geospatial mapping techniques, this study aims to predict at-risk sites for agricultural microplastic contamination within the study area, the Murray-Darling Basin (MDB) agricultural region of New South Wales (NSW), Australia. This study focuses on three key pollutant transport pathways: the transportation of microplastics from agricultural sites via rainfall-runoff, the retention of microplastics in agricultural soils, and the transportation of microplastics into groundwater. To the best of the authors' knowledge, this applied methodology is a first in the field of microplastic research and represents a foundation for further risk assessment studies to build upon, as the literature base expands globally and within Australia.

This study employs Quantum Geographic Information Systems version 3.34.6 (QGIS), which is a widely used, free, and open-source geospatial software that enables the input, processing, analysis, and output of a wide range of map data and information [34].

2. Materials and Methods

The methodology of this study began with a scope definition and the selection of a geographical study area. Following this, the global literature base was reviewed to determine possible risk factors that promote the three key transport pathways of microplastics as an agricultural pollutant: the transportation of microplastics from agricultural sites via rainfall-runoff, the retention of microplastics in agricultural soils, and the transportation of microplastics into groundwater. Input data were then sourced to represent each of these risk factors within the study area, for each transport pathway. The study then combined the effects of these risk factors through geospatial modelling overlay techniques by standardising multivariate data sources for results-mapping production.

2.1. Study Area

The chosen study extent encompassed a portion of the Murray-Darling Basin (MDB) in the Australian state of New South Wales (NSW). NSW was chosen due to the quality and availability of spatial input data from the NSW State Government. The MDB encompasses more than 1 million square kilometres, including drainage catchments of the Murray and Darling Rivers, Australia's first and third longest rivers, respectively. The MDB produces approximately 40% of Australia's produce and is home to about 2.4 million people, including members of over 50 First Nations groups [35]. Additionally, the region is ecologically important, with 35 endangered species, 16 international wetlands, and 120 varieties of waterbirds [35]. Figure 1 illustrates the broad range of land uses within the MDB.

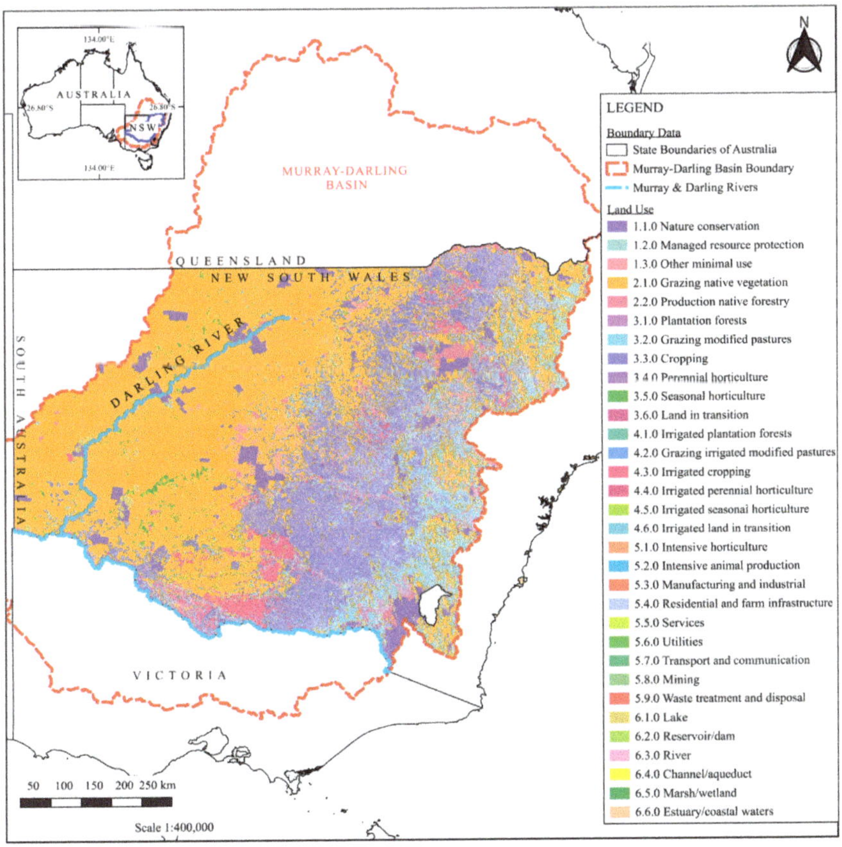

Figure 1. NSW MDB and land use. Mapping produced by using publicly available input data [36–41].

2.2. Land Use

Due to the limited categorisation of which agricultural land types are most applied with biosolids, and the lack of data toward this end in Australia, this study assumed uniform distributions of biosolid application rates across the study area and the selected agricultural land use types. Based on a review of global biosolid usage literature, two land use types were selected for the study: broadacre cropping and horticulture. Data on land use was obtained from the NSW government [40] and an overall study area polygon was drawn based on this land use. More than 90% of the chosen land use types found in the NSW MDB were captured within the study extent polygon and its boundaries were aligned to enclosed local government areas (LGAs). The study contained 50 LGAs, with boundaries and names redacted for reporting purposes. Figure 2 shows the study land use types and the study extent. Study extent polygons were divided into the three zones and aligned to internal LGA boundaries for analysis and presentation.

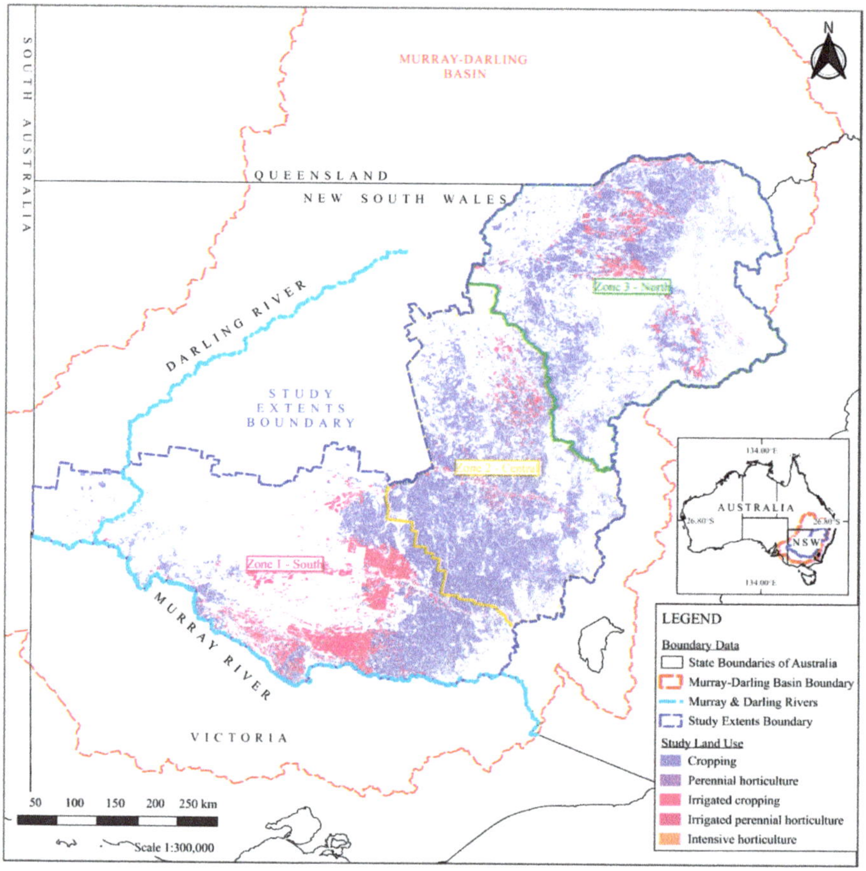

Figure 2. Study area extent and land use data. Mapping produced by using publicly available input data [37–41].

2.3. *Risk Factors for Each Transport Pathway and Representative Input Data for Each Risk Factor*

Potential risk factors for each of the three transport pathways were determined from the current global literature base on agricultural microplastic pollution as described below. Relevant spatial data were then sourced, or generated, for the mapping of each risk factor.

2.3.1. Rainfall-Runoff of Microplastics

In agricultural soils, microplastics are subject to rainfall-runoff cycles that ultimately transport them to downstream freshwater environments such as rivers, creeks, lakes, and further aquatic environments [18]. There are many negative effects of microplastics in aquatic environments: disrupting metabolic processes in corals [42], reducing chlorophyll absorption [43], and transporting other adsorbed pollutants such as perfluoroalkyl and polyfluoroalkyl substances (PFAS) [44]. Several key risk factors have been identified in the currently available literature, which may encourage the runoff of microplastics from applied biosolids.

Clay soils: Clay soils are the least permeable soil profiles, which reduce microplastic retention in soil during rainfall-runoff cycles. Applied biosolids are at risk of increasing runoff from poorly draining clay soils [45,46]. Figure 3 shows a modelled percentage of clay within the upper layers of soils (0–200 cm depth), within the study extent.

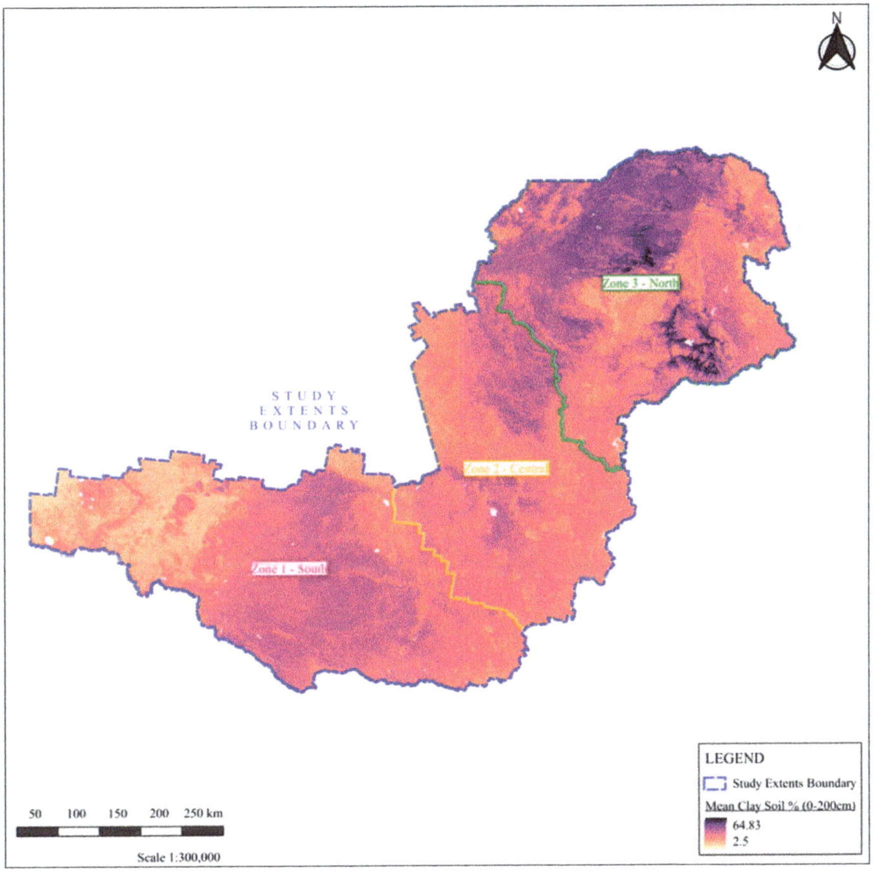

Figure 3. Mean clay soil % (0–200 cm depth). Mapping produced by using publicly available input data [46].

High average annual rainfall: In Ontario, Canada, a region that experiences a relatively high average annual rainfall of up to 1000 mm/year, one study showed that—from a large amount of biosolids applied—greater than 99% of the microplastics contained in the biosolids were unaccounted for just one year after application [47]. This could be explained by a large rainfall event shortly after the biosolid fertiliser was applied, resulting in potential mass runoff and microplastic loss [47]. Similar results were found in Hampshire, England, a study area that receives a similarly high average of 800 mm/year precipitation; researchers found that microplastic concentrations were higher during the summer when rainfall was lower, suggesting runoff may be responsible for microplastic transport [31]. In the Shaanxi province of China, researchers tested different climatic condition effects on agricultural microplastic retention with similar rates of biosolids application and land use. There was a retention of 2490 microplastics kg^{-1} in the arid northern zone versus 1700 microplastics kg^{-1} in the far southern region (which receives 1400 mm of rainfall each year) [48].

Figure 4 shows the average annual rainfall, sourced from publicly available data from the Australian Bureau of Meteorology for the period 1991–2020 [49] across the study extent.

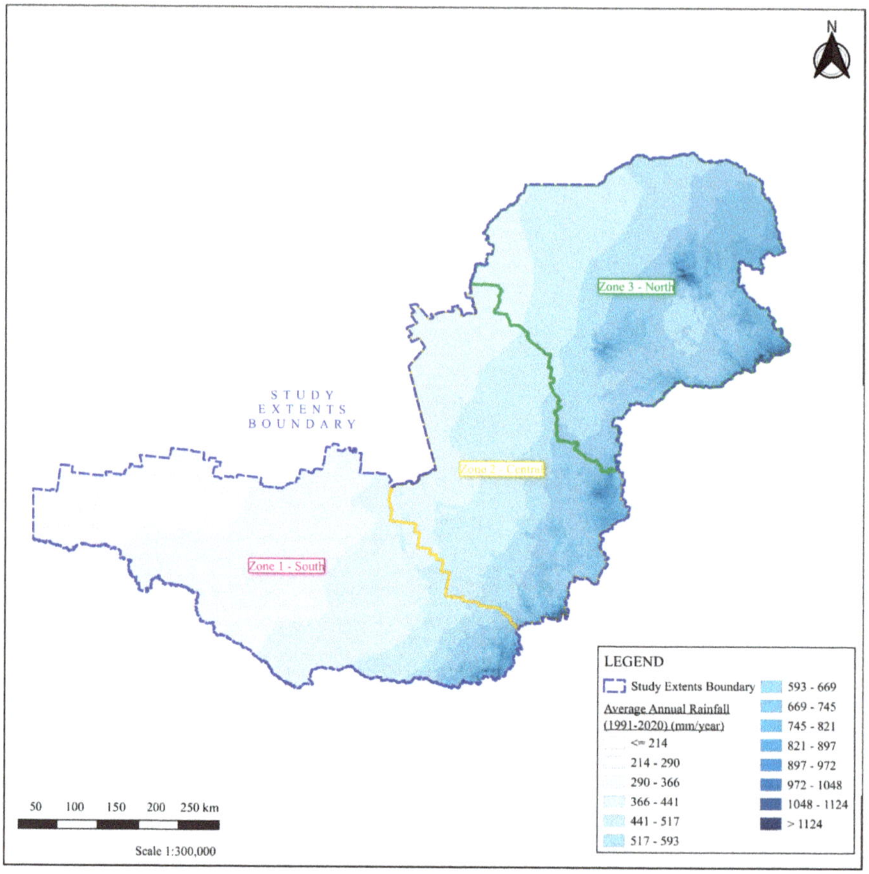

Figure 4. Average annual rainfall (1991–2020). Mapping produced by using publicly available input data [49].

Proximity to nearby waterways: The proximity of an agricultural land use parcel to a waterway plays a key role in ranking the risk of agricultural runoff entering downstream waterways. A publicly available hydrograph was obtained [38] for the entire state extent, in the form of vector line data with an attribute for the watercourse name. As shown in Figure 5a, a centroid for each study land use polygon was created and the QGIS function 'Shortest line between features' was used to find the waterway string closest to each land use polygon. Figure 5b shows the resultant lines generated by this function, with distance measurements appended as attributes to the land use parcels. Once the distance attribute was assigned to each land parcel, a rasterisation average grid was placed over the study land use extent, using the average value (distance to the nearest waterway) for each

100 × 100 m area to map this risk factor across the study extent. The results of this raster average can be seen in Figure 6.

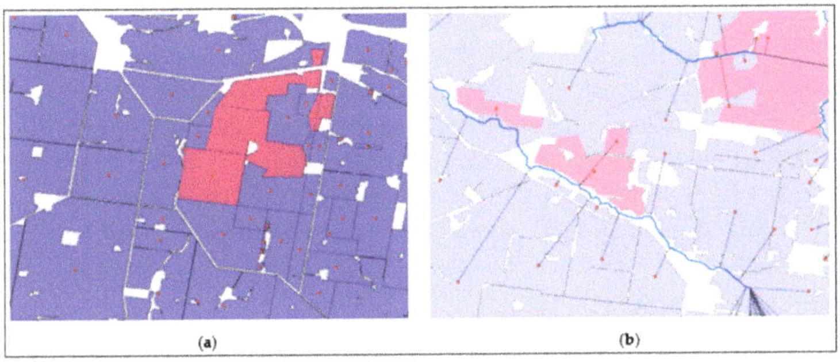

Figure 5. (**a**) Study land use parcels and centroids (red dots) [40]. (**b**) Shortest distance from the study land use centroid to the nearest waterway. Mapping produced by using publicly available input data [38].

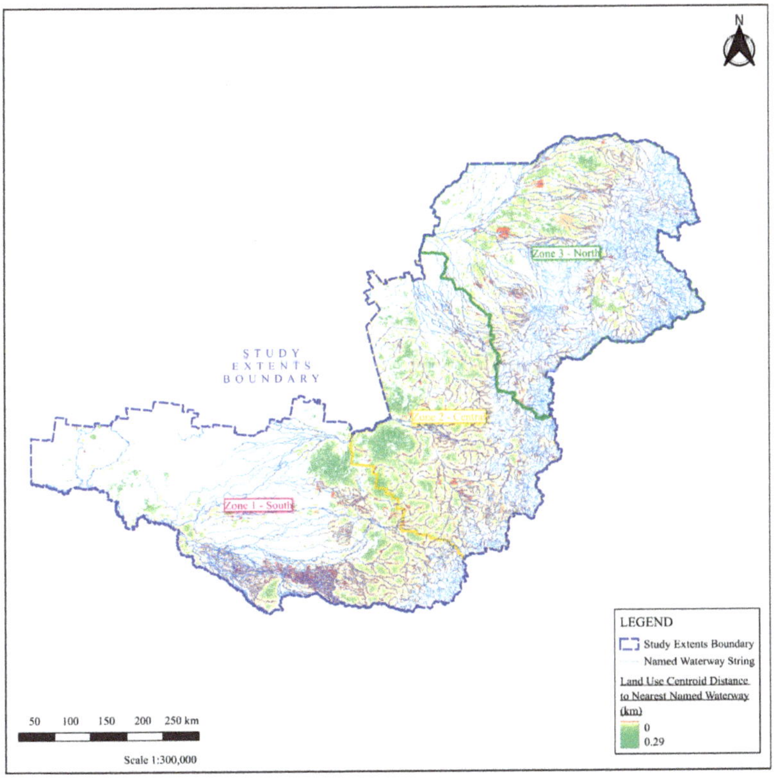

Figure 6. Land use centroid to the nearest named waterway. Mapping produced by using publicly available input data [38,40].

2.3.2. Retention of Microplastics in Soil Profiles

Agricultural soils that retain microplastics are more likely to promote reduced soil fertility and soil microbial community, leading to poor soil quality and nutrient cycling [14], reduced crop yields and plant growth [16], increased soil–water evaporation [15], and decreased livestock fertility and transport to food supply [17]. The following risk factors were identified in this study as contributing to retention:

Aridity: A study conducted in 31 agricultural fields in Melipilla, Chile, found that fields treated with 40 tonnes ha^{-1} of biosolids retained a residual concentration of 2000 microplastics kg^{-1} two years after application [29]. Similar conclusions were reached by another study in a semi-arid Spanish climate [30]. The aforementioned study in China also found that microplastic retention and concentrations were higher in the more arid northern zone, as compared to the other more temperate areas in the study [48]. Using spatial data sourced from the NSW Government [50], the 'aridity index' was used to represent aridity in the study extent. The aridity index dataset used measures the long-term balance between rainfall and net radiation [51]. The data across the study can be seen in Figure 7.

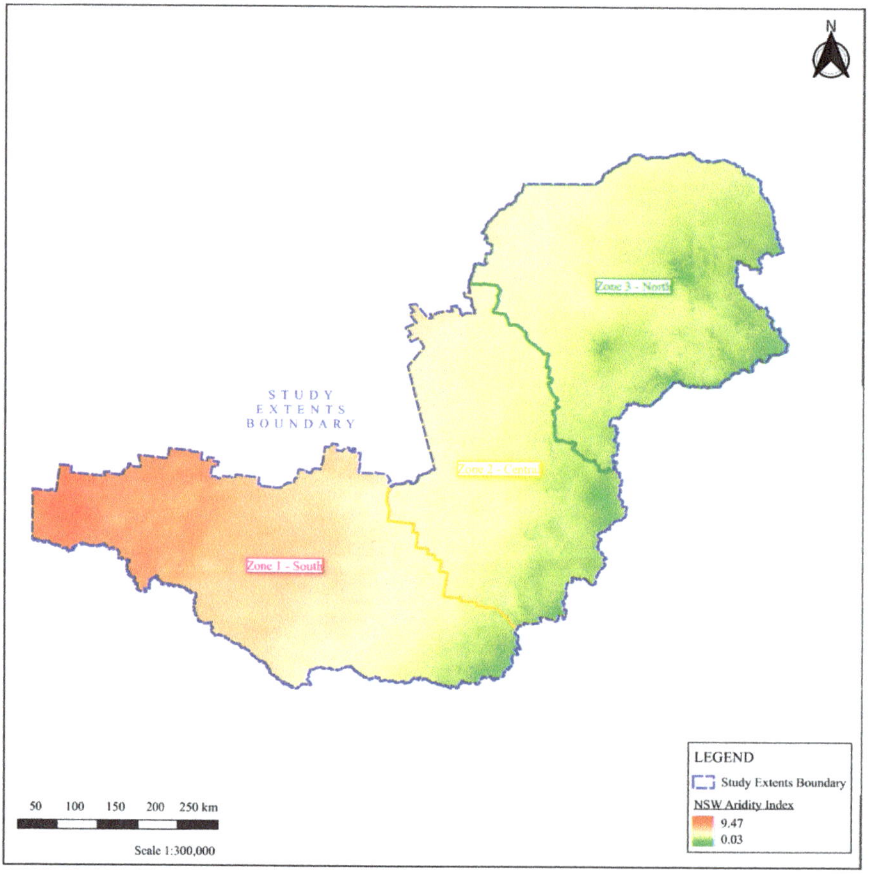

Figure 7. Aridity index. Mapping produced by using publicly available input data [50].

Horticultural land use: The use of horticultural land in agricultural soils appears to be another key variable. In the study previously mentioned [48], researchers found the highest levels of microplastic concentration in fruit-tree fields and the lowest levels in rice fields. In another study focusing on fruit orchard plantation sites amended with biosolids, a high concentration was detectable 15 years after biosolid application [52]. In France, a study covering cropping, vineyards, and fruit orchards found median concentration values of 258 microplastics kg^{-1} with maximum values ranging from 1290 microplastics kg^{-1} in crops to 3096 microplastics kg^{-1} in vineyards and fruit orchards [53].

Figure 8 shows the horticultural land use for the study extent. The study area contains over 89,000 hectares of this land use type, despite appearing sparse in the figure.

High-ionic strength soils: Several laboratory studies have shown that soil–water solutions with high-ionic strengths reduce the vertical transportation potential of microplastics, thereby increasing their retention potential in upper soil horizons [54–56]. As direct soil–water ionic strength measurements are not currently available within the study extent, salinity was used as a substitute measure, often being used to represent the ionic strength of water [57].

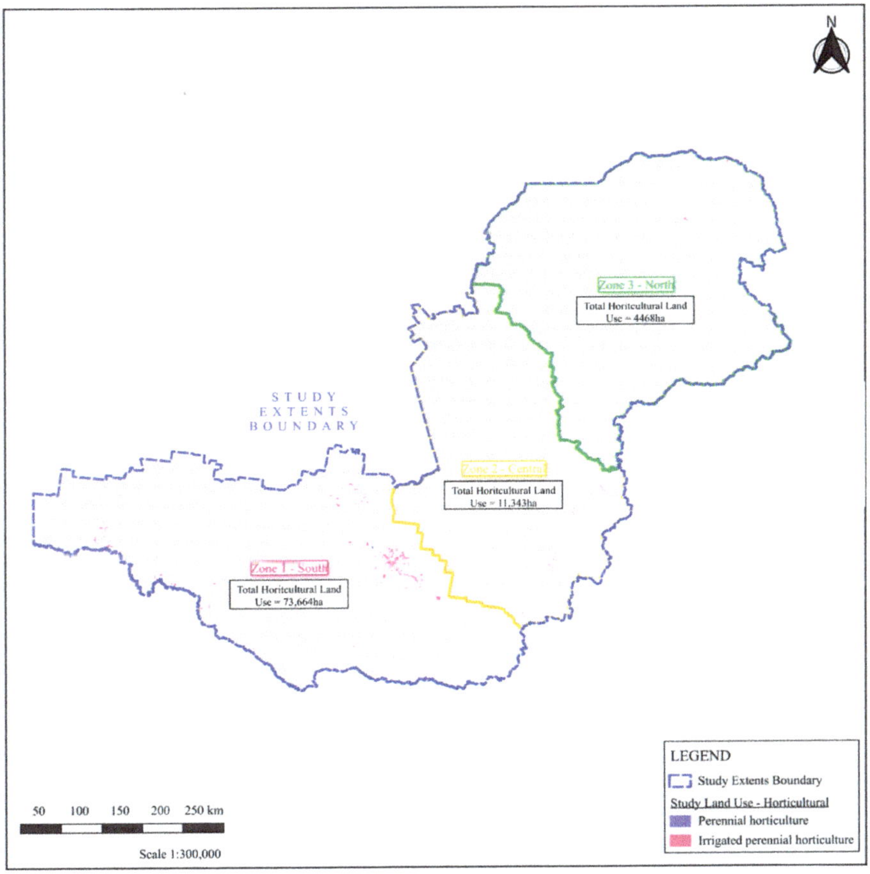

Figure 8. Horticultural land use. Mapping produced by using publicly available input data [58].

Input data were sourced from the NSW government, which provides a state-wide mapping layer for salinity hazard rankings [58]. Data were input into the project QGIS environment as vector polygons and salinity risk attributes were used to represent high- and low-salinity values. As shown in Figure 9, the original data were qualitative and replaced with numerical classes for mapping.

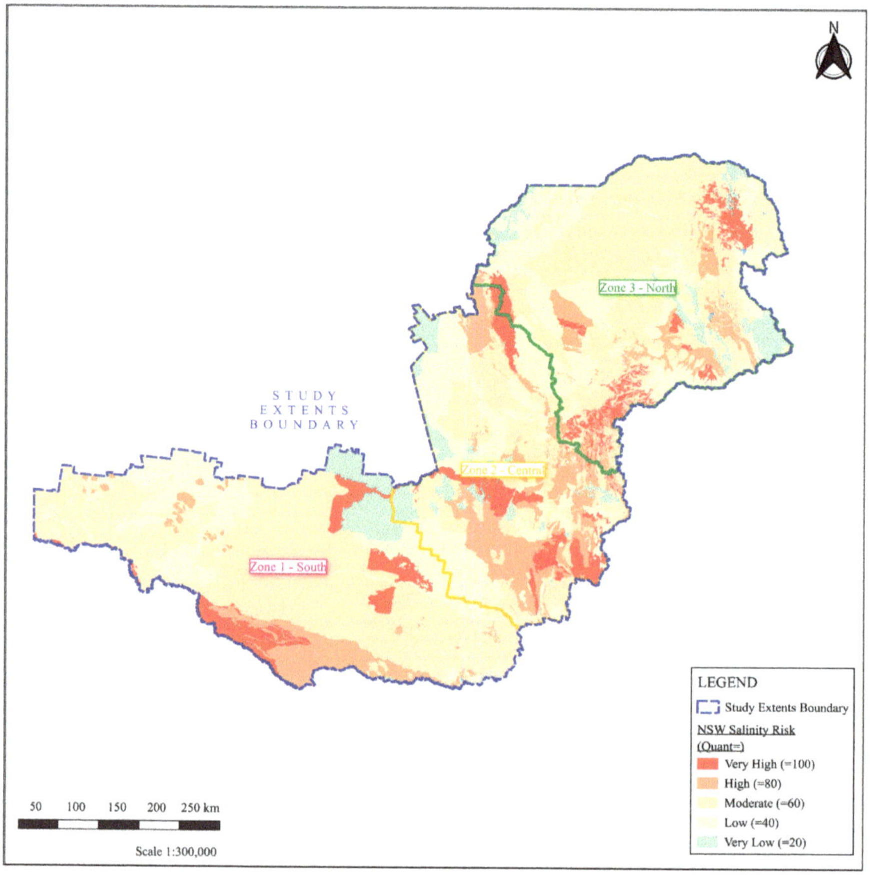

Figure 9. Salinity (used as a surrogate for ionic strength). Mapping produced by using publicly available input data [40].

2.3.3. Groundwater Infiltration of Microplastics

Whilst research is limited, recent studies have detected elevated microplastics in groundwater near agricultural activities that may have been treated with biosolids [19,59]. In a groundwater sample in the Australian state of Victoria, substantially high microplastic concentrations were found, attributed to the proximity to agricultural land that may have been fertilized with biosolid [33]. Microplastics in groundwater may reach bores used for potable water or irrigation. Eventually, they can reach more downstream waterways through groundwater aquifers. Microplastics can also adsorb pollutants such as PFAS [44] and heavy metals like cadmium [20], which can then also be transported through soils to groundwater.

Low groundwater levels: Low-lying groundwater is obviously a concern to any potential groundwater contamination. This has been acknowledged by the NSW Department of Primary Industries, in relation to biosolid amendments [45]. Groundwater mapping data were obtained as publicly available point data from the NSW government [60].

Low-ionic strength soils: As previously mentioned, high-ionic strength soil–water matrices may inhibit the vertical transportation of microplastics. In this study, it was assumed that low-ionic strength soil–water matrices might transport microplastics deeper within their profile, providing a greater risk for groundwater infiltration within low-lying areas.

High sandy soils %/hydraulic conductivity: In 2004, the NSW Department of Primary Industries identified that sandy soils were at a 'severe' risk of transporting biosolid contaminants to groundwater [45]. The risk identified is consistent with the findings of multiple studies. According to one study, microplastics transported through sandy soils of varying sizes could potentially penetrate to depths that, over time, could expose these microplastics to subterranean fauna or aquifer systems [61]. It has also been suggested that environments prone to drought cycles may increase penetration depths linearly with alternating wet/dry cycles, with the study's model predicting microplastic penetration depths of up to 7 m [61]. According to another study, copper content in sandy soils decreased with microplastic accumulation, suggesting potential copper leaching occurred as hydraulic conductivity passed water through the sandy soil [62]. In a sandy-loam soil profile containing high-density PE microplastics ranging from 48 μm to 2.0 mm, Ref. [20] found reduced soil adsorption of cadmium, as well as increased soil desorption.

Figure 10 shows the modelled percentage of sand within the upper soil layers (0–200 cm depth) from the Terrestrial Ecosystem Research Network (TERN) [46].

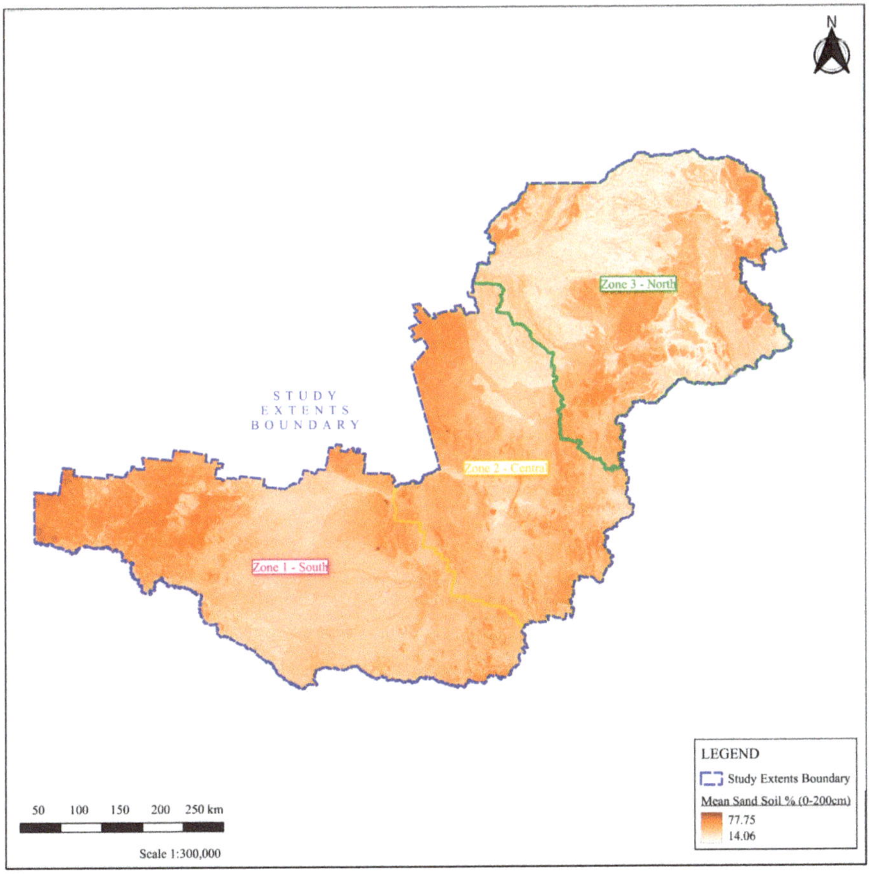

Figure 10. Mean sand soil % (0–200 cm depth). Mapping produced by using publicly available input data [46].

Table S1 in the Supplementary document provides a summary of the three study transport pathways, the risk factors that have been determined to promote these risks, and the spatial data used to represent these factors for the mapping portion of this study.

2.4. Geospatial Mapping Overlay Process for Each Transport Pathway

2.4.1. Data Reclassification into Standardised Quantiles

The different input data sources were multivariate, so QGIS was used for standardisation. First, the raster input data (raster squares of different sizes) were rescaled into 100 × 100 m squares across the study area. Then, each raster was reclassified into 100 quantile distribution classes based on the average values within these squares. Statistical quantiles were ranked according to risk factor, with higher numbers indicating greater risk (for example, the 99th percentile of annual rainfall value in the study extent was given a quantitative raster value of 99). An example of quantile reclassification, displaying the upper and lower quantiles of average annual rainfall, can be seen in Table 1 below.

Table 1. QGIS quantile reclassification example data, average annual rainfall (mm/year and corresponding quantiles 1–100).

Min (mm/Year)	Max (mm/Year)	Quantile
0.00	232.21	1
232.21	250.93	2
250.93	254.83	3
...	...	...
800.08	835.69	98
835.69	886.61	99
886.61	100,000.00	100

It should be noted that although most risk factors can be reclassified into 1–100 quantiles, a different approach was taken for the risk factors of horticultural land use and ionic strength. In the case of horticultural land use, a binary approach was used, with squares containing land use assigned a value of 100 (i.e., contributing most to the risk factors), and squares without land use assigned a value of 0. The input data for ionic strength were available as part of a qualitative set of attributes, ranking salinity risk through five categories: very low, low, moderate, high, and very high. In order to quantify ionic strength, n values (20, 40, 60, 80, 100) were assigned to the high- and low-ionic strength rasters, corresponding to the five original risk categories.

2.4.2. QGIS Raster Calculator to Combine the Effects of Risk Factors for Each Transport Pathway

The raster calculator is a tool within QGIS that enables the algebraic addition of raster squares that have the same geographical location. Figure 11 illustrates the process that occurs during raster calculator addition. To produce a raster map output that is representative of the risk index across the study extent, raster addition was applied to each transport pathway to sum the reclassified risk factor rasters. Each risk factor for each transport pathway was added to the other risk factors in that pathway. As shown in Figure 12, the study risk factors were reclassified into quantile distributions, and the raster calculations were added for each transport pathway. Based on the ascending ranking of all processed quantile risk factor data, a higher sum resulting from the addition process indicates a raster square (100 × 100-meter geographical area) at a higher risk. For each transport pathway, multiple risk factors were identified. This resultant number was normalised by dividing the sum of the raster addition by the number of risk factors for that transport pathway, resulting in an integer between 1 and 100 for any given results raster.

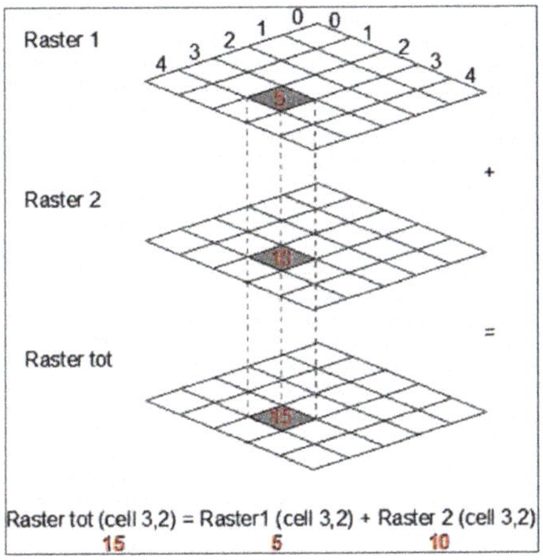

Figure 11. Raster calculator addition process. Image reproduced from [63], under CC BY 4.0 license.

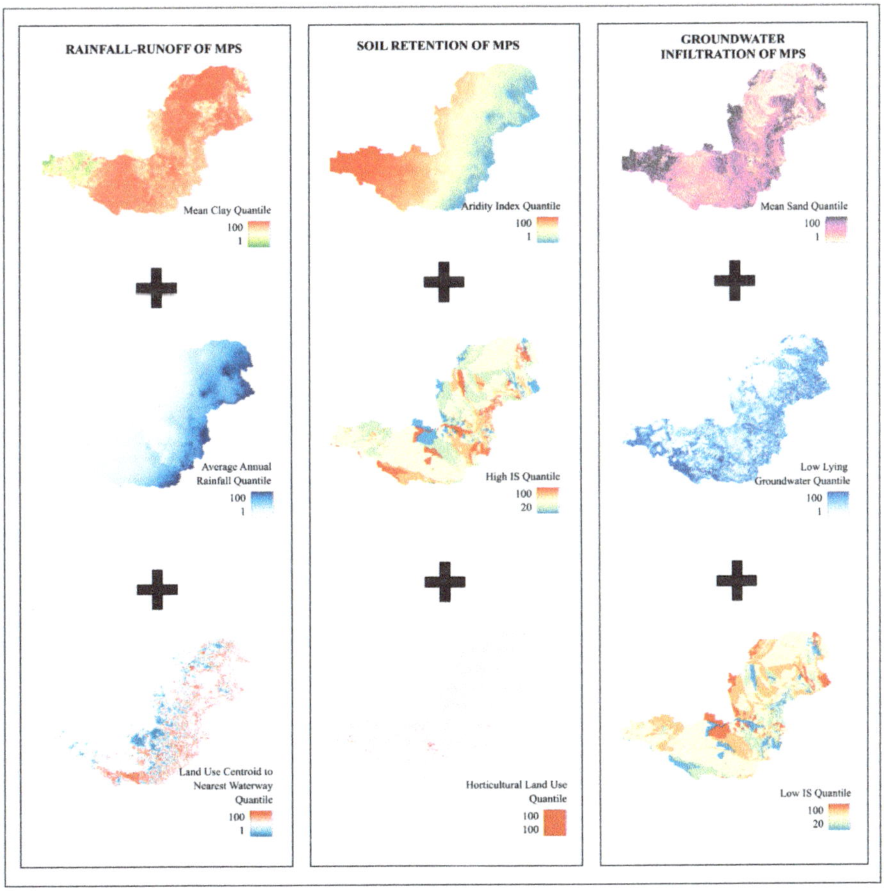

Figure 12. Quantile reclassification and raster addition of risk factors for each transport pathway.

Results rasters were 'clipped' to the extent of the study land use within the study areas. Zonal statistics were generated at both a high level for comparison of the three zones and at a more granular level by way of local government areas within the study area.

QGIS zonal statistics were also used to average the results rasters for each contained LGA polygon. The results for each of the transport pathways were ranked alongside the contributing study land use area within each LGA. The weighted mean ranking was developed by taking into consideration the risk associated with each transport pathway, as well as the contributing land use area of each LGA.

3. Results

3.1. Rainfall-Runoff of Microplastics

The risk index mapped for rainfall-runoff from microplastics across the study extent is shown in Figure 13. A general trend is observed toward the eastern areas of the study being at a higher risk for this transport pathway. The central zone displays the highest level of general risk, as indicated by its mean risk index value of 30.49. A low-risk profile is displayed in the southern zone with a mean value of 15.80; however, there is an isolated high-risk pocket adjacent to the eastern extent of the Murray River within the southern zone. The northern zone, with a mean risk index value of 24.13, displays less concentrated areas of risk than the southern zone.

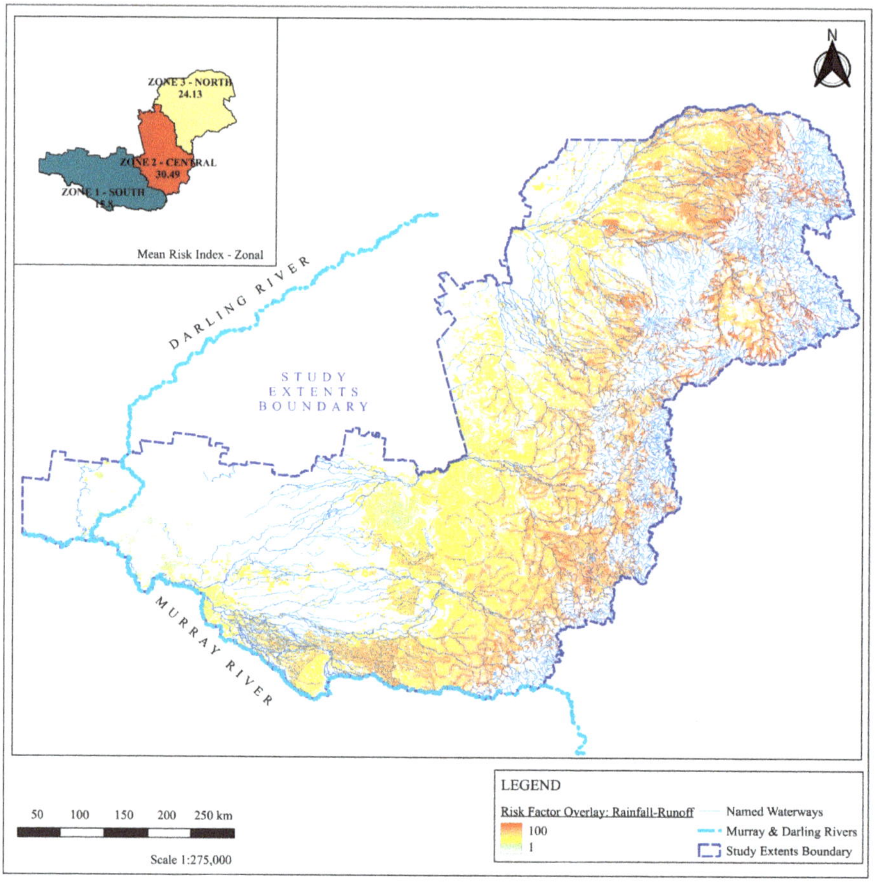

Figure 13. Results—rainfall-runoff of microplastics.

3.2. Retention of Microplastics in Soil Profiles

Based on the results raster presented in Figure 14, areas with the greatest risk of retaining microplastics in their soils can be identified. In general, there is a greater risk along the south boundary of the study area, adjacent to the Murray River, as well as smaller pockets of very high-risk zones where horticultural land use is present. The southern zone is at the highest risk, with a risk index value of 45.04. Both the central and northern zones are at similar levels of risk, with mean values of 33.64 and 29.61.

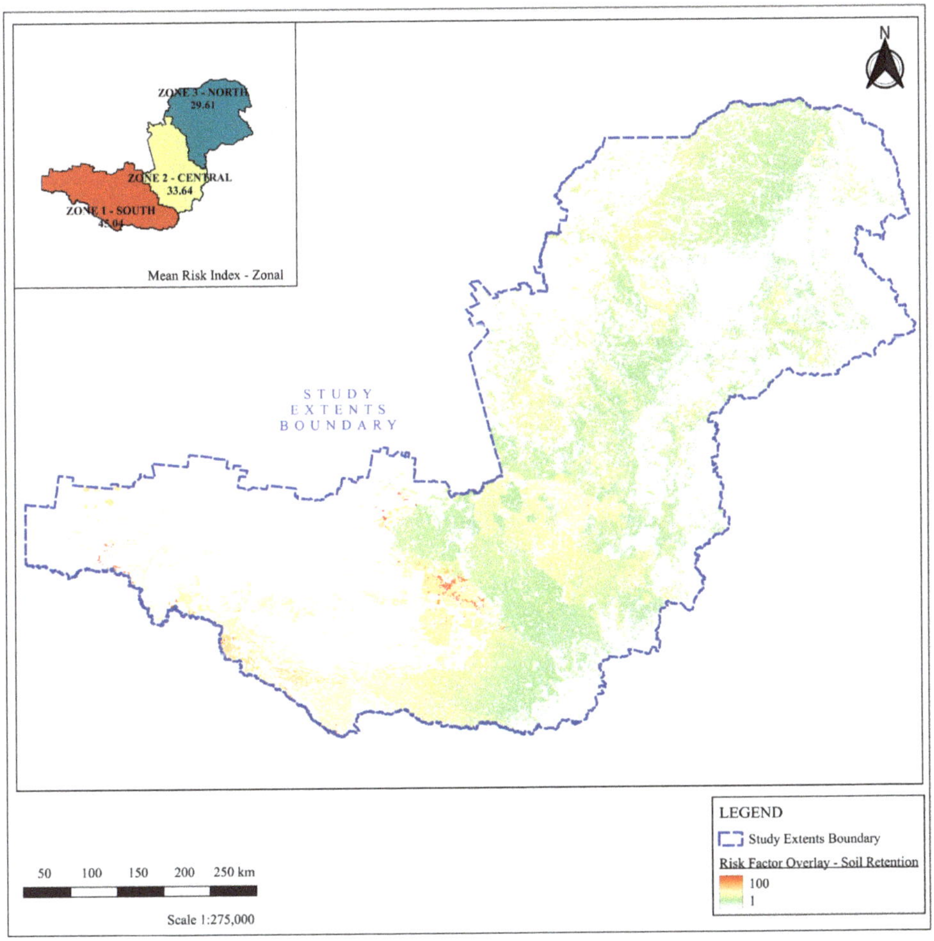

Figure 14. Results—soil retention of microplastics.

3.3. Groundwater Infiltration of Microplastics

The results of the risk mapping for groundwater infiltration of microplastics are shown in Figure 15. Based on the results, high risk is generally observed in the central-west and central-east zones of the study area, with smaller pockets found in the south-western zone. Overall, the southern zone has the highest mean risk value of 56.26, closely followed by the central zone, which has a mean value of 55.00. In addition to the zonal statistics outlined above for each transport pathway, a generalised statistical analysis was performed for each of the three transport pathways over the entire study area. Table 2 provides a statistical overview of this analysis.

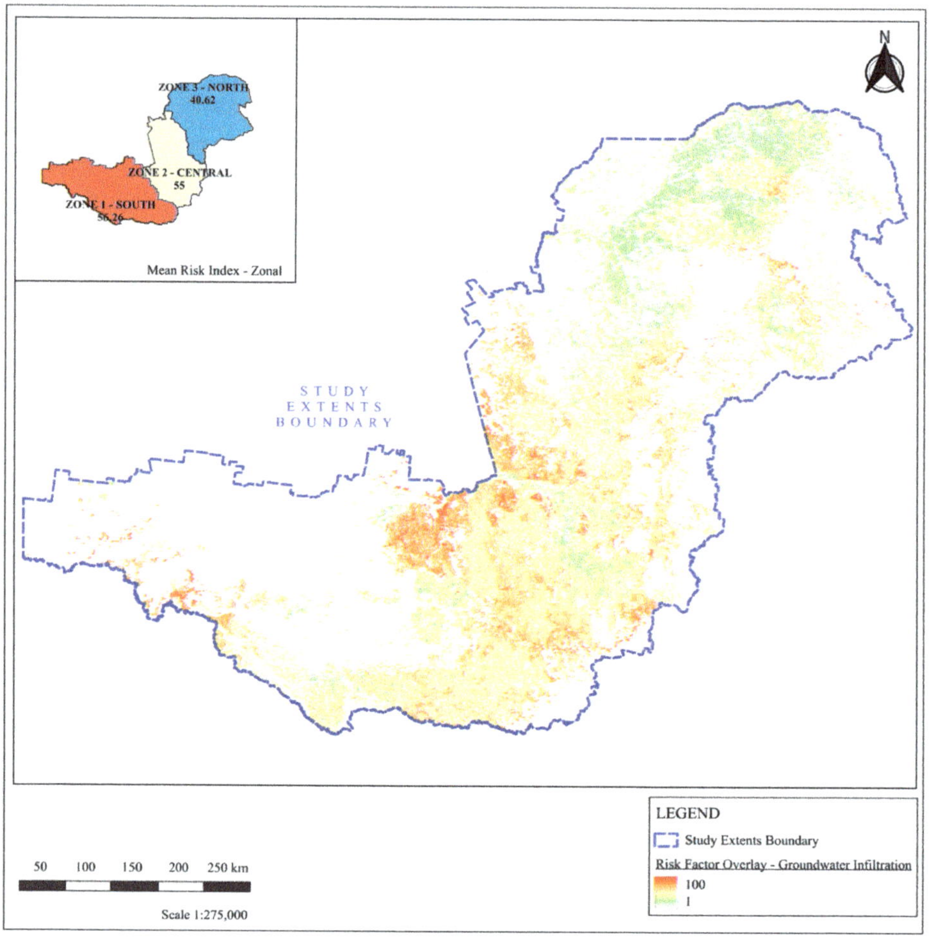

Figure 15. Results—groundwater infiltration of microplastics.

Table 2. Risk factor overlay results raster zonal statistics.

Transport Pathway	Min Value	Max Value	Range	Mean Value	Standard Deviation
Rainfall-runoff of microplastics	0.431	97.667	97.236	9.661	23.687
Soil retention of microplastics	7.333	99.333	92.000	35.581	9.588
Groundwater infiltration of microplastics	10.120	98.713	88.593	50.683	15.038

3.4. Local Government Area Results

The list of local government areas (LGAs) that are most at risk within the study extent can be found in Table 3. In the results presentation, the names of LGAs are redacted.

Table 3. Risk factor overlay results by LGA/ranking. Colour graduation from green (low risk) to red (high risk) are used in the rankings.

LGA	Study Zone	Study Land Use Area (1)	Rainfall-Runoff Risk (2)	Soil Retention Risk (3)	Ground Water Risk (4)	Mean Rank (1,2,3,4)
A	1-South	6	22.5	2	4	8.625
B	3-North	1	11	22	11.5	11.375
C	3-North	3	18.5	15	9	11.375
D	1-South	5	25	10	7.5	11.875
E	2-Central	4	22	14	9	12.25
F	2-Central	2	22.5	18	10	13.125
G	3-North	8	17.5	17	12.5	13.75
H	1-South	9	22.5	13	11	13.875
I	2-Central	15	20	12	13.5	15.125
J	1-South	17	26	8	12.5	15.875
K	2-Central	16	23	11	13.5	15.875
L	2-Central	11	11.5	24	17.5	16
M	2-Central	10	27	16	13	16.5
N	3-North	7	10.5	33	20	17.625
O	2-Central	12	19	23	17.5	17.875
P	1-South	13	27	19	16	18.75
Q	2-Central	14	21.5	25	19.5	20
R	1-South	29	28.5	7	18	20.625
S	3-North	18	14	28	23	20.75
T	1-South	21	26	21	21	22.25
U	1-South	33	27.5	9	21	22.625
V	2-Central	23	27.5	20	21.5	23
W	1-South	36	37	1	18.5	23.125
X	3-North	20	18.5	31	25.5	23.75
Y	3-North	22	15	32	27	24
Z	1-South	35	40.5	5	20	25.125
AA	2-Central	26	23	26	26	25.25
AB	3-North	27	21	27	27	25.5
AC	1-South	41	39	3	22	26.25
AD	1-South	19	21	37	28	26.25
AE	1-South	39	43	4	21.5	26.875
AF	2-Central	25	29	30	27.5	27.875
AG	3-North	24	21.5	36	30	27.875
AH	1-South	42	42	6	24	28.5
AI	2-Central	31	27.5	29	30	29.375
AJ	2-Central	32	22.5	35	33.5	30.75
AK	1-South	28	23	40	34	31.25
AL	3-North	30	19.5	41	35.5	31.5
AM	2-Central	34	22.5	38	36	32.625
AN	2-Central	37	26.5	34	35.5	33.25
AO	3-North	38	20	39	38.5	33.875
AP	3-North	40	22	43	41.5	36.625
AQ	2-Central	43	24.5	44	43.5	38.75
AR	1-South	46	26.5	42	44	39.625
AS	2-Central	44	24.5	46	45	39.875
AT	3-North	47	24	45	46	40.5
AU	3-North	45	24	47	46	40.5

4. Discussion

This GIS-based geospatial modelling study examined three key pathways for the transport of agricultural microplastics as generated by biosolid applications in the study area in the Australian state of New South Wales (NSW). Whilst the study focused on this discrete Australian area, we believe that the applicability of the methodology represents a foundation for similar studies across various global scales. This approach could contribute to the understanding of microplastic pollution risks from biosolid application.

Since there are limited research studies on the risk assessments of particular locations, actual field concentrations, and transport behaviours of agricultural microplastics in Australia, this study developed a set of risk factors that contribute to the three transport pathways by using emerging trends from the global literature base. Using publicly available high-quality spatial input data, these risk factors were mapped and subsequently overlaid to determine the risks associated with each method of transport.

According to study results, regarding the rainfall-runoff of microplastics, a greater risk was observed in the eastern regions of the NSW Murray-Darling Basin (MDB), where average annual rainfall is high, and aridity is low. Additionally, a localised region of high risk was observed in the southern portion of the study area, adjacent to the Murray River, where agricultural properties are in relatively close proximity to waterways (mostly irrigation channels draining to natural streams, creeks, and rivers).

The southern extent of the New South Wales Murray-Darling Basin is more likely to retain microplastics in soil profiles, with high-risk areas generally found in small pockets of horticultural land use, indicating the importance of this risk factor. From the mapping, medium- to high-risk areas of agricultural land adjacent to the Murray River appeared once again, in a region that is relatively arid and contains soils of high potential ionic strength.

According to the risk mapping for groundwater infiltration of microplastics, the western and southern zones of the study area were generally at the highest relative risk, with large areas at medium to high risk. High-risk areas were again found near the Murray River in concentrations of agricultural land, representing an emerging trend from all three transport pathway risk assessments.

While the study results may be limited by its broad area and assumptions made in lieu of present data in an emerging area of research, the methodology used in this study represents a novel foundational framework for the development of similar future risk assessments. By incorporating future research into the field concentrations and transport behaviours of agricultural microplastics, such risk assessments can present more high-resolution mapping, highlighting to policymakers and the agricultural industry areas of the agricultural landscape that are appropriate for biosolid amendment, and those that should be avoided to prevent the risk of microplastic pollution and downstream effects.

5. Conclusions

Globally, plastic production and consumption are on the rise and microplastic particles (microplastics), broken-down forms of these plastics, are emerging as major pollutants. The use of biosolids, which are wastewater treatment plant byproducts sold to the agriculture industry as green alternatives to inorganic fertilizers, is causing microplastics to reach agricultural settings in huge quantities. It has been shown that microplastics may cause significant environmental and ecological harm in agricultural settings, with certain risk factors increasing the likelihood that microplastics will accumulate in soils, run off into nearby waterways, or penetrate groundwater once they are introduced to agricultural settings.

This study used a GIS-based methodology to analyse a portion of the Murray-Darling Basin in New South Wales, Australia, overlaying risk factors associated with three key pathways of microplastic pollution: rainfall-runoff, soil retention, and groundwater infiltration. Through the standardisation of multivariate input data, this study was able to identify particular areas of risk for each transport pathway and map them clearly. Identifying these areas of greatest risk will allow targeted future field investigations within Australia, and the novelty of the methodology applied will provide a foundation for global studies to undertake similar investigative mapping.

This study was limited by the lack of research in the field of agricultural microplastic pollution, and future studies should verify the accuracy of mapping through field investigations and empirical studies.

A similar GIS methodology to that used in this study is applicable globally, at various scales of study, with a multitude of possibilities for input data and risk factor overlay, as the literature base develops.

Within Australia, it is recommended that further investigations be conducted to determine the actual areas that are applied with biosolids, the levels of microplastic concentrations in different types of agricultural land use, and the concentrations of microplastics in downstream waterways and groundwater near agricultural land use.

Supplementary Materials: The following supporting information can be downloaded at https://www.mdpi.com/article/10.3390/agronomy15010047/s1, Table S1: Risk factors, categories, sources, and spatial data and sources used.

Author Contributions: K.M.: conceptualization, data curation, investigation, writing—original draft preparation. K.N.: funding acquisition, writing—review and editing. All authors have read and agreed to the published version of the manuscript.

Funding: This research received no external funding. Funding for the APC will be provided by the Author's ADEB fund.

Data Availability Statement: The raw data supporting the conclusions of this article will be made available by the authors on request.

Acknowledgments: This study is supported by the University of Southern Queensland, Australia.

Conflicts of Interest: The authors declare no conflicts of interest.

References

1. Geyer, R.; Jambeck, J.R.; Law, K.L. Production, Use, and Fate of All Plastics Ever Made. *Sci. Adv.* **2017**, *3*, e1700782. [CrossRef]
2. Frias, J.P.G.L.; Nash, R. Microplastics: Finding a Consensus on the Definition. *Mar. Pollut. Bull.* **2019**, *138*, 145–147. [CrossRef] [PubMed]
3. O'Brien, S.; Rauert, C.; Ribeiro, F.; Okoffo, E.D.; Burrows, S.D.; O'Brien, J.W.; Wang, X.; Wright, S.L.; Thomas, K.V. There's Something in the Air: A Review of Sources, Prevalence and Behaviour of microplastics in the Atmosphere. *Sci. Total Environ.* **2023**, *874*, 162193. [CrossRef] [PubMed]
4. Gambino, I.; Bagordo, F.; Grassi, T.; Panico, A.; De Donno, A. Occurrence of microplastics in Tap and Bottled Water: Current Knowledge. *Int. J. Environ.* **2022**, *19*, 5283. [CrossRef]
5. Abbasi, S.; Soltani, N.; Keshavarzi, B.; Moore, F.; Turner, A.; Hassanaghaei, M. Microplastics in Different Tissues of Fish and Prawn from the Musa Estuary, Persian Gulf. *Chemosphere* **2018**, *205*, 80–87. [CrossRef] [PubMed]
6. Leslie, H.A.; van Velzen, M.J.M.; Brandsma, S.H.; Vethaak, A.D.; Garcia-Vallejo, J.J.; Lamoree, M.H. Discovery and Quantification of Plastic Particle Pollution in Human Blood. *Environ. Int.* **2022**, *163*, 107199. [CrossRef]
7. Napper, I.E.; Davies, B.F.R.; Clifford, H.; Elvin, S.; Koldewey, H.J.; Mayewski, P.A.; Miner, K.R.; Potocki, M.; Elmore, A.C.; Gajurel, A.P.; et al. Reaching New Heights in Plastic Pollution—Preliminary Findings of microplastics on Mount Everest. *One Earth* **2020**, *3*, 621–630. [CrossRef]
8. Kang, H.-M.; Byeon, E.; Jeong, H.; Kim, M.-S.; Chen, Q.; Lee, J.-S. Different Effects of Nano- and microplastics on Oxidative Status and Gut Microbiota in the Marine Medaka *Oryzias Melastigma*. *J. Hazard. Mater.* **2021**, *405*, 124207. [CrossRef]
9. Liu, S.; Li, H.; Wang, J.; Wu, B.; Guo, X. Polystyrene microplastics Aggravate Inflammatory Damage in Mice with Intestinal Immune Imbalance. *Sci. Total Environ.* **2022**, *833*, 155198. [CrossRef]
10. Jin, H.; Yang, C.; Jiang, C.; Li, L.; Pan, M.; Li, D.; Han, X.; Ding, J. Evaluation of Neurotoxicity in BALB/c Mice Following Chronic Exposure to Polystyrene microplastics. *Environ. Health Perspect.* **2022**, *130*, 107002. [CrossRef]
11. Vandenberg, L.N.; Luthi, D.; Quinerly, D. Plastic Bodies in a Plastic World: Multi-Disciplinary Approaches to Study Endocrine Disrupting Chemicals. *J. Clean. Prod.* **2017**, *140*, 373–385. [CrossRef]

12. Park, E.-J.; Han, J.-S.; Park, E.-J.; Seong, E.; Lee, G.-H.; Kim, D.-W.; Son, H.-Y.; Han, H.-Y.; Lee, B.-S. Repeated-Oral Dose Toxicity of Polyethylene microplastics and the Possible Implications on Reproduction and Development of the next Generation. *Toxicol. Lett.* **2020**, *324*, 75–85. [CrossRef] [PubMed]
13. Wright, S.L.; Kelly, F.J. Plastic and Human Health: A Micro Issue? *Env. Sci. Technol.* **2017**, *51*, 6634–6647. [CrossRef] [PubMed]
14. Qian, H.; Zhang, M.; Liu, G.; Lu, T.; Qu, Q.; Du, B.; Pan, X. Effects of Soil Residual Plastic Film on Soil Microbial Community Structure and Fertility. *Water Air Soil Pollut.* **2018**, *229*, 261. [CrossRef]
15. Wan, Y.; Wu, C.; Xue, Q.; Hui, X. Effects of Plastic Contamination on Water Evaporation and Desiccation Cracking in Soil. *Sci. Total Environ.* **2019**, *654*, 576–582. [CrossRef]
16. Wu, X.; Liu, Y.; Yin, S.; Xiao, K.; Xiong, Q.; Bian, S.; Liang, S.; Hou, H.; Hu, J.; Yang, J. Metabolomics Revealing the Response of Rice (*Oryza Sativa*, L.) Exposed to Polystyrene microplastics. *Environ. Pollut.* **2020**, *266*, 115159. [CrossRef]
17. Urli, S.; Pause, F.C.; Crociati, M.; Baufeld, A.; Monaci, M.; Stradaioli, G. Impact of microplastics and Nanoplastics on Livestock Health: An Emerging Risk for Reproductive Efficiency. *Animals* **2023**, *13*, 1132. [CrossRef]
18. Wang, J.; Bucci, K.; Helm, P.A.; Hoellein, T.; Hoffman, M.J.; Rooney, R.; Rochman, C.M. Runoff and Discharge Pathways of microplastics into Freshwater Ecosystems: A Systematic Review and Meta-Analysis. *FACETS* **2022**, *7*, 1473–1492. [CrossRef]
19. Abel, J. Fluctuations of Microplastic Concentrations in Groundwater and a Rural Creek Relating to Seasonal Farming Practices. In Proceedings of the 2024 Emerging Contaminants in the Environment Conference (ECEC24), Virtual, 24–25 April 2024.
20. Zhang, S.; Han, B.; Sun, Y.; Wang, F. Microplastics Influence the Adsorption and Desorption Characteristics of Cd in an Agricultural Soil. *J. Hazard. Mater.* **2020**, *388*, 121775. [CrossRef]
21. Marchuk, S.; Tait, S.; Sinha, P.; Harris, P.; Antille, D.L.; McCabe, B.K. Biosolids-Derived Fertilisers: A Review of Challenges and Opportunities. *Sci. Total Environ.* **2023**, *875*, 162555. [CrossRef]
22. ANZBP. Australian Biosolids Statistics. Available online: https://www.biosolids.com.au/guidelines/australian-biosolids-statistics/ (accessed on 9 March 2024).
23. Ziajahromi, S.; Slynkova, N.; Dwyer, J.; Griffith, M.; Fernandes, M.; Jaeger, J.E.; Leusch, F.D.L. Comprehensive Assessment of microplastics in Australian Biosolids: Abundance, Seasonal Variation and Potential Transport to Agroecosystems. *Water Res.* **2024**, *250*, 121071. [CrossRef] [PubMed]
24. Mohajerani, A.; Karabatak, B. Microplastics and Pollutants in Biosolids Have Contaminated Agricultural Soils: An Analytical Study and a Proposal to Cease the Use of Biosolids in Farmlands and Utilise Them in Sustainable Bricks. *Waste Manag.* **2020**, *107*, 252–265. [CrossRef]
25. Harley-Nyang, D.; Memon, F.A.; Jones, N.; Galloway, T. Investigation and Analysis of microplastics in Sewage Sludge and Biosolids: A Case Study from One Wastewater Treatment Works in the UK. *Sci. Total Environ.* **2022**, *823*, 153735. [CrossRef]
26. Okoffo, E.D.; Tscharke, B.J.; O'Brien, J.W.; O'Brien, S.; Ribeiro, F.; Burrows, S.D.; Choi, P.M.; Wang, X.; Mueller, J.F.; Thomas, K.V. Release of Plastics to Australian Land from Biosolids End-Use. *Environ. Sci. Technol.* **2020**, *54*, 15132–15141. [CrossRef]
27. Zhang, L.; Xie, Y.; Liu, J.; Zhong, S.; Qian, Y.; Gao, P. An Overlooked Entry Pathway of microplastics into Agricultural Soils from Application of Sludge-Based Fertilizers. *Environ. Sci. Technol.* **2020**, *54*, 4248–4255. [CrossRef]
28. van den Berg, P.; Huerta-Lwanga, E.; Corradini, F.; Geissen, V. Sewage Sludge Application as a Vehicle for microplastics in Eastern Spanish Agricultural Soils. *Environ. Pollut.* **2020**, *261*, 114198. [CrossRef] [PubMed]
29. Corradini, F.; Meza, P.; Eguiluz, R.; Casado, F.; Huerta-Lwanga, E.; Geissen, V. Evidence of Microplastic Accumulation in Agricultural Soils from Sewage Sludge Disposal. *Sci. Total Environ.* **2019**, *671*, 411–420. [CrossRef] [PubMed]
30. Schell, T.; Hurley, R.; Buenaventura, N.T.; Mauri, P.V.; Nizzetto, L.; Rico, A.; Vighi, M. Fate of microplastics in Agricultural Soils Amended with Sewage Sludge: Is Surface Water Runoff a Relevant Environmental Pathway? *Environ. Pollut.* **2022**, *293*, 118520. [CrossRef]
31. Radford, F.; Horton, A.; Hudson, M.; Shaw, P.; Williams, I. Agricultural Soils and microplastics: Are Biosolids the Problem? *Front. Soil Sci.* **2023**, *2*, 941837. [CrossRef]
32. Ziajahromi, S.; Lu, H.-C.; Dwyer, J.; Fernandes, M.; Griffith, M.; Leusch, F.D.L. Transport and Accumulation of microplastics from Biosolids to Australian Agricultural Soils: Detection of microplastics Down to 1 µm. *Environ. Sci. Technol.* **2024**, *58*, 17048–17057. [CrossRef]
33. Samandra, S.; Johnston, J.M.; Jaeger, J.E.; Symons, B.; Xie, S.; Currell, M.; Ellis, A.V.; Clarke, B.O. Microplastic Contamination of an Unconfined Groundwater Aquifer in Victoria, Australia. *Sci. Total Environ.* **2022**, *802*, 149727. [CrossRef] [PubMed]
34. QGIS. QGIS Spatial without Compromise. Available online: https://www.qgis.org/ (accessed on 1 September 2024).
35. Murray-Darling Basin Authority. The Basin. Available online: https://www.mdba.gov.au/basin (accessed on 16 November 2024).
36. Australian Bureau of Statistics. Digital Boundary Files: State Boundaries 2019. Available online: https://www.abs.gov.au/statistics/standards/australian-statistical-geography-standard-asgs-edition-3/jul2021-jun2026/access-and-downloads/digital-boundary-files#cite-window1 (accessed on 3 August 2024).

37. Department of Industry, Science and Resources. NSW Local Government Areas—Geoscape Administrative Boundaries Datasets. Available online: https://data.gov.au/dataset/ds-dga-f6a00643-1842-48cd-9c2f-df23a3a1dc1e/details?q= (accessed on 4 August 2024).
38. NSW Government—Spatial Services NSW. Water Theme—Hydro Line. Available online: https://portal.spatial.nsw.gov.au/portal/home/item.html?id=7b0e959effd749c788d304a4179abf8a (accessed on 18 August 2024).
39. Murray-Darling Basin Authority. Murray Darling Basin Boundary—Water Act 2007. Available online: https://data.gov.au/data/dataset/murray-darling-basin-boundary (accessed on 3 August 2024).
40. State Government of NSW. NSW Department of Climate Change, Energy, the Environment and Water NSW Landuse 2017—SEED. Available online: https://datasets.seed.nsw.gov.au/dataset/nsw-landuse-2017 (accessed on 3 August 2024).
41. Australian Bureau of Statistics. Digital Boundary Files: Australia Boundary 2019. Available online: https://www.abs.gov.au/statistics/standards/australian-statistical-geography-standard-asgs-edition-3/jul2021-jun2026/access-and-downloads/digital-boundary-files#cite-window1 (accessed on 3 August 2024).
42. Reichert, J.; Schellenberg, J.; Schubert, P.; Wilke, T. Responses of Reef Building Corals to Microplastic Exposure. *Environ. Pollut.* **2018**, *237*, 955–960. [CrossRef] [PubMed]
43. Nerland, I.L.; Halsband, C.; Allan, I.; Thomas, K.V. *Microplastics in Marine Environments: Occurrence, Distribution and Effects*; Norwegian Institute for Water Research: Oslo, Norway, 2024; p. 71.
44. Llorca, M.; Schirinzi, G.; Martínez, M.; Barceló, D.; Farré, M. Adsorption of Perfluoroalkyl Substances on microplastics under Environmental Conditions. *Environ. Pollut.* **2018**, *235*, 680–691. [CrossRef]
45. NSW Department of Primary Industries. Landform and Soil Requirements for Biosolids and Effluent Reuse. Available online: https://www.dpi.nsw.gov.au/__data/assets/pdf_file/0004/187627/biosolid-effluent-release.pdf (accessed on 17 August 2024).
46. Terrestrial Ecosystem Research Network (TERN). Soil and Landscape Grid of Australia. Available online: https://esoil.io/TERNLandscapes/Public/Pages/SLGA/ProductDetails-SoilAttributes.html#DataAcknowledgements (accessed on 17 August 2024).
47. Crossman, J.; Hurley, R.R.; Futter, M.; Nizzetto, L. Transfer and Transport of microplastics from Biosolids to Agricultural Soils and the Wider Environment. *Sci. Total Environ.* **2020**, *724*, 138334. [CrossRef] [PubMed]
48. Ding, L.; Zhang, S.; Wang, X.; Yang, X.; Zhang, C.; Qi, Y.; Guo, X. The Occurrence and Distribution Characteristics of microplastics in the Agricultural Soils of Shaanxi Province, in North-Western China. *Sci. Total Environ.* **2020**, *720*, 137525. [CrossRef]
49. Bureau of Meteorology. Rainfall—Climate Averages Maps. Available online: http://www.bom.gov.au/climate/maps/averages/rainfall/ (accessed on 10 August 2024).
50. State Government of NSW and NSW Department of Climate Change, Energy, the Environment and Water. NSW Aridity Index—High Resolution (30 Meter)—SEED. Available online: https://datasets.seed.nsw.gov.au/dataset/nsw-aridity-index-high-resolution-30-meter (accessed on 24 August 2024).
51. Nyman, P.; Sherwin, C.; Langhans, C.; Lane, P.; Sheridan, G. Downscaling Regional Climate Data to Calculate the Radiative Index of Dryness in Complex Terrain. *Aust. Meteorol. Oceanogr. J.* **2014**, *64*, 109–122. [CrossRef]
52. Zubris, K.A.V.; Richards, B.K. Synthetic Fibers as an Indicator of Land Application of Sludge. *Environ. Pollut.* **2005**, *138*, 201–211. [CrossRef]
53. Palazot, M.; Soccalingame, L.; Froger, C.; Jolivet, C.; Bispo, A.; Kedzierski, M.; Bruzaud, S. First National Reference of Microplastic Contamination of French Soils. *Sci. Total Environ.* **2024**, *918*, 170564. [CrossRef] [PubMed]
54. Chu, X.; Li, T.; Li, Z.; Yan, A.; Shen, C. Transport of Microplastic Particles in Saturated Porous Media. *Water* **2019**, *11*, 2474. [CrossRef]
55. Li, M.; Zhang, X.; Yi, K.; He, L.; Han, P.; Tong, M. Transport and Deposition of Microplastic Particles in Saturated Porous Media: Co-Effects of Clay Particles and Natural Organic Matter. *Environ. Pollut.* **2021**, *287*, 117585. [CrossRef]
56. Wang, Y.; Xu, L.; Chen, H.; Zhang, M. Retention and Transport Behavior of Microplastic Particles in Water-Saturated Porous Media. *Sci. Total Environ.* **2022**, *808*, 152154. [CrossRef] [PubMed]
57. US EPA. Ionic Strength. Available online: https://www.epa.gov/caddis/ionic-strength (accessed on 1 September 2024).
58. State Government of NSW and NSW Department of Climate Change, Energy, the Environment and Water. NSW State-Wide Hydrogeological Landscapes 2020 (First Edition)—SEED. Available online: https://datasets.seed.nsw.gov.au/dataset/nsw-state-wide-hydrogeological-landscapes-2020-first-edition (accessed on 1 September 2024).
59. Jeong, E.; Kim, Y.-I.; Lee, J.-Y.; Raza, M. Microplastic Contamination in Groundwater of Rural Area, Eastern Part of Korea. *Sci. Total Environ.* **2023**, *895*, 165006. [CrossRef]
60. Bureau of Meteorology. National Groundwater Information System. Available online: https://data.gov.au/data/dataset/national-groundwater-information-system (accessed on 24 August 2024).
61. O'Connor, D.; Pan, S.; Shen, Z.; Song, Y.; Jin, Y.; Wu, W.-M.; Hou, D. Microplastics Undergo Accelerated Vertical Migration in Sand Soil Due to Small Size and Wet-Dry Cycles. *Environ. Pollut.* **2019**, *249*, 527–534. [CrossRef]

62. Lincmaierová, K.; Botyanszká, L.; Lichner, L.; Toková, L.; Zafeiriou, I.; Bondarev, D.; Horák, J.; Šurda, P. Assessing Microplastic Induced Changes in Sandy Soil Properties and Crop Growth. *Agriengineering* **2023**, *5*, 1555–1567. [CrossRef]
63. Blanda, V.; Blanda, M.; Russa, F.; Scimeca, R.; Scimeca, S.; D'Agostino, R.; Auteri, M.; Torina, A. Geo-Statistical Analysis of *Culicoides* spp. Distribution and Abundance in Sicily, Italy. *Parasites Vectors* **2018**, *11*, 78. [CrossRef]

agronomy

MDPI

Article

Circular Approach in Development of Microbial Biostimulants Using Winery Wastewater

Tatjana Dujković, Ivana Danilov *, Vanja Vlajkov, Marina Savić, Zdravko Šumić, Aleksandar Jokić and Jovana Grahovac

Faculty of Technology Novi Sad, University of Novi Sad, Bulevar cara Lazara 1, 21000 Novi Sad, Serbia; tatjana.dujkovic@uns.ac.rs (T.D.); vanja.vlajkov@uns.ac.rs (V.V.); msavic.savic5@gmail.com (M.S.); sumic@uns.ac.rs (Z.Š.); jokic@uns.ac.rs (A.J.); johana@uns.ac.rs (J.G.)

* Correspondence: ivana.pajcin@uns.ac.rs

Abstract: Sustainable development requires implementation of eco-friendly practices and a circular approach in both agricultural and industrial systems. This study evaluated winery flotation wastewater (WFW) as a cultivation substrate for *Bacillus* sp. 10/R isolated from grapevine rhizosphere for sustainable biostimulant production. The bacterial isolate was characterized by 16S rRNA sequencing and biochemical tests, showing the highest similarity with *Bacillus mojavensis* and *Bacillus halotolerans*. Plant growth-promoting traits were assessed via assays for hydrolytic enzymes, ACC (1-aminocyclopropane-1-carboxylate) deaminase, and IAA (indole acetic acid) production, as well as for phosphate solubilization. The isolate was cultivated in WFW, including monitoring of biomass growth, enzymatic activity, and substrate composition changes. The resulting cultivation broths based on WFW (WFW-CB) and nutrient broth (NB-CB) were tested as barley seed treatment at five dosages, using sterile media and water as controls. The results have displayed strong pectinase (EAI–enzyme activity index 2.79) and cellulase activity (2.33), moderate xylanase (1.75) and ACC deaminase activity (growth zone 54.67 ± 0.58 mm), and moderate IAA production (9.66 µg/mL). Biomass content has increased by two log units within 48 h (up to 9.06 log CFU/mL), with stable pectinase activity (~2.2 U/mL). Germination assays revealed that 10% WFW-CB and 50% WFW enhanced germination indices and biomass, whereas undiluted WFW and WFW-CB inhibited germination. These results indicate that WFW is a suitable substrate for *Bacillus* sp. 10/R cultivation, linking industrial wastewater valorization with plant biostimulant production in a circular economy framework.

Keywords: *Bacillus* sp. 10/R; winery flotation wastewater; plant growth promotion; hydrolytic enzymes; IAA; ACC deaminase; phosphate solubilization; barley; germination; seed treatment

1. Introduction

Although the use of chemical fertilizers and pesticides has significantly enhanced agricultural productivity over the past five decades, their excessive application has also contributed to environmental issues, including groundwater contamination, soil degradation, and biodiversity loss [1–3]. Therefore, it is essential to develop innovative, environmentally sustainable agricultural practices that can maintain high productivity while mitigating the adverse impacts of climate change and environmental degradation due to chemical pollution [4]. Green biotechnological products represent one of the promising solutions in this field. In this context, plant growth-promoting (PGP) bacteria have emerged as key players in the development of microbial-based products intended for sustainable agricultural use [5,6]. These beneficial microorganisms exhibit a variety of direct and indirect mechanisms, including nitrogen fixation, nutrient solubilization, phytohormone production, synthesis of hydrolytic enzymes, antibiotics and siderophores, as well as induction of systemic resistance in plants [7,8]. Owing to these diverse activities, PGP bacteria-based products can be effectively utilized as biofertilizers, biostimulants, biocontrol agents, and soil quality enhancers, providing a promising tool to improve crop productivity within sustainable agricultural systems [9,10].

Among PGP microorganisms, including wide range of bacterial and fungal species such as *Azotobacter*, *Azospirillum*, *Rhizobium*, *Pseudomonas*, *Trichoderma*, etc., *Bacillus* species stand out as the most studied and utilized in sustainable agriculture due to their diverse functional traits and adaptability to various environmental conditions. These rhizobacteria are recognized for their ability to enhance plant growth and resilience to abiotic stresses, while also providing biological control potential against plant pathogens [11,12]. A key advantage of *Bacillus* spp. is their ability to form endospores, which ensures high environmental stability and viability during formulation, storage and field application, making them particularly suitable for large-scale use in agricultural settings [13,14]. Application of various formulations of *Bacillus*-based plant biostimulants has been investigated across the wide range of crops and testing conditions (including seed germination experiments, growth chambers, greenhouse and field experiments), indicating their contributions to improved plant growth and crop quality, resilience and advanced immune responses of plants under abiotic and biotic stresses [15–17].

However, despite the promising potential of microbial-based bioagents, including *Bacillus* spp., their wider application as biofertilizers in agricultural production systems is still limited, mainly due to the high costs associated with biotechnological production processes. In this regard, the utilization of alternative, complex media derived from industrial effluents represents a promising and underexplored approach for the cost-effective production of microbial biocontrol agents/plant biostimulants [18]. Such media would not only contribute to the economic feasibility of bioprocesses but also align with the principles of circular economy and industrial symbiosis by valorizing waste streams while reducing freshwater consumption [19,20].

The wine industry is considered one of the most significant socio-economic activities in Europe [21]. According to the International Organization of Vine (OIV), the global vineyard surface area in 2024 was reported at 7.1 million hectares, while global wine production reached approximately 226 million hectoliters [22]. Processing 1000 kg of grapes typically results in 750 L of wine, alongside 1650 L of wastewater and about 200 kg of solid residues, which present a significant environmental burden due to their high organic load and the presence of volatile organic compounds [23]. However, the rich organic composition of these waste streams also indicates their potential for valorization through biotechnological approaches, particularly microbial bioconversion, offering opportunities for the development of sustainable processes within the wine industry [24]. The utilization of winery waste for the cultivation of *Bacillus* spp. provides an opportunity to reduce environmental pollution while simultaneously producing biofertilizers and plant growth-promoting agents, aligning with the circular economy principles [14].

The aim of this study was to evaluate the potential of *Bacillus* sp. 10/R, isolated from the grapevine rhizosphere, as a plant growth-promoting (PGP) agent. Specifically, the study focused on investigating the production of hydrolytic enzymes, indole-3-acetic acid (IAA), ACC deaminase, and phosphate solubilization as the prominent PGP traits. In addition, the nutritional potential of winery flotation wastewater (WFW) was assessed by its application as a growth substrate for cultivation of the aforementioned bacterial isolate. Throughout the cultivation period, biomass accumulation and enzyme activities were monitored. Finally, the cultivation broth, applied at different concentrations as seed treatment, was used to evaluate its effect on barley seed germination. The main hypothesis of the study thus was that the WFW as an industrial effluent could be applied as a bacterial growth substrate in production of *Bacillus*-based plant biostimulants. The proposed industrial symbiosis framework would contribute to circular utilization of winery waste in a subsequent industrial sector producing bio-based alternatives for agrochemicals aimed primarily for applications in organic agriculture, therefore improving sustainability indices of both industrial and agricultural sector.

2. Materials and Methods

2.1. Microorganisms

In this study, *Bacillus* sp. 10/R, isolated from the rhizosphere of Merlot grapevine (*Vitis vinifera* cv. Merlot), was used as the biostimulant active component. For the isolation procedure, 1 g of the collected rhizosphere sample was resuspended in 9 mL of sterile saline and incubated at 28 °C for 15 min with continuous agitation on a laboratory rotary shaker (KS 4000i control, IKA® Werke, Staufen im Breisgau, Germany) at 150 rpm. After homogenization, selective isolation of spore-forming bacteria was achieved by subjecting the suspension to heat treatment at 100 °C for 7 min. Serial tenfold dilutions (10^{-1} to 10^{-3}) were prepared, and 500 µL of each dilution was spread onto HiChrome Bacillus agar (HiMedia Laboratories, Mumbai, India) and incubated at 28 °C for 48 h. Colonies with morphological characteristics typical for *Bacillus* spp. were subcultured on fresh selective media until pure cultures were obtained [25]. The pure isolate was maintained on nutrient agar slants at 4 °C, as well as a glycerol stock culture at −80 °C in the culture collection of the Laboratory for Biochemical Engineering, Faculty of Technology Novi Sad, University of Novi Sad, Serbia.

2.2. Genomic DNA Extraction and 16S rRNA Sequencing

Bacillus sp. 10/R was cultured in 20 mL of nutrient broth (HiMedia Laboratories, Mumbai, India) overnight at 28 °C with continuous agitation at 150 rpm on a laboratory rotary shaker (KS 4000i control, IKA® Werke, Staufen im Breisgau, Germany). Cells were harvested by centrifugation at 5000 rpm for 15 min, and the resulting pellet was resuspended in 10 mL of lysis buffer containing 0.3 M sucrose, 25 mM EDTA, 25 mM Tris-HCl, and RNase (2 U), adjusted to pH 7.5. The suspension was incubated with lysozyme (10 mg) at 37 °C for 30 min, followed by the addition of proteinase K (5 mg, Sigma Aldrich–Merck, Darmstadt, Germany) and 1 mL of 10% (*w*/*v*) SDS, and incubated at 55 °C for 90 min. Subsequently, 15 mL of chloroform and 3.6 mL of 5 M NaCl were added, and the mixture was subjected to end-over-end rotation for 20 min. Cell debris were removed by centrifugation at 5000 rpm for 20 min, and DNA was precipitated from the supernatant using isopropanol (1:1, *v*/*v*). The DNA pellet was washed with 1 mL of 70% (*v*/*v*) ethanol, air-dried, and dissolved in DNA- and RNA-free water (Thermo Fischer Scientific, Waltham, MA, USA) [26].

The 16S rRNA gene was amplified from the extracted genomic DNA using universal bacterial primers 27F (5′-AGAGTTTGATCCTGGCTCAG-3′) and 1492R (5′-GGTTACCTTG TTACGACTT-3′) in a Surecycler 8800 Thermocycler (Agilent Technologies, Santa Clara, CA, USA). Amplified DNA fragments were bidirectionally sequenced (Macrogen Europe, Amsterdam, The Netherlands) with the same primers as in the amplification step. The obtained sequences were identified using the blastn feature of BLAST (Basic Local Arrangement Search Tool—https://blast.ncbi.nlm.nih.gov/, accessed on 16 July 2025) and evolutionary analyses were performed using MEGA12 software. The phylogenetic tree was generated using the maximum likelihood method and the 16S rRNA sequence of *Escherichia coli* NBRC 102203 was applied as an outgroup.

2.3. Biochemical Characterization of Bacillus sp. 10/R

Biochemical characterization of *Bacillus* sp. 10/R was performed using the commercial API test kits: API 50 CH, API 20 E, and API ZYM (bioMérieux, Marcy-l'Étoile, France), whose application was aimed at deciphering the ability of the bacterial isolate to consume various nutrient sources, produce several metabolites relevant for biochemical distinction among bacterial species, as well as to better understand enzymatic profile of the investigated strain, respectively. All tests were carried out following the manufacturer's instructions. Briefly, bacterial cells were suspended in sterile saline solution to the recommended turbidity and inoculated into the respective API strips or galleries, containing appropriate substrates for the aforementioned biochemical reactions. The strips were incubated at 28 °C for 24 to 48 h, depending on the test requirements. After incubation, the results were recorded based on color changes or enzymatic activity indicators according to the API reading guidelines.

2.4. Screening of Plant-Growth Promotion Traits

2.4.1. Screening of Enzyme Production on Agar Plates

The enzymatic activity of *Bacillus* sp. 10/R isolate was screened using semi-solid media containing specific substrates for the target enzymes (pectinase, cellulase, xylanase, protease), important for complex nutrient release in agricultural applications. The incubation conditions were similar in each enzyme activity screening experiment: 120 h at 28 °C. All media were sterilized in an autoclave at 121 °C and 2.1 bar for 20 min, poured into Petri dishes, and allowed to solidify. Subsequently, plates were point-inoculated at the center using a sterile inoculating pin from an actively growing culture of the strain *Bacillus* sp. 10/R. After the incubation period, the diameters of the growth zones and the halo zones around the colonies were measured. For certain assays aimed at specific enzymes, appropriate reagent or indicator solutions were applied onto the surface of the media to enhance the visualization of the halo zones where needed, as explained in the following sections.

The Enzymatic Activity Index (EAI) for each tested enzyme was calculated according to the following formula [9]:

$$EAI = \frac{\text{diameter of bacterial growth} + \text{diameter of halo zone}}{\text{diameter of bacterial growth}} \quad (1)$$

Pectinase Activity

The pectinolytic activity of *Bacillus* sp. 10/R was assessed on a medium containing (g/L): $NaNO_3$ 1, KCl 1, K_2HPO_4 1, $MgSO_4{\cdot}7H_2O$ 1, yeast extract 0.5, pectin 10, and agar 15 [27]. The pH of the medium was adjusted to 7.00 ± 0.02 prior to the addition of agar. Halo zones indicating pectinolytic activity were visualized by flooding the plates with Gram's iodine solution for 5 min, followed by rinsing with distilled water and the measurement of the corresponding diameters of bacterial growth and halo zones.

Cellulase and Xylanase Activity

The cellulase and xylanase activity of *Bacillus* sp. 10/R isolate was assessed on medium containing (g/L): CMC or xylan 5, glucose 1, yeast extract 0.5, KCl 1, K_2HPO_4 1, $NaNO_3$ 1, $MgSO_4{\cdot}7H_2O$ 0.5, agar 17 [28]. For improved visualization of the clear zones surrounding microbial growth, the plates were flooded with Congo red solution (0.5% *w*/*v*) for 15 min after incubation, followed by rinsing with 1 M NaCl solution for 15 min and the measurement of the corresponding diameters of bacterial growth and halo zones.

Protease Activity

The proteolytic activity of the investigated *Bacillus* sp. 10/R was evaluated on modified skim milk agar containing (g/L): skim milk powder (Centrohem, Stara Pazova, Serbia) 28, tryptone 5, yeast extract 2.5, glucose 1, and agar 15 [29]. The pH of the medium was adjusted to 7.00 ± 0.02 before agar was added. After the incubation under the previously described conditions, the halo zones and bacterial growth diameters were measured.

2.4.2. ACC Deaminase Activity

The ability of *Bacillus* sp. 10/R to produce ACC deaminase was assessed using agar-solidified DF minimal medium supplemented with 3 mM 1-aminocyclopropane-1-carboxylate (ACC—Sigma Aldrich–Merck, Darmstadt, Germany) as the sole nitrogen source, following the procedure described by Penrose and Glick (2003) [30]. Briefly, bacterial isolates were streaked onto DF minimal medium plates containing ACC and incubated at 28 °C for up to 7 days. Growth on this medium, which lacks alternative nitrogen sources, was taken as an indicator of ACC deaminase activity, since only bacteria capable of utilizing ACC as a nitrogen source are able to grow under these conditions.

2.4.3. Phosphate Solubilization Assay

Phosphate solubilization by *Bacillus* sp. BioSol021 was evaluated using Pikovskaya's agar medium (Himedia Laboratories, Mumbai, India). The plates were point-inoculated at the center by transferring a small amount of biomass from an actively growing culture using a sterile inoculating pin. The plates were incubated at 28 °C for 120 h. Following incubation, the diameters of the colonies and the surrounding halo zones were measured, and the phosphate solubilization index (PSI) was calculated according to the following formula [31].

$$PSI = \frac{\text{diameter of bacterial growth} + \text{diameter of halo zone}}{\text{diameter of bacterial growth}} \quad (2)$$

2.4.4. Indole Acetic Acid Production

For indole-3-acetic acid (IAA) production assay *Bacillus* sp. 10/R was cultivated in nutrient broth medium (HiMedia Laboratories, Mumbai, India) supplemented with L-tryptophan (1.02 g/L) as a precursor of IAA. The cultivation was carried out in Erlenmeyer flasks at 28 °C for 48 h on a laboratory rotary shaker at 150 rpm under spontaneous aeration. The determination of IAA was performed by colorimetric method with slight modifications according to Syed-Ab-Rahman et al. (2018) [32]. Briefly, 1 mL of the culture supernatant was mixed with 2 mL of Salkowski reagent (1.2% [*w*/*v*] $FeCl_3$ in 7.9 M H_2SO_4) and incubated in the dark at room temperature for 30 min. The development of a pink coloration indicated the production of IAA. After incubation, the absorbance was measured at 535 nm using a UV–Vis spectrophotometer (UV 1800, Shimadzu, Kyoto, Japan). Distilled water served as a blank, and the IAA concentration in the samples was determined using a calibration curve prepared with an indole-3-acetic acid standard (Sigma-Aldrich–Merck, Darmstadt, Germany).

2.5. Cultivation Media

Inoculum of *Bacillus* sp. 10/R was prepared using liquid commercial medium nutrient broth (HiMedia Laboratories, Mumbai, India). The cultivation of the producing microorganisms was performed using a medium based on winery flotation wastewater (WFW). The initial quality parameters of the WFW were determined as described in the following section. Prior to use, the wastewater was diluted with distilled water in a 1:2 (*v*/*v*) ratio, and the initial pH of the cultivation medium was adjusted to 6.5 ± 0.2. The media were sterilized by autoclaving at 121 °C for 20 min under a pressure of 2.1 bar.

2.6. Physicochemical Characterization of WFW

The physicochemical characterization of WFW was conducted prior to medium inoculation and after 48 h of cultivation. To determine the basic nutritional profile of the WFW medium and the 48-h cultivation broth, several methods and techniques were used to assess dry matter, sugar content, protein and cellulose content, as well as alcohol content. The total solids of samples was determined gravimetrically, following the method described in the Official Gazette of the SFRY, No. 29/83 [33]. Soluble solids were measured refractometrically using an Abbe universal refractometer (Carl Zeiss, Oberkochen, Germany), with direct readings taken from the refractometer scale. The total sugar content in the samples was analyzed using the Luff-Schoorl method [33], while reducing sugars were quantified volumetrically according to the procedure outlined in the same source. Cellulose (crude fiber) content was determined using the Kirschner–Ganakova method, as described by Vračar (2001) [34]. The protein content was determined by the Kjeldahl method [35]. Alcohol (ethanol) content was determined by distillation, followed by oxidation with potassium dichromate in the presence of sulfuric acid. The excess dichromate was titrated with ammonium ferrous sulfate using iron-ortho-phenanthroline as an indicator, according to the Official Gazette of the SFRY, No. 29/83 [33].

2.7. Inoculum Preparation and Cultivation Media and Conditions

The inoculum of *Bacillus* sp. 10/R was prepared by transferring a loopful of the previously refreshed culture into sterile nutrient broth, followed by cultivation at 28 °C for 24 h on a laboratory rotary shaker at 150 rpm. The previously prepared and sterilized WFW-based medium was inoculated with 10% (*v*/*v*) of the inoculum. Cultivation was carried out at 28 °C for 96 h on a laboratory rotary shaker at 175 rpm under spontaneous aeration. Samples of the cultivation broth were collected at 24-h intervals for the determination of biomass content and enzymatic activity.

2.8. Biomass Content Determination

The droplet plate method was applied to assess biomass content of *Bacillus* sp. 10/R. Serial dilutions of the bacterial suspension were prepared by adding 0.1 mL of sample into 0.9 mL of sterile saline, vortexing for 8 s, and repeating to achieve a 10-fold dilution series. For plating, agar plates were divided into four quadrants, each labeled for a different dilution. After vortexing, 10 μL of each dilution was added as four evenly spaced drops (total volume of 40 μL) onto the designated area of the agar plates.

The drops were allowed to absorb (~15–20 min on pre-dried agar), after which the plates were inverted and incubated 28 °C for 17–20 h. Colonies within drops containing 3–30 colonies were counted, and the viable cell concentration was calculated as colony-forming units per milliliter (CFU/mL) using the formula [36]

$$\text{CFU} = \frac{\text{average number of colonies per drop}}{0.01\ \text{mL}} \times \text{dilution factor} \tag{3}$$

2.9. Quantitative Determination of Enzyme Activity in Liquid Culture

Pectinase activity was determined using a modified dinitrosalicylic acid (DNS) method as described by Mohandas et al. [27]. Briefly, culture broth of *Bacillus* sp. 10/R was centrifuged at 10,000 rpm for 10 min, and the supernatant was used as the crude enzyme extract. The reaction mixture containing 0.5 mL of culture supernatant and 0.5 mL of 1% (*w*/*v*) pectin (Sigma Aldrich–Merck, Darmstadt, Germany) as substrate was incubated at 50 °C for 30 min in a water bath. After incubation, 1.5 mL of DNS reagent was added, and the mixture was heated in a boiling water bath for 5 min. The samples were then cooled, diluted appropriately, and the absorbance was measured at 540 nm using a spectrophotometer (UV-1800, Shimadzu, Kyoto, Japan). The cultivation medium (sterilized WFW) without the bacteria was used as a control. Mono-D-galacturonic acid was used to prepare the standard calibration curve, and one unit of pectinase activity was defined as the amount of enzyme required to release 1 μmol of galacturonic acid per mL per minute under the assay conditions. The pectinase activity was calculated using the following formula:

$$\text{Pectinase activity} = \frac{\text{m(monogalacturonic acid}(\mu\text{g}))}{\text{v(sample volume)} \times \text{t(incubation time)} \times 212.15} \quad (4)$$

Quantitative determination of endo-cellulase and endo-xylanase activities in the culture supernatants was performed using the Megazyme assay kits (K-CellG5, K-XylX6) according to the manufacturer's instructions (Megazyme, Wicklow, Ireland). The original extracts were preprared by diluting supernatant samples in sodium phosphate buffer. For enzyme assays, 0.10 mL of substrate solution (CellG5 for cellulase, XylX6 for xylanase) was mixed with 0.1 mL of the diluted samples (original extracts) and incubated for 10 min at 40 °C. The reactions were stopped with 3.0 mL of the stopping reagent (2% (w/v) Tris buffer (pH 10)), and absorbance was measured at 400 nm (UV-1800, Shimadzu, Kyoto, Japan). Reagent blanks were prepared for both cellulase and xylanase by mixing the stopping reagent, substrate solution, and the diluted enzymes. Enzyme activity was expressed as CellG5 units/mL for cellulase and XylX6 units/mL for xylanase. One CellG5 unit is defined as the amount of enzyme required to release 1 μmol of 4-nitrophenol from CellG5 per minute, while one XylX6 unit is defined as the amount of enzyme required to release 1 μmol of 4-nitrophenol from XylX6 per minute under the assay conditions.

2.10. Seed Germination Assay

For seed germination assay barley (*Hordeum vulgare*, Nonius variety) seeds were used, which were commercially supplied from Institute of Field and Vegetable Crops, Novi Sad, Serbia. The seeds were surface sterilized using the chlorine bleach solution (6% (v/v), 1 min) and thoroughly washed using the sterile distilled water for 5 min [37]. After drying, ten seeds were placed in 90 mm diameter Petri dishes containing filter paper. The moisture required for germination was provided by adding 2 mL of tap water onto the filter paper. Each dish with seeds was treated with 1 mL of cultivation broth based on WFW or nutrient broth (NB), applied either undiluted or diluted 2, 4, 10 and 20 folds. WFW and NB in the same volume and dilutions as the treatments were used as controls, as well as tap water. The Petri dishes were sealed with parafilm to prevent moisture loss and incubated in the dark for the first 48 h at 25 °C, followed by incubation under ambient light with natural day-night cycles for the next 5 days at the same temperature. Each day, the germination percentage (GP, %) for each treatment and control was determined by counting the number of germinated and non-germinated seeds. The minimal root length of 1 mm was used for evaluation of germination test as positive. These data were also used to calculate the mean germination time (MGT, days; Equation (5)), where n_i represents the number of seeds germinated on day d_i, and N is the total number of seeds germinated after 7 days. Additionally, the germination index (GI, seeds/day; Equation (6)) was calculated starting from day 3 of the germination period. At the end of the 7-day germination period, the following seedling parameters were measured: fresh mass (FM, g), dry mass (DM, g; after drying at 80 °C to constant mass), total length (TL, cm), root length (RL, cm), and shoot length (SL, cm). Based on these measurements, seedling vigour indices I and II (SVI-I and SVI-II; Equations (7) and (8), respectively [38]) were calculated.

$$\text{MGT} = \frac{\sum(n_i \times d_i)}{N} \quad (5)$$

$$GI = \sum \left(\frac{n_3}{d_3}\right) + \ldots \left(\frac{n_7}{d_7}\right) \quad (6)$$

$$SVI - I = GP \times TL \quad (7)$$

$$SVI - II = GP \times DM \quad (8)$$

2.11. Statistical Analyses

The experimental data were analyzed using TIBCO Statistica® 14.1.0 software (TIBCO Software Inc., part of Cloud Software Group, San Ramon, CA, USA) to evaluate statistical differences using the One-Way ANOVA, followed by Duncan's multiple range post hoc test, taking into account 95% confidence intervals.

3. Results

3.1. Identification of the Producing Microorganism

The identification of producing microorganism was performed using 16S rRNA gene sequencing. The results showed the highest similarity of the producing microorganism with *Bacillus mojavensis* and *Bacillus halotolerans* based on BLAST analysis, with sequence similarity values around 98%. Phylogenetic analysis based on 16S rRNA sequences was performed to determine the relationship of the isolate with the closely related *Bacillus* species. The constructed phylogenetic tree (Figure 1) confirmed the BLAST results, clearly showing the closest relationship of the isolate with *Bacillus mojavensis* and *Bacillus halotolerans* strains. The 16S rRNA sequence of *Bacillus* sp. 10/R has been deposited in the GenBank database under the accession number PX210502.

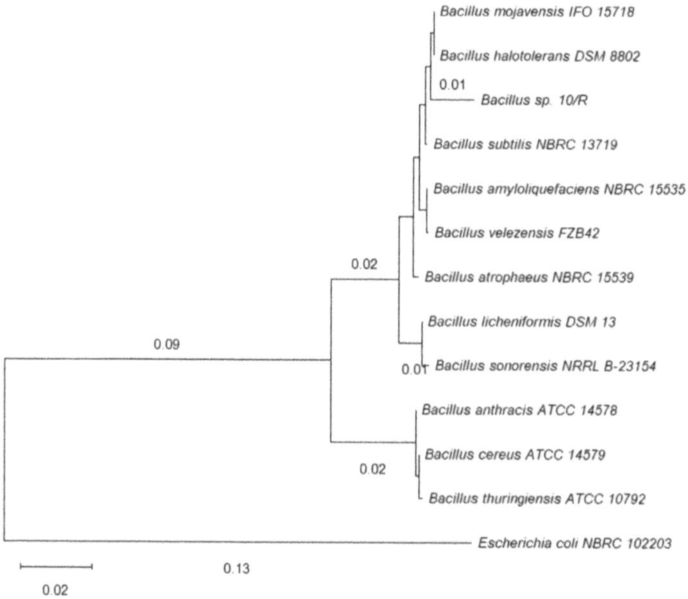

Figure 1. Phylogenetic tree based on 16S rRNA sequences of the isolate *Bacillus* sp. 10/R and closely related *Bacillus* species, constructed using the Neighbor-Joining method and the Kimura 2-parameter model in MEGA12. Branch lengths are proportional to evolutionary distances, and the tree is drawn to scale. Positions containing gaps or missing data were eliminated, yielding 1017 nucleotide positions.

3.2. Biochemical Profiling of the Strain Bacillus sp. 10/R

In order to investigate the biochemical characteristics of the *Bacillus* sp. 10/R isolate, API test kits were used. The results obtained after 48 h of incubation of the inoculated API 50 CH strips, which include 50 biochemical tests for the study of bacterial carbohydrate metabolism, are shown in Table 1.

Table 1. The results of API 50 CH biochemical tests for the isolate *Bacillus* sp. 10/R.

API 50 CH Test Strip Results					
Glycerol	+	D-mannitol	+	D-raffinose	+
Erythritol	−	D-sorbitol	+	Starch (amidon)	+
D-arabinose	−	methyl-α-D-mannopyranoside	−	Glycogen	+
L-arabinose	+	methyl-α-D-glycopyranoside	+	Xylitol	−
D-ribose	+	*N*-acetylglucosamide	−	Gentiobiose	−
D-xylose	+	Amygdalin	+	D-turanose	−
L-xylose	−	Arbutin	+	L-lyxose	−
D-adonitol	−	Esculin	+	D-tagatose	+
Methyl-β-D-xylopyranoside	−	Salicin	+	D-fucose	−
D-galactose	−	D-cellobiose	+	L-fucose	−
D-glucose	+	D-maltose	+	D-arabitol	−
D-fructose	+	D-lactose	−	L-arabitol	−
D-mannose	+	D-melibiose	−	Potassium gluconate	+
L-sorbose	−	D-saccharose	+	Potassium 2-ketoglyconate	−
L-rhamnose	−	D-trehalose	+	Potassium 5-ketoglyconate	−
Dulcitol	−	Inulin	+		
Inositol	−	D-melezitose	−		

"+" designates positive rection results, while "−" designates negative reaction result.

The second API kit used for the biochemical characterization of *Bacillus* sp. 10/R was API 20 E. Since the last 10 tests included in this kit are already covered by the API 50 CH test strip, Table 2 shows the results of the remaining 10 tests from the API 20E kit.

Table 2. The results of API 20 E biochemical tests for the isolate *Bacillus* sp. 10/R.

API 20 E Test Strip Results									
ONPG	+	ADH	+	LDC	−	ODC	−	CIT	+
H_2S	−	URE	−	TDA	−	IND	+	VP	+

ONPG–β-galactosidase; ADH–arginine dihydrolase; LDC–lysine decarboxylase; ODC–ornithine decarboxylase; CIT–citrate utilization; H_2S–H_2S production; URE–urease; TDA–tryptophane deaminase; IND–indole production; VP–acetoin production; "+" designates positive rection results, while "−" designates negative reaction result.

The results of the API ZYM tests are presented in Table 3. API ZYM is a semi-quantitative micromethod used for the investigation of enzymatic activities.

Table 3. The results of API ZYM biochemical tests for the isolate *Bacillus* sp. 10/R.

API ZYM Test Strip Results			
Control		Acid phosphatase	+
Alkaline phosphatase	+	Naphthol-AS-BI-phosphohydrolase	+
Esterase (C 4)	+	α-galactosidase	−
Esterase Lipase (C 8)	+	β-galactosidase	−
Lipase (C 14)	−	β-glucuronidase	−
Leucine arylamidase	+	α-glucosidase	+
Valine arylamidase	+	β-glucosidase	−
Cystine arylamidase	−	*N*-acetyl-β-glucosaminidase	−
Trypsin	−	α-mannosidase	−
α-chymotrypsin	−	α-fucosidase	−

"+" designates positive rection results, while "−" designates negative reaction result.

3.3. Plant Growth Promoting Traits of the Strain Bacillus sp. 10/R

In order to assess the potential of the investigated *Bacillus* isolate as a bioinoculant for plants, several traits considered crucial for plant growth promotion (PGP) were evaluated. These included enzymatic activity of the hydrolytic enzymes (cellulase, xylanase, pectinase and protease), indole-3-acetic acid (IAA) production, phosphate solubilization, and ACC deaminase production. The results of the enzymatic activity assays are shown in Figure 2, presenting growth and halo zone diameters, as well as the enzymatic activity index (EAI). The lowest EAI value was recorded for protease activity, where no halo zone appeared around the bacterial growth. On the other hand, the highest EAI was observed for pectinase production, with a value of 2.79, indicating a strong capacity for the extracellular production of this enzyme. The enzymatic activity indices for cellulase and xylanase were 2.33 and 1.75, respectively.

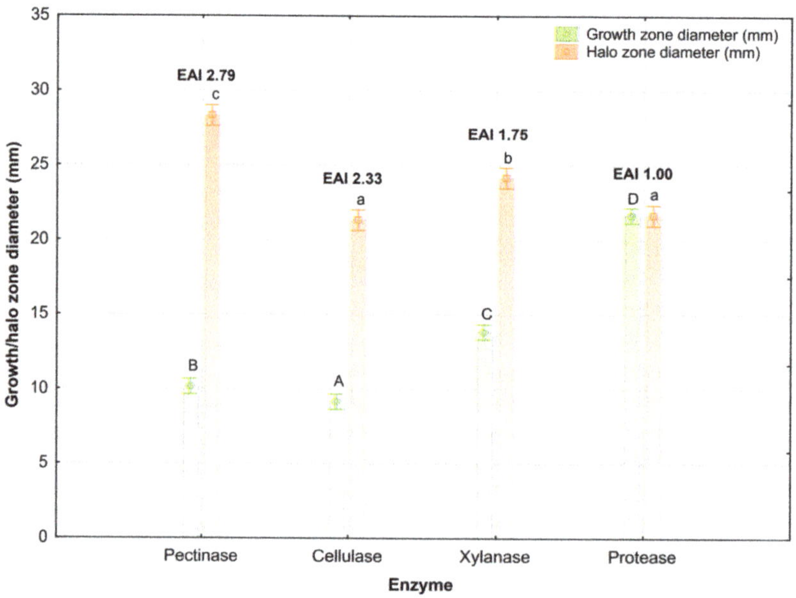

Figure 2. Enzymatic activity of *Bacillus* sp. 10/R. Letters (a–d and A–D) represent different levels of statistical significance based on Duncan's multiple range test, which has been performed after the separate One-way ANOVAs (analyses of variance) for two presented data series (growth zone diameter and halo zone diameter). Values marked with the same letter are at the same level of significance (95% confidence interval). EAI–enzymatic activity index.

The ability of the *Bacillus* sp. 10/R isolate to produce ACC deaminase was confirmed by its growth on DF minimal medium supplemented with 1-aminocyclopropane-1-carboxylic acid (ACC) as the sole nitrogen source. A growth zone of 54.67 ± 0.58 mm was observed, indicating the strain's capability to utilize ACC. In contrast, phosphate solubilization activity was relatively low, with a phosphate solubilization index (PSI) of only 1.13 ± 0.07. The results of these assays are presented in Table 4.

Table 4. Plant growth promoting traits of *Bacillus* sp. 10/R in terms of ACC deaminase production and phosphate solubilization.

Parameter	Growth Zone Diameter (mm)	Halo Zone Diameter (mm)	PSI
ACC deaminase production	54.67 ± 0.58	/	/
Phosphate solubilization	11.17 ± 0.29	12.67 ± 1.15	1.13 ± 0.07

PSI—phosphate solubilization index, ACC—1-aminocyclopropane-1-carboxylic acid.

The ability of *Bacillus* sp. 10/R to produce indole-3-acetic acid (IAA), a key phytohormone involved in the regulation of plant growth and root development, was evaluated using a colorimetric method based on reaction with Salkowski's reagent. The assay was performed using the cultivation broth supernatant after incubation under the optimized conditions. The concentration of IAA detected in the supernatant was 9.66 μg/mL, indicating a moderate level of auxin production by the isolate *Bacillus* sp. 10/R.

3.4. Basic Nutritional Profile of WFW Medium and Cultivation Broth

The characterization of the WFW medium was performed by analyzing total and soluble solids, sugar content, protein, cellulose, and alcohol content in order to assess its potential as a substrate for microbial bioconversion. The same parameters were evaluated after 48 h of *Bacillus* sp. 10/R cultivation to determine changes in the nutritional profile caused by bacterial activity. The obtained results are presented in Table 5.

Table 5. The basic nutritional and pH value profile of WFW medium before and after cultivation.

Parameter	Before Cultivation	After Cultivation
Total solids (%)	7.96 ± 0.06	6.99 ± 0.13
Soluble solids (°Bx)	6.42 ± 0.35	6.00 ± 0.00
Total sugar content (%)	6.61 ± 1.35	2.68 ± 0.06
Reducing sugar content (%)	4.28 ± 0.47	/
Protein content (%)	0.64 ± 0.06	0.29 ± 0.02
Cellulose content (%)	0.01 ± 0.01	0.00 ± 0.00
pH value	3.54 ± 0.27	5.95 ± 0.25

3.5. Bacterial Growth in WFW Medium During 96 h of Cultivation

To investigate bacterial survival and growth in the used winery waste-based medium, biomass content of *Bacillus* sp. 10/R was measured every 24 h of cultivation. The drop plate method was employed for biomass determination, and the obtained results are presented in Figure 3. The biomass increased by approximately 2 log units after 48 h of cultivation and remained close to that level over the following 48 h. Considering that the maximum bacterial content was observed at 48 h, this time point was selected as the final cultivation time for subsequent analyses.

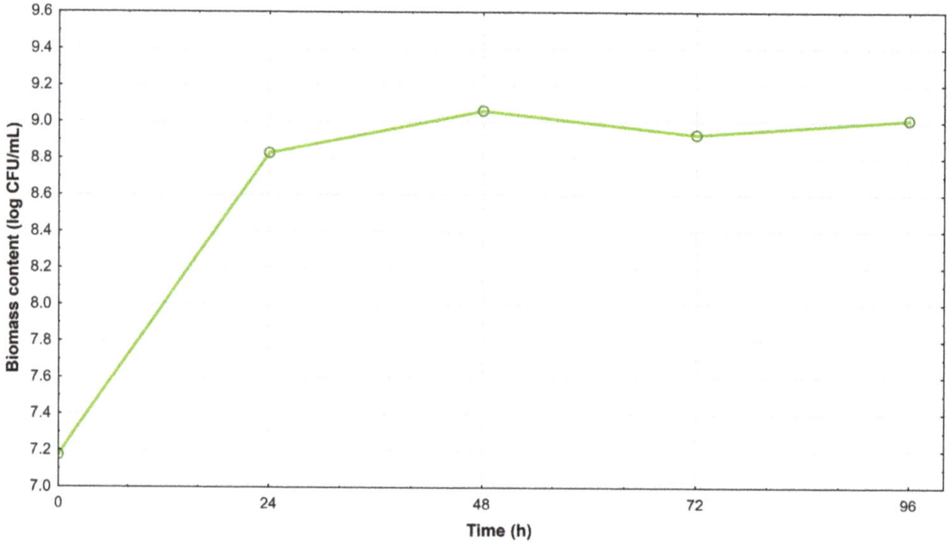

Figure 3. Growth dynamics of *Bacillus* sp. 10/R in WFW-based medium during 96 h of cultivation.

3.6. Enzyme Activity in Liquid WFW-Based Culture

Throughout the 48 h-cultivation of *Bacillus* sp. 10/R on WFW-based medium, enzymatic activity was measured by the DNS method for pectinase activity, while cellulase and xylanase activity was determined using CellG5 method and XylX6 method using the commercial enzyme assay kits. The obtained results in terms of enzymatic activity are presented in Table 6.

Table 6. Pectinase, cellulase, and xylanase activities of *Bacillus* sp. 10/R cultivated using the winery flotation wastewater medium, monitored over 48 h of cultivation.

Time of Cultivation (h)	Pectinase Activity (U/mL)	Cellulase Activity (CellG5 U/mL)	Xylanase Activity (XylX6 U/mL)
12	2.19	0.00	0.00
24	2.19	0.01	0.00
36	2.19	0.00	0.00
48	2.20	0.08	0.00

U/mL = (μMol/(mL × min)).

3.7. Seed Germination

The efficiency of *Bacillus* sp. 10/R cultivation broth in promoting barley seed germination was evaluated using the cultivation broths prepared on WFW and nutrient broth (NB) at different dilutions. These treatments were compared with the corresponding dilutions of WFW and NB alone, as well as with tap water, which served as the negative control. Germination rate was recorded daily during a 7-day experiment. After 7 days, root and shoot lengths of each seedling were measured, together with fresh mass. Dry mass was determined after drying at 80 °C to constant weight. The results for germination rate, seedling length, and fresh and dry mass are presented in Figure 4. The experimental data obtained after 7 days were further used to calculate mean germination time (MGT), germination index (GI), and seedling vigor indices I and II (SVI-I and SVI-II), as shown in Figure 5. In both figures, cultivation broths are designated as CB.

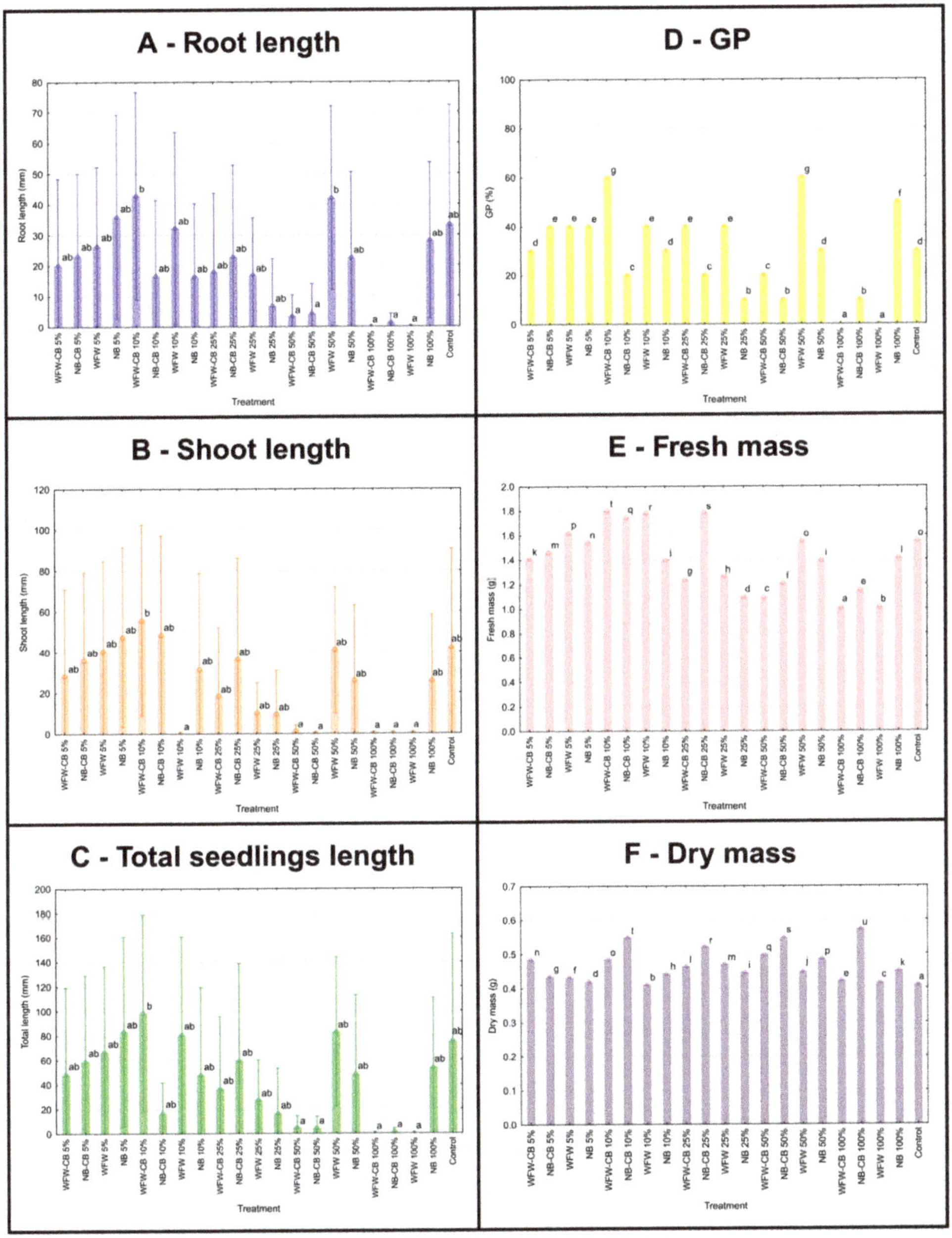

Figure 4. Barley growth and germination parameters after treatment using cultivation broths of *Bacillus* sp. 10/R based on winery flotation wastewater (WFW-CB) and nutrient broth (NB) in various concentrations (5%, 10%, 25%, 50%, 100%, *v*/*v*) after 7 day-incubation. Letter designations represent different levels of statistical significance based on Duncan's multiple range test. Values marked with the same letter are at the same level of significance (95% confidence interval).

As illustrated in Figures 4 and 5, the concentration of cultivation broths and media strongly influenced germination rate as well as seedling length and biomass. The WFW-based cultivation broth applied at 10% concentration had a particularly pronounced effect on shoot elongation, resulting in the highest total seedling length (98.4 mm) among all treatments. The highest germination percentage (60%) was recorded for seeds treated with 10% WFW-CB and 50% WFW, whereas no germination occurred in treatments with 100% WFW and 100% WFW-CB. Regarding the plant biomass synthesis, fresh mass was highest in seeds treated with 10% WFW-CB, while dry mass was greatest in seedlings treated with 10% NB-CB. Overall, all treatments with cultivation broths yielded higher dry mass compared to media alone and the tap water control.

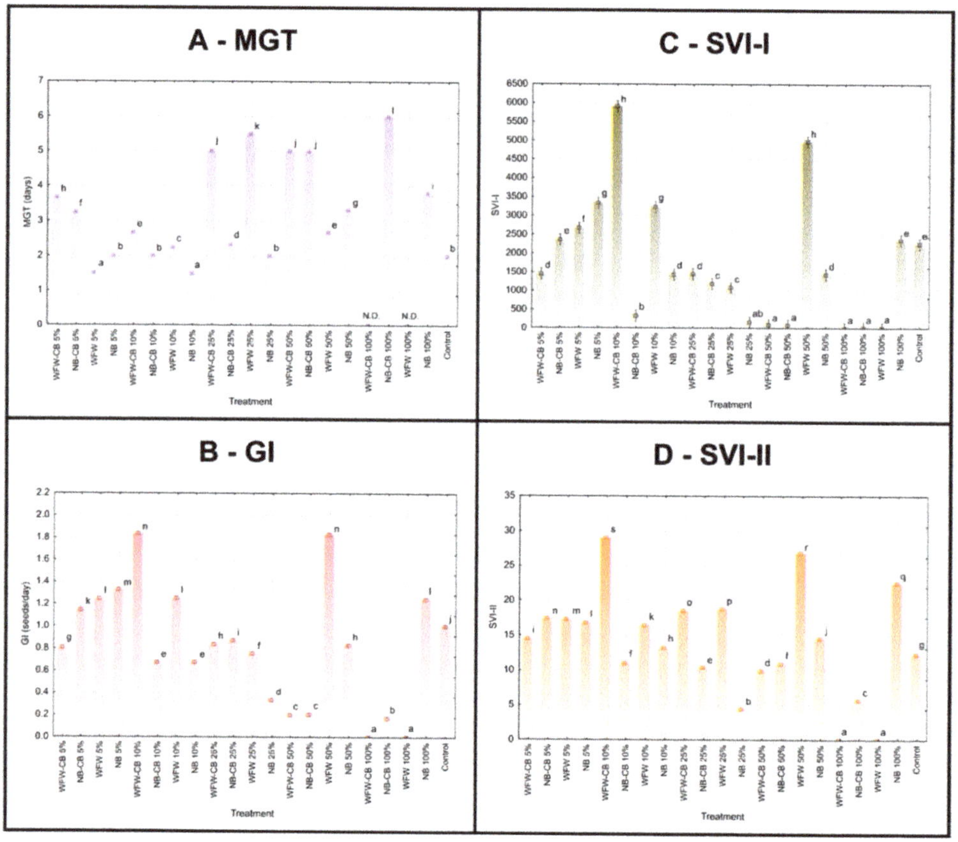

Figure 5. Mean germination time (MGT), germination index (GI), and seedling vigor indices I and II (SVI-I and SVI-II) of barley seeds/seedlings after treatment using cultivation broths of *Bacillus* sp. 10/R based on winery flotation wastewater (WFW-CB) and nutrient broth (NB) in various concentrations (5%, 10%, 25%, 50%, 100%, *v*/*v*) after 7 day-incubation. Letter designations represent different levels of statistical significance based on Duncan's multiple range test. Values marked with the same letter are at the same level of significance (95% confidence interval).

For MGT (Figure 5), no values could be determined for seeds treated with 100% WFW-CB and 100% WFW due to the absence of germination. The highest GI value was observed in seeds treated with 10% WFW-CB (1.84 seeds/day), followed closely by 50% WFW (1.83 seeds/day). Among the treatments where germination occurred, the lowest GI values were recorded for 100% NB-CB (0.17), 50% WFW-CB (0.20), and 50% NB-CB (0.20).

4. Discussion

In recent years, microorganisms used as bioinoculants and PGP microorganisms have garnered increasing attention due to their potential to enhance sustainable agriculture. Their ability to stimulate plant growth and improve plant health is attributed to a variety of direct and indirect mechanisms [39–42]. These include direct PGP mechanism facilitating nutrient acquisition and phytohormones production (auxins, gibberellins, abscisic acid, and cytokinins), as well indirect PGP mechanisms targeting biocontrol of competing organisms and plant pathogens, as well as production of compounds involved in quorum sensing and plant systemic resistance induction [43,44]. The most commonly studied and applied PGP microorganisms include species from the genera *Bacillus*, *Pseudomonas*, *Azospirillum*, *Azotobacter*, and *Rhizobium*, as well as beneficial fungi such as *Trichoderma* [45,46]. The genus *Bacillus* stands out as one of the most promising bacterial genera due to its wide range of beneficial activities and is among the most commonly used for plant growth promotion. Numerous species within this genus, such as *B. velezensis*, *B. subtilis*, *B. amyloliquefaciens*, *B. macerans*, *B. circulans*, *B. azotofixans*, and *B. coagulans*, have been well-documented for their ability to enhance plant growth through various mechanisms [47,48].

4.1. PGP Traits of B. mojavensis/B. halotolerans Strains

The grapevine rhizosphere isolate *Bacillus* sp. 10/R used in this study was molecularly identified using 16S rRNA gene sequencing and showed the highest similarity with *Bacillus mojavensis* and *Bacillus halotolerans*, with sequence identity values around 98%. Although this level of similarity suggests a close relationship with these species, it does not reach the $\geq$98.65% threshold generally accepted for confident species-level identification based on 16S rRNA. Therefore, the isolate was designated as *Bacillus* sp., most closely related to the *B. mojavensis*/*B. halotolerans* clade. To achieve precise taxonomic resolution, further studies such as multilocus sequence analysis or whole-genome sequencing would be required. Previous studies have highlighted the potential of *B. mojavensis* as a PGP species with multiple beneficial traits. For instance, Prajakta et al. (2019) [49] demonstrated that *B. mojavensis* produces volatile metabolites with antifungal activity against *Rhizoctonia solani*, along with siderophores, IAA, chitinases, and phosphate-solubilizing ability, confirming its dual role in biocontrol and plant growth promotion. Ghazala et al. (2023) [50] showed that *B. mojavensis* strain PGPB I4 enhanced wheat growth under salinity stress by improving physiological parameters, chlorophyll content, and antioxidant activity, whereas Sdiri et al. (2024) [51] reported that *B. mojavensis* I4 (BmI4) improved potato growth and tolerance to salinity through similar mechanisms, including enhanced membrane stability and modulation of antioxidant enzymes. In addition, Danish et al. (2022) [52] showed that *B. mojavensis* BZ-13 alleviated silver nanoparticle-induced toxicity in *Withania somnifera*, improving plant biomass, photosynthetic efficiency, and secondary metabolite accumulation while reducing oxidative stress and nanoparticle uptake. Several studies have highlighted the potential of *Bacillus halotolerans* as a PGP agent under salinity stress. Kapadla et al. (2022) [53] reported that the strain AD9, isolated from the saline coastal soil, exhibited multiple PGP traits at high salt concentrations (15–20%), including high levels of ammonia production and phosphate solubilization. Inoculation with AD9 significantly improved rice growth under salinity stress, resulting in increased root length, plant dry weight, and tiller number compared to the control. Similarly, Oliva et al. (2023) [54] characterized halotolerant strains including *Bacillus* sp. M21 and M23, closely related to *B. halotolerans*, and demonstrated that their PGP activities, such as IAA production and ammonium release, were strongly influenced by the present NaCl levels.

4.2. Biochemical Profile of *Bacillus sp.* 10/R

The biochemical characterization of *Bacillus* sp. 10/R using API 50 CH, API 20 E, and API ZYM systems revealed a versatile metabolic profile. The isolate was capable of fermenting a broad range of carbohydrates, including glucose, maltose, arabinose, salicin, cellobiose, and trehalose, indicating strong saccharolytic potential, while failing to utilize sugars such as galactose, turanose, lyxose, and dulcitol. API 20 E results showed positive reactions for β-galactosidase, arginine dihydrolase, citrate utilization, indole production and acetoine production, demonstrating the strain's ability to metabolize diverse substrates and produce key metabolic intermediates, consistent with other *Bacillus* species. The enzymatic profiling using API ZYM indicated high activity of esterases (C4 and C8), alkaline phosphatase, leucine arylamidase, valine arylamidase, acid phosphatase, and naphthol-AS-BI-phosphohydrolase, reflecting the strain's potential to degrade a variety of substrates, including proteins and phosphoesters. In contrast, activities of lipase (C14), cystine arylamidase, trypsin, α-chymotrypsin, β-glucosidase, and other glycoside hydrolases were absent or weak, suggesting selective enzymatic capabilities. Overall, these results demonstrate that *Bacillus* sp. 10/R possesses a diverse enzymatic repertoire and carbohydrate utilization profile, which may contribute to its adaptability and potential for biotechnological applications such as plant growth promotion and biocontrol.

4.3. Enzymatic Activity of *Bacillus sp.* 10/R

The remarkable adaptability of *Bacillus* strains to diverse environmental conditions is one of the key advantages of this genus, enabling cultivation on a wide range of substrates, including nutrient-rich industrial waste streams [55]. The production of hydrolytic enzymes provides these bacteria with the ability to access and utilize nutrients from complex substrates. Screening for hydrolytic enzyme activity among *Bacillus* isolates is therefore a common step in the assessment of their PGP potential, since such enzymatic activity not only enhances the utilization of nutrients available in the soil by plants and other beneficial soil microorganisms, but also contributes to the improved degradation of crop residues and direct antagonism against phytopathogens through cell wall degradation mediated by these enzymes [9]. Cellulase, xylanase, pectinase, and protease activities were evaluated as the key enzymatic functions involved in degradation of the predominant organic substrates typically present in soil. Cellulases target cellulose originating from plant residues, xylanases hydrolyze hemicellulose components such as xylan from lignocellulosic material, pectinases degrade pectin-rich components derived from fruit and vegetable tissues, while proteases break down proteins released from microbial biomass and other organic inputs. The selection of these enzymes was based on their central role in mineralizing the major organic fractions in soil and facilitating nutrient cycling [56,57].

Using a semi-quantitative agar plate assay, enzymatic activity was expressed as the enzymatic activity index (EAI) to provide a comparative measure of hydrolytic potential, based on the extracellular enzymatic activity, as shown in Figure 1. The investigated isolate *Bacillus* sp. 10/R exhibited the highest extracellular pectinase activity (EAI = 2.79), indicating strong capacity for hydrolyzing pectin-rich substrates and suggesting adaptation to environments enriched with fruit- and vegetable-derived residues, which is also beneficial in case of using grape-based winery effluents for cultivation of the aforementioned strain, as in this study. According to Kabir and Tasmim (2019) [58], a halo zone diameter of ≥15 mm indicates very good pectinolytic activity, which is consistent with the results obtained in our study. Among 95 isolates from coffee pulp investigated by Oumer and Abate (2018) [59], 31.58% exhibited pectinolytic activity as evidenced by clear zones around bacterial growth, while 10 isolates were considered good pectinase producers with an EAI ≥ 2.00. Cellulase activity (EAI = 2.33) was also notable, reflecting the potential to degrade cellulose-containing plant material, whereas xylanase activity was comparatively lower (EAI = 1.75), which may indicate a limited hemicellulolytic capacity under the applied experimental conditions. In a study by Castaldi et al. (2021) [60], among 22 *Bacillus* isolates, the highest proportion of strains produced xylanase and protease activities (90%), while 80% exhibited cellulolytic activity. Furthermore, cellulase activity was investigated among 398 isolates from various environments and detected in 6.5% of them, with enzymatic indexes ranging between 0.34 and 5.2 [61]. Xylanase EAI achieved by *Bacillus* sp. BioSol021 was 2.85, which is significantly higher than the result achieved by *Bacillus* sp. 10/R [9]. In contrast, protease activity showed an EAI of 1.00, as no clear hydrolysis zone surrounding the growth area was observed. Notably, the largest growth zone (21.67 $\pm$ 0.58 mm) was recorded when the isolate was cultivated on a protein-based medium supplemented with skim milk as the nutrient source. Although no surrounding clear zone of extracellular protein hydrolysis was detected, the pronounced colony growth suggests that the isolate was able to effectively utilize the available amino acid-based nutrients subsequently with their enzymatic release from milk proteins. This may imply the presence of cell-bound or intracellular proteolytic enzymes, the uptake of small peptides and amino acids naturally present in skim milk, or an alternative metabolic adaptation that supports growth without extensive extracellular protein degradation.

4.4. *Direct and Indirect PGP Mechanisms Exhibited by Bacillus sp. 10/R*

Abiotic and biotic stress significantly reduce crop growth and productivity by altering plant morphological and physiological traits, often through increased ethylene production, which inhibits root development. Plant growth-promoting microorganisms can alleviate these negative effects, with ACC deaminase production representing a key mechanism of stress mitigation. This enzyme hydrolyzes ACC, the ethylene precursor, into ammonia and α-ketobutyrate, thereby reducing ethylene levels under stress conditions and promoting root growth and overall plant development [62,63]. Diverse microbial taxa, including *Acinetobacter*, *Aeromonas*, *Burkholderia*, *Bacillus*, *Enterobacter*, *Hallobacillus*, *Sphingomonas* and *Streptomyces*, have been identified as efficient ACC deaminase producers. Exploiting such microbes offers a promising strategy to improve plant tolerance to environmental stresses, including drought [64]. In the present study, the ability of *Bacillus* sp. 10/R to grow on ACC-supplemented medium devoid of other nitrogen sources, reaching a colony diameter of 54.67 $\pm$ 0.58 mm, indicates substantial ACC deaminase activity, further supporting its potential role in enhancing plant stress tolerance. After nitrogen, phosphorus represents the second most essential macronutrient required for optimal plant growth and development [65]. Phosphorus is essential for crop development, particularly during early vegetative growth, seedling establishment, and maturation, where it plays a key role in enhancing seed quality and yield through its involvement in various morphological and physiological processes. Phosphate-solubilizing microorganisms facilitate the release of phosphorus from both inorganic and organic soil pools via solubilization and mineralization, thereby acting as PGP agents. This activity not only improves plant growth and grain yield but also reduces the need for chemical fertilizers [66]. Candales et al. (2017) [67] investigated the PGP potential of *Bacillus* isolates and found that, among 15 tested strains, only two, *Bacillus subtilis* GIBI 200 and *Bacillus pumilus* GIBI 206, were able to solubilize phosphorus. In the study by Kumari et al. (2024) [68], five out of seven bacterial isolates demonstrated phosphate solubilization on Pikovskaya medium, with PSI values ranging from 2.3 to 4.5. Similarly, PGP *Bacillus* strains examined by Chandra et al. (2021) [69] exhibited PSI values between 2.5 and 3.6. In contrast, the results obtained in our study indicate a low phosphate-solubilizing ability of *Bacillus* sp. 10/R compared to the aforementioned studies, with over 50% lower PSI values (1.13 $\pm$ 0.07).

IAA is primarily responsible for root elongation by stimulating cell division and tissue differentiation, which subsequently increases plant fresh mass. Therefore, its production is considered one of the main PGP traits to be assessed when evaluating the potential active components of biostimulants [70]. The production of IAA as a plant growth-promoting trait has been widely investigated in numerous studies, showing variable results. Goud et al. (2025) [71] reported that, out of 80 bacterial isolates, 67 were capable of producing IAA and its derivatives, indicating that IAA production is a common feature among many bacteria. In their study, IAA was synthesized predominantly via the tryptophan-dependent pathway, with concentrations ranging from 3.8 to 70.0 μg/mL. Vlajkov et al. (2023) [9] reported an IAA concentration of 15 μg/mL for *Bacillus* sp. BioSol021, while in the study by Chandra et al. (2021) [69], IAA concentrations ranged from 2.49 to 5.47 μg/mL, with the highest level produced by *Bacillus subtilis* strain GIBI 200. In the present study, *Bacillus* sp. 10/R produced IAA at a concentration of 9.66 μg/mL, which demonstrates a substantial capacity for IAA synthesis.

4.5. Winery Flotation Wastewater as a Substrate for Bacillus sp. 10/R Growth

Winery wastewater (WWW) represents a major effluent stream of the winemaking industry, generated mainly from cleaning and processing operations, and is characterized by high volumes, strong seasonal variability, and an acidic pH value. Its composition includes elevated concentrations of readily biodegradable organic matter (sugars, ethanol, organic acids), suspended solids, and polyphenolic compounds, alongside moderate levels of nitrogen, phosphorus, and salts [72,73]. The reported COD (chemical oxygen demand) values typically range between 320 and 49,000 mg/L, while BOD_5 (biological oxygen demand) averages around 6500 mg/L, highlighting its high organic strength and biodegradability [73], However, the low N and P relative to organic load often necessitates nutrient supplementation for microbial processes. Despite these challenges, the nutrient-rich nature of WWW and its biodegradability make it a suitable substrate for microbial bioconversion processes, including anaerobic digestion, where it can serve as a source for bioenergy and value-added products such as biofertilizers [72,74].

To assess the potential of WFW as a substrate for the cultivation of *Bacillus* sp. 10/R, its basic nutritional profile was investigated, including total and soluble solids, total and reducing sugar content, protein content, and alcohol content. The same analyses were performed at the end of the cultivation period to monitor nutrient consumption by the bacteria. The results presented in Table 5 correspond to WFW diluted three times, which was the form used for the cultivation of *Bacillus* sp. 10/R. The initial pH value of the prepared medium was 3.54 ± 0.27, representing an acidic environment, and was subsequently adjusted to 6.5 ± 0.2 prior to inoculation. It has been gradually dropping to 5.95 ± 0.25 by the end of the cultivation, thus suggesting the production of acidic bacterial metabolites during the cultivation course. The cultivation was monitored over time, and although samples were collected beyond 48 h, this time point was chosen as the endpoint for compositional analyses, as it corresponded to the maximum biomass content and peak viable cell density. Prior to inoculation, the medium contained 7.96 ± 0.06% total solids and 6.42 ± 0.35 °Bx soluble solids, reflecting a high load of dissolved organic matter. The total sugar content was 6.61 ± 1.35%, of which 4.28 ± 0.47% were reducing sugars, indicating a substantial proportion of readily utilizable monosaccharides and disaccharides. The protein content of 0.64 ± 0.06% suggested the presence of organic nitrogen sources, while cellulose content was negligible (0.01 ± 0.01%), as expected for a liquid waste stream derived from fruit processing, considering that the major part of insoluble carbohydrates remains in the solid fraction after grape processing [21]. After 48 h of cultivation, total solids decreased to 6.99 ± 0.13%, and soluble solids to 6.00 ± 0.00 °Bx. Total sugars declined to 2.68 ± 0.06%, whereas reducing sugars were completely depleted, indicating the rapid utilisation of simple carbohydrates by the bacteria. Protein content dropped to 0.29 ± 0.02%, corresponding to a ~55% reduction, suggesting significant assimilation of organic nitrogen during growth. Cellulose was undetectable at the end of cultivation. The results demonstrate that the medium initially provided sufficient amounts of readily available carbon and nitrogen sources to support *Bacillus* sp. 10/R growth, but that growth cessation was likely associated with the depletion of utilizable reducing sugars and organic nitrogen, rather than the exhaustion of total carbohydrate content.

Enzyme activity of *Bacillus* sp. 10/R cultivated on WFW medium was monitored over 48 h, at four time points (12, 24, 36 and 48 h). Pectinase activity was determined using the DNS method, while cellulase and xylanase activities were assessed using the CellG5 and XylX6 commercial assay kits, respectively. The culture supernatants were obtained by centrifugation of cultivation broths at 10,000 rpm for 10 min and used for enzyme activity measurements. Pectinase activity remained relatively constant during cultivation, with values of 2.19 U/mL at the first three time points and 2.20 U/mL at 48 h. In the study by Mohandas et al. (2018) [27], using the same method to assess pectinase activity of *Bacillus sonorensis* MPTD1, maximum activity reached 2.43 (μM/mL)/min. Among four soil-derived *Bacillus* isolates (*B. firmus*, *B. coagulans*, *B. endophyticus* and *B. vietnamensis*), *B. firmus* exhibited the highest pectinase activity, recorded as 0.0053 IU/mL/min [75]. Utami et al. (2022) [76] reported that pectinase activity of *Bacillus* sp. 2P11, monitored over 96 h, peaked at 72 h with a value of 0.391 U/mL, suggesting the strong pectinolytic activity of the strain *Bacillus* sp. 10/R used in this study.

In contrast to pectinase activity, no xylanase activity was detected under the tested conditions, while cellulase activity was low, with values of 0.01 and 0.08 CellG5 U/mL measured at 24 h and 48 h, respectively. The absence of detectable cellulase and xylanase activities in the supernatants of *Bacillus* sp. 10/R cultivated on WFW medium can be explained by the composition of the substrate. Previous analysis of WFW showed that the cellulose content was extremely low (0.01%, calculated value), which likely limited induction of cellulolytic activity. Since xylan is typically present in close association with cellulose in lignocellulosic materials, its concentration in WFW can also be assumed to be negligible. Given that production of hydrolytic enzymes such as cellulases and xylanases in *Bacillus* spp. is strongly regulated by substrate availability and is usually induced only in the presence of polysaccharides or their degradation products, the absence of sufficient inducers in WFW likely resulted in the lack of detectable enzyme activity under the tested conditions [77–80]. The substrate dependence of enzyme synthesis was further confirmed in this study by agar plate assays, where *Bacillus* sp. 10/R exhibited clear cellulase and xylanase activity on media supplemented with CMC and xylan, respectively. These findings indicate that the lack of activity in liquid cultures with WFW was not due to the inability of the strain to produce these enzymes, but rather to the absence of appropriate inducers.

4.6. Seed Treatment of Barley Using Circular Biostimulant Based on Bacillus sp. 10/R and Winery Flotation Wastewater

This study has investigated the effects of a seed treatment based on the plant biostimulant produced using *Bacillus* sp. 10/R and WFW on barley seed germination and the initial seedling growth. A wide range of studies have demonstrated that *Bacillus* isolates can significantly enhance seed germination and early seedling growth in various plant species [67,81–84]. Among them there are some investigations about *Bacillus* beneficial effects on barley seeds. For instance, specific *Bacillus* strains, such as BMG1 and BMG2, have been shown to increase the germination percentage of barley seeds and promote substantial elongation of both roots and shoots. The isolate BMG1 was found to be exceptionally effective, increasing the shoot length of barley by 2.23 times compared to the untreated control, while the isolate PAZE-6 demonstrated the most significant effect on root growth, increasing the root length of barley by up to 60% [85]. Moreover, research by Teker Yıldız and Acar (2025) [86] indicates that *Bacillus* strains can alleviate abiotic stresses, with *Bacillus thuringiensis* and *Bacillus cereus* helping to improve germination and growth under saline conditions The inoculation of *Bacillus cereus* led to a 17% increase in root length in barley seedlings compared to stressed but untreated plants. *Bacillus thuringiensis* was even more effective at increasing root length [80]. Co-inoculated phosphorus-releasing bacteria (*Bacillus megaterium* var. *phosphaticum*, *Arthrobacter agilis*) and nitrogen-fixing bacteria (*Azospirillum lipoferum* Br17, *Azotobacter chroococcum)* resulted in a grain yield of up to 4.5 t/ha (tons per hectare) in spring barley. This was a significant increase compared to the control, which produced 3.37 t/ha [87]. The effect of *Bacillus* sp. 10/R cultivation broths on barley germination revealed a clear dependence on both the type of medium and the applied concentration. As shown in Figure 4, undiluted WFW and WFW-CB completely inhibited seed germination, which points to the presence of inhibitory compounds in winery wastewater that are not sufficiently neutralized or metabolized even after microbial growth. Such effects are frequently reported for wastewater rich in polyphenols and other secondary metabolites that may exhibit phytotoxic properties at high concentrations [88]. In the study by Mosse et al. (2010) [88], the effects of undiluted winery wastewater on the germination and early growth of several crop species, including lucerne, millet, phalaris and barley were assessed, and it was demonstrated that high concentrations of polyphenols and other secondary metabolites in the wastewater completely inhibited seed germination and reduced seedling growth. However, partial dilution of the media drastically altered their impact in our study. The highest germination rate, reaching 60%, was observed with 10% WFW-CB and 50% WFW, suggesting that at sub-inhibitory concentrations certain components of WFW, together with microbial metabolites, may act as stimulants of germination. This dual effect indicates that WFW serves not only as a growth substrate for bacteria but also as a potential source of bioactive molecules, which at appropriate concentrations can enhance early plant development. With respect to biomass, fresh mass was highest in seedlings treated with 10% WFW-CB, while for dry mass the greatest value was recorded with undiluted NB-CB, followed by 10% WFW-CB. Importantly, in all cases cultivation broths promoted higher dry mass compared to the negative control (tap water), confirming that bacterial metabolites contributed positively to seedling development, likely through the combined action of hydrolytic enzymes, phytohormone-like compounds, and nutrient mobilization. In summary, the increase in the investigated barley germination and growth parameters indicated suitability of 10% WFW-CB as a potential plant biostimulant, as well as suitability of *Bacillus* sp. 10/R as a biostimulant active component, with a necessity to further investigate possible mechanisms of barley growth promotion in more detail taking into account plant responses and large scale in vivo trials.

5. Conclusions

The results presented in this study reveal that the isolate *Bacillus* sp. 10/R from the grapevine rhizosphere, taxonomically closely related to *Bacillus mojavensis* and *Bacillus halotolerans*, could be considered as a potential plant growth promoting agent, with proven abilities to synthesize hydrolytic enzymes facilitating plant and soil microflora nutrition, compounds modulating plant immune responses and plant growth hormone auxin. Furthermore, an environment-friendly proof of concept was reported when it comes to the biotechnological production process for obtaining a circular microbial biostimulant. In this study, we tested barley seed treatment, which was based on the aforementioned bacterial strain used as an active component and its ability to effectively utilize nutrients present in winery flotation wastewater, whose suitability as a substrate for bacterial growth was also proven. Further research should precisely elucidate the plant responses regarding the suggested plant growth promotion mechanisms, taking into account wider range of crops, as well larger-scale testing in greenhouse and field trials. The need to investigate other possible application methods beyond seed treatment, as well as defining the optimal dosage of the biostimulant, considering the potential plant growth inhibitory effects observed in this study and the necessity to understand its effects to the plant–environment ecosystem, has also been foreseen. The proposed bioprocess solution should also be further analyzed for its techno-economical feasibility to determine the cost–benefit ratio for wineries as waste substrate generators, producers of microbial biostimulants and farmers as the end-users, thus developing both industrial and agricultural symbiosis for food production with reduced environmental footprint.

Author Contributions: Conceptualization, T.D. and I.D.; methodology, J.G., I.D., V.V. and Z.Š.; software, T.D.; validation, T.D., I.D., Z.Š. and M.S.; formal analysis, T.D. and M.S.; investigation, T.D. and M.S.; resources, J.G. and Z.Š.; data curation, T.D., I.D., A.J. and V.V.; writing—original draft preparation, T.D. and I.D.; writing—review and editing, J.G. and V.V.; visualization, T.D.; supervision, J.G.; project administration, J.G. and Z.Š.; funding acquisition, J.G. All authors have read and agreed to the published version of the manuscript.

Funding: This research was funded by the programs 451-03-136/2025-03/200134 and 451-03-137/2025-03/200134 of the Ministry of Science, Technological Development and Innovations of the Republic of Serbia and by the Provincial Secretariat for Higher Education and Scientific Research of the Autonomous Province of Vojvodina in the framework of the project "Development of industrial symbiosis in the AP Vojvodina through valorization of fruit processing by-products using green technologies" (project number 003075720 2024 09418 003 000 000 001 04 004).

Data Availability Statement: The original contributions presented in this study are included in the article. Further inquiries can be directed to the corresponding author(s).

Acknowledgments: During the preparation of this manuscript/study, the authors used Gemini Flash 2.5 for the purposes of grammar checking. The authors have reviewed and edited the output and take full responsibility for the content of this publication.

Conflicts of Interest: The authors declare no conflicts of interest.

Abbreviations

The following abbreviations are used in this manuscript:

PGP	plant-growth promotion
ACC	1-aminocyclopropane-1-carboxylic acid
WFW	winery flotation wastewater
IAA	indole-3-acetic acid
NB	nutrient broth
EAI	enzymatic activity index
CB	cultivation broth

References

1. Elumalai, P.; Gao, X.; Parthipan, P.; Luo, J.; Cui, J. Agrochemical pollution: A serious threat to environmental health. *Curr. Opin. Environ. Sci. Health* **2025**, *43*, 100597. [CrossRef]
2. Ramakrishna, W.; Yadav, R.; Li, K. Plant growth promoting bacteria in agriculture: Two sides of a coin. *Appl. Soil Ecol.* **2019**, *138*, 10–18. [CrossRef]
3. Asghar, W.; Kondo, S.; Iguchi, R.; Mahmood, A.; Kataoka, R. Agricultural utilization of unused resources: Liquid food waste material as a new source of plant growth-promoting microbes. *Agronomy* **2020**, *10*, 954. [CrossRef]

4. Orozco-Mosqueda, M.d.C.; Flores, A.; Rojas-Sánchez, B.; Urtis-Flores, C.A.; Morales-Cedeño, L.R.; Valencia-Marin, M.F.; Chávez-Avila, S.; Rojas-Solis, D.; Santoyo, G. Plant Growth-Promoting Bacteria as Bioinoculants: Attributes and Challenges for Sustainable Crop Improvement. *Agronomy* **2021**, *11*, 1167. [CrossRef]
5. Basu, A.; Prasad, P.; Das, S.N.; Kalam, S.; Sayyed, R.Z.; Reddy, M.S.; El Enshasy, H. Plant Growth Promoting Rhizobacteria (PGPR) as Green Bioinoculants: Recent Developments, Constraints, and Prospects. *Sustainability* **2021**, *13*, 1140. [CrossRef]
6. Kumar, V.V. Biofertilizers and Biopesticides in Sustainable Agriculture. In *Role of Rhizospheric Microbes in Soil*; Meena, V., Ed.; Springer: Singapore, 2018; Volume 1, pp. 377–398. [CrossRef]
7. Ansabayeva, A.; Makhambetov, M.; Rebouh, N.Y.; Abdelkader, M.; Saudy, H.S.; Hassan, K.M.; Nasser, M.A.; Ali, M.A.A.; Ebrahim, M. Plant growth-promoting microbes for resilient farming systems: Mitigating environmental stressors and boosting crops productivity—A review. *Horticulturae* **2025**, *11*, 260. [CrossRef]
8. Etesami, H. The dual nature of plant growth-promoting bacteria: Benefits, risks, and pathways to sustainable deployment. *Curr. Res. Microb. Sci.* **2025**, *9*, 100421. [CrossRef]
9. Vlajkov, V.; Pajčin, I.; Vučetić, S.; Anđelić, S.; Loc, M.; Grahovac, M.; Grahovac, J. *Bacillus*-Loaded Biochar as Soil Amendment for Improved Germination of Maize Seeds. *Plants* **2023**, *12*, 1024. [CrossRef]
10. Souza, R.D.; Ambrosini, A.; Passaglia, L.M. Plant growth-promoting bacteria as inoculants in agricultural soils. *Genet. Mol. Biol.* **2015**, *38*, 401–419. [CrossRef] [PubMed]
11. Alina, S.O.; Constantinscu, F.; Petruta, C.C. Biodiversity of *Bacillus subtilis* group and beneficial traits of *Bacillus* species useful in plant protection. *Rom. Biotechnol. Lett.* **2015**, *20*, 10737–10750.
12. Azeem, M.; Javed, S.; Zahoor, A.F. *Bacillus* species as potential plant growth promoting Rhizobacteria for drought stress resilience. *Russ. J. Plant Physiol.* **2023**, *70*, 59. [CrossRef]
13. Ngalimat, M.S.; Mohd Hata, E.; Zulperi, D.; Ismail, S.I.; Ismail, M.R.; Mohd Zainudin, N.A.I.; Saidi, N.B.; Yusof, M.T. Plant Growth-Promoting Bacteria as an Emerging Tool to Manage Bacterial Rice Pathogens. *Microorganisms* **2021**, *9*, 682. [CrossRef]
14. Dmitrović, S.; Pajčin, I.; Vlajkov, V.; Grahovac, M.; Jokić, A.; Grahovac, J. Dairy and Wine Industry Effluents as Alternative Media for the Production of *Bacillus*-Based Biocontrol Agents. *Bioengineering* **2022**, *9*, 663. [CrossRef] [PubMed]
15. Slatni, T.; Zorrig, W.; Razzegui, A.; Hernández, J.A.; Barba-Espín, G.; Hamed, K.B.; Díaz-Vivancos, P. Halophilic *Bacillus* improve barley growth on calcareous soil via enhanced photosynthetic performance and metabolomic re-programing. *J. Plant Physiol.* **2025**, *309*, 154495. [CrossRef]
16. Akhtyamova, Z.; Arkhipova, T.; Sharipova, G.; Ivanov, R.; Nuzhnaya, T.; Kudoyarova, G.; Veselov, D. The effect of plant growth-promoting bacteria *Bacillus subtilis* IB-22 on the hydraulic conductivity and abundance of PIP2 aquaporins in the roots of an abscisic acid-deficient barley mutant. *Int. J. Mol. Sci.* **2024**, *25*, 10706. [CrossRef] [PubMed]
17. Ajayi, O.; Grover, S.; Yimer, B.; Vinje, M.; Mahalingam, R. Bacterial seed endophytes promote barley growth and inhibits *Fusarium graminearum* in vitro. *BMC Res. Notes* **2024**, *17*, 289. [CrossRef] [PubMed]
18. Fenibo, E.O.; Ijoma, G.N.; Matambo, T. Biopesticides in sustainable agriculture: A critical sustainable development driver governed by green chemistry principles. *Front. Sustain. Food Syst.* **2021**, *5*, 619058. [CrossRef]
19. Neves, A.; Godina, R.; Azevedo, S.G.; Matias, J.C. A comprehensive review of industrial symbiosis. *J. Clean. Prod.* **2020**, *247*, 119113. [CrossRef]
20. Yi, H.; Li, M.; Huo, X.; Zeng, G.; Lai, C.; Huang, D.; An, Z.; Qin, L.; Liu, X.; Li, B.; et al. Recent development of advanced biotechnology for wastewater treatment. *Crit. Rev. Biotechnol.* **2020**, *40*, 99–118. [CrossRef]
21. Rodrigues, R.P.; Gando-Ferreira, L.M.; Quina, M.J. Increasing Value of Winery Residues through Integrated Biorefinery Processes: A Review. *Molecules* **2022**, *27*, 4709. [CrossRef]
22. Statistical Report on World Vitiviniculture. 2024. Available online: https://www.oiv.int/what-we-do/statistics (accessed on 16 July 2025).
23. Oliveira, M.; Duarte, E. Integrated approach to winery waste: Waste generation and data consolidation. *Front. Environ. Sci. Eng.* **2016**, *10*, 168–176. [CrossRef]
24. Sousa, A.C.; Dias, C.; Martins, A.R.; Gomes, A.G.; Santos, C.A. Using winery effluents for cultivating microalgae as bio-additives for vineyards. *J. Appl. Phycol.* **2025**, *37*, 1619–1632. [CrossRef]
25. Vlajkov, V.; Grahovac, M.; Budakov, D.; Loc, M.; Pajčin, I.; Milić, D.; Grahovac, J. Distribution, genetic diversity and biocontrol of aflatoxigenic *Aspergillus flavus* in Serbian maize fields. *Toxins* **2021**, *13*, 687. [CrossRef]
26. Nikodinovic, J.; Barrow, K.D.; Chuck, J.A. High yield preparation of genomic DNA from *Streptomyces*. *Biotechniques* **2003**, *35*, 932–936. [CrossRef] [PubMed]
27. Mohandas, A.; Raveendran, S.; Parameswaran, B.; Abraham, A.; Athira, R.S.; Mathew, A.K.; Pandey, A. Production of pectinase from *Bacillus sonorensis* MPTD1. *Food Technol. Biotechnol.* **2018**, *56*, 110. [CrossRef] [PubMed]
28. Amore, A.; Parameswaran, B.; Kumar, R.; Birolo, L.; Vinciguerra, R.; Marcolongo, L.; Faraco, V. Application of a new xylanase activity from *Bacillus amyloliquefaciens* XR44A in brewer's spent grain saccharification. *J. Chem. Technol. Biotechnol.* **2015**, *90*, 573–581. [CrossRef]

29. Adinarayana, K.; Ellaiah, P.; Prasad, D.S. Purification and partial characterization of thermostable serine alkaline protease from a newly isolated *Bacillus subtilis* PE-11. *AAPS PharmSciTech* **2003**, *4*, 440–448. [CrossRef]
30. Penrose, D.M.; Glick, B.R. Methods for isolating and characterizing ACC deaminase-containing plant growth-promoting rhizobacteria. *Physiol. Plant.* **2003**, *118*, 10–15. [CrossRef] [PubMed]
31. Gupta, S.; Pandey, S. ACC deaminase producing bacteria with multifarious plant growth promoting traits alleviates salinity stress in French bean (*Phaseolus vulgaris*) plants. *Front. Microbiol.* **2019**, *10*, 1506. [CrossRef]
32. Syed Ab Rahman, S.F.; Singh, E.; Pieterse, C.M.J.; Schenk, P.M. Emerging microbial biocontrol strategies for plant pathogens. *Plant Sci.* **2018**, *267*, 102–111. [CrossRef]
33. SFRY. Ordinance on methods of taking samples and performing chemical and physical analyzes for the purpose of quality control of fruit and vegetable products. *Off. Gaz. SFRY* **1983**, 29.
34. Vračar, O.L. *Manual for Quality Control of Fresh and Processed Fruits and Vegetables and Mushrooms and Refreshing Soft Drinks*; University of Novi Sad, Faculty of Technology: Novi Sad, Serbia, 2001.
35. Marcó, A.; Rubio, R.; Compañó, R.; Casals, I. Comparison of the Kjeldahl method and a combustion method for total nitrogen determination in animal feed. *Talanta* **2002**, *57*, 1019–1026. [CrossRef]
36. Montana State University. Drop Plate Method. Available online: https://www.cs.montana.edu/webworks/projects/stevesbook/contents/chapters/chapter011/section008/blue/page002.html (accessed on 11 July 2025).
37. Vlajkov, V.; Pajčin, I.; Loc, M.; Budakov, D.; Dodić, J.; Grahovac, M.; Grahovac, J. The Effect of Cultivation Conditions on Antifungal and Maize Seed Germination Activity of *Bacillus*-Based Biocontrol Agent. *Bioengineering* **2022**, *9*, 797. [CrossRef]
38. Walia, M.K.; Mohammed, Y.A.; Franck, W.L.; Chen, C. Evaluation of early seedling development of Chickpea and its relation to seed yield. *Agrosystems Geosci. Environ.* **2020**, *3*, e20005. [CrossRef]
39. El-Saadony, M.T.; Saad, A.M.; Soliman, S.M.; Salem, H.M.; Ahmed, A.I.; Mahmood, M.; El-Tahan, A.M.; Ebrahim, A.A.M.; Abd El-Mageed, T.A.; Negm, S.H.; et al. Plant growth-promoting microorganisms as biocontrol agents of plant diseases: Mechanisms, challenges and future perspectives. *Front. Plant Sci.* **2022**, *13*, 923880. [CrossRef]
40. Hakim, S.; Naqqash, T.; Nawaz, M.S.; Laraib, I.; Siddique, M.J.; Zia, R.; Mirza, M.S.; Imran, A. Rhizosphere Engineering with Plant Growth-Promoting Microorganisms for Agriculture and Ecological Sustainability. *Front. Sustain. Food Syst.* **2021**, *5*, 617157. [CrossRef]
41. Pandey, A.; Tripathi, A.; Srivastava, P.; Choudhary, K.K.; Dikshit, A. Plant growth-promoting microorganisms in sustainable agriculture. In *Role of Plant Growth Promoting Microorganisms in Sustainable Agriculture and Nanotechnology*; Kumar, A., Singh, A.K., Choudhary, K.K., Eds.; Woodhead Publishing: Cambridge, UK, 2019; pp. 1–19. [CrossRef]
42. Cantabella, D.; Dolcet-Sanjuan, R.; Teixidó, N. Using plant growth-promoting microorganisms (PGPMs) to improve plant development under in vitro culture conditions. *Planta* **2022**, *255*, 117. [CrossRef] [PubMed]
43. Taheri, P.; Puopolo, G.; Santoyo, G. Plant growth-promoting microorganisms: New insights and the way forward. *Microbiol. Res.* **2025**, *297*, 128168. [CrossRef] [PubMed]
44. Singh, A.A.; Singh, A.K. Role of bacterial quorum sensing in plant growth promotion. *World. J. Microbiol. Biotechnol.* **2025**, *41*, 18. [CrossRef]
45. Dhawi, F. The role of plant growth-promoting microorganisms (PGPMs) and their feasibility in hydroponics and vertical farming. *Metabolites* **2023**, *13*, 247. [CrossRef]
46. Naamala, J.; Smith, D.L. Relevance of Plant Growth Promoting Microorganisms and Their Derived Compounds, in the Face of Climate Change. *Agronomy* **2020**, *10*, 1179. [CrossRef]
47. Khan, A.R.; Mustafa, A.; Hyder, S.; Valipour, M.; Rizvi, Z.F.; Gondal, A.S.; Yousuf, Z.; Iqbal, R.; Daraz, U. *Bacillus* spp. as Bioagents: Uses and Application for Sustainable Agriculture. *Biology* **2022**, *11*, 1763. [CrossRef]
48. Sansinenea, E. *Bacillus* spp.: As Plant Growth-Promoting Bacteria. In *Secondary Metabolites of Plant Growth Promoting Rhizomicroorganisms*; Singh, H., Keswani, C., Reddy, M., Sansinenea, E., García-Estrada, C., Eds.; Springer: Singapore, 2019. [CrossRef]
49. Prajakta, B.M.; Suvarna, P.P.; Raghvendra, S.P.; Alok, R.R. Potential biocontrol and superlative plant growth promoting activity of indigenous *Bacillus mojavensis* PB-35(R11) of soybean (*Glycine max*) rhizosphere. *SN Appl. Sci.* **2019**, *1*, 1143. [CrossRef]
50. Ghazala, I.; Chiab, N.; Saidi, M.N.; Gargouri-Bouzid, R. The plant growth-promoting bacteria strain *Bacillus mojavensis* I4 enhanced salt stress tolerance in durum wheat. *Curr. Microbiol.* **2023**, *80*, 178. [CrossRef] [PubMed]
51. Sdiri Ghidawi, J.; Ghazala, I.; Haddar, A.; Bouazizi, O.; Gargouri-Bouzid, R.; Nouri-Ellouz, O. Effect of the Plant Growth-Promoting Bacteria Strain *Bacillus Mojavensis* I4 on Potato Growth, Physiology, Tuber Yield, and Quality Under Salt Stress Conditions. *Potato Res.* **2025**, *68*, 2205–2227. [CrossRef]
52. Danish, M.; Shahid, M.; Zeyad, M.T.; Bukhari, N.A.; Al-Khattaf, F.S.; Hatamleh, A.A.; Ali, S. *Bacillus mojavensis*, a metal-tolerant plant growth-promoting bacterium, improves growth, photosynthetic attributes, gas exchange parameters, and Alkalo-Polyphenol Contents in Silver Nanoparticle (Ag-NP)-Treated *Withania somnifera* L. (Ashwagandha). *ACS Omega* **2022**, *7*, 13878–13893. [CrossRef]

53. Kapadia, C.; Patel, N.; Rana, A.; Vaidya, H.; Alfarraj, S.; Ansari, M.J.; Gafur, A.; Poczai, P.; Sayyed, R.Z. Evaluation of Plant Growth-Promoting and Salinity Ameliorating Potential of Halophilic Bacteria Isolated from Saline Soil. *Front. Plant Sci.* **2022**, *13* 946217. [CrossRef]
54. Oliva, G.; Di Stasio, L.; Vigliotta, G.; Guarino, F.; Cicatelli, A.; Castiglione, S. Exploring the Potential of Four Novel Halotolerant Bacterial Strains as Plant-Growth-Promoting Rhizobacteria (PGPR) under Saline Conditions. *Appl. Sci.* **2023**, *13*, 4320. [CrossRef]
55. Barros, F.F.; Simiqueli, A.P.; de Andrade, C.J.; Pastore, G.M. Production of Enzymes from Agroindustrial Wastes by Biosurfactant Producing Strains of *Bacillus subtilis*. *Biotechnol. Res. Int.* **2013**, *2013*, 103960. [CrossRef] [PubMed]
56. Shrestha, S.; Kognou, A.L.M.; Zhang, J.; Qin, W. Different Facets of Lignocellulosic Biomass Including Pectin and Its Perspectives *Waste Biomass Valorization* **2021**, *12*, 4805–4823. [CrossRef]
57. Sharma, V.; Tsai, M.-L.; Nargotra, P.; Chen, C.-W.; Kuo, C.-H.; Sun, P.-P.; Dong, C.-D. Agro-Industrial Food Waste as a Low-Cost Substrate for Sustainable Production of Industrial Enzymes: A Critical Review. *Catalysts* **2022**, *12*, 1373. [CrossRef]
58. Kabir, M.S.; Tasmim, T. Isolation of pectinase producing bacteria from the rhizosphere of Andrographis paniculata nees and 16S rRNA gene sequence comparison of some potential strains. *Adv. Microbiol.* **2019**, *9*, 1–13. [CrossRef]
59. Oumer, O.J.; Abate, D. Screening and molecular identification of pectinase producing microbes from coffee pulp. *Biomed. Res. Int* **2018**, *2018*, 2961767. [CrossRef]
60. Castaldi, S.; Petrillo, C.; Donadio, G.; Piaz, F.D.; Cimmino, A.; Masi, M.; Evidente, A.; Isticato, R. Plant Growth Promotion Function of *Bacillus* sp. Strains Isolated from Salt-Pan Rhizosphere and Their Biocontrol Potential against *Macrophomina phaseolina* *Int. J. Mol. Sci.* **2021**, *22*, 3324. [CrossRef]
61. Balla, A.; Silini, A.; Cherif-Silini, H.; Bouket, A.C.; Boudechicha, A.; Luptakova, L.; Alenezi, F.N.; Belbahri, L. Screening of Cellulolytic Bacteria from Various Ecosystems and Their Cellulases Production under Multi-Stress Conditions. *Catalysts* **2022**, *12*, 769. [CrossRef]
62. Belimov, A.A.; Hontzeas, N.; Safronova, V.I.; Demchinskaya, S.V.; Piluzza, G.; Bullitta, S.; Glick, B.R. Cadmium-tolerant plant growth-promoting bacteria associated with the roots of Indian mustard (*Brassica juncea* L. Czern.). *Soil Biol. Biochem.* **2005**, *37*, 241–250. [CrossRef]
63. Saleem, A.R.; Brunetti, C.; Khalid, A.; Della Rocca, G.; Raio, A.; Emiliani, G.; De Carlo, A.; Mahmood, T.; Centritto, M. Drought response of *Mucuna pruriens* (L.) DC. inoculated with ACC deaminase and IAA producing rhizobacteria. *PLoS ONE* **2018**, *13*, e0191218. [CrossRef] [PubMed]
64. Kour, D.; Khan, S.S.; Kour, H.; Kaur, T.; Devi, R.; Rai, A.K.; Yadav, A.N. ACC deaminase producing phytomicrobiomes for amelioration of abiotic stresses in plants for agricultural sustainability. *J. Plant Growth Regul.* **2024**, *43*, 963–985. [CrossRef]
65. Ahmad, I.; Ahmad, M.; Hussain, A.; Jamil, M. Integrated use of phosphate-solubilizing *Bacillus subtilis* strain IA6 and zinc-solubilizing *Bacillus* sp. strain IA16: A promising approach for improving cotton growth. *Folia Microbiol.* **2021**, *66*, 115–125. [CrossRef]
66. Afzal, A.; Bahader, S.; Ul Hassan, T.; Naz, I.; Din, A.U. Rock phosphate solubilization by plant growth-promoting *Bacillus velezensis* and its impact on wheat growth and yield. *Geomicrobiol. J.* **2023**, *40*, 131–142. [CrossRef]
67. Cendales, T.C.; González, C.A.R.; Cuásquer, C.P.V.; Alzate, O.A.T.; Rodríguez, A.H. Efecto de *Bacillus* sobre la germinación y crecimiento de plántulas de tomate (*Solanum lycopersicum* L.). *Acta Biol. Colomb.* **2017**, *22*, 37. [CrossRef]
68. Kumari, S.; Kumar, P.; Kiran, S.; Kumari, S.; Singh, A. Characterization of culture condition dependent, growth responses of phosphate solubilizing bacteria (*Bacillus subtilis* DR2) on plant growth promotion of *Hordeum vulgare*. *Vegetos* **2024**, *37*, 266–276. [CrossRef]
69. Chandra, P.; Khobra, R.; Sundha, P.; Sharma, R.K.; Jasrotia, P.; Chandra, A.; Singh, D.P.; Singh, G.P. Plant growth promoting *Bacillus*-based bio formulations improve wheat rhizosphere biological activity, nutrient uptake and growth of the plant. *Acta Physiol. Plant.* **2021**, *43*, 139. [CrossRef]
70. Danilov, I.; Vlajkov, V.; Šumić, Z.; Milić, A.; Horecki, A.T.; Dujković, T.; Živanović, N.; Simin, N.; Lesjak, M.; Grahovac, J. Valorization of Strawberry Juice Production Wastewater: Possibilities for Polyphenols Recovery and Plant Biostimulant Production. *Foods* **2024**, *13*, 3224. [CrossRef]
71. Goud, M.S.; Sharma, S.K.; Kharbikar, L.L.; Prasanna, R.; Sangwan, S.; Dahuja, A.; Dixit, A. *Bacillus* species consortium with tryptophan-dependent and-independent pathways mediated production of IAA and its derivatives modulates soil biological properties, growth and yield of wheat. *Plant Soil* **2025**, *508*, 71–97. [CrossRef]
72. Latessa, S.H.; Hanley, L.; Tao, W. Characteristics and practical treatment technologies of winery wastewater: A review for wastewater management at small wineries. *J. Environ. Manag.* **2023**, *342*, 118343. [CrossRef] [PubMed]
73. Ioannou, L.A.; Puma, G.L.; Fatta-Kassinos, D. Treatment of winery wastewater by physicochemical, biological and advanced processes: A review. *J. Hazard. Mater.* **2015**, *286*, 343–368. [CrossRef]
74. Ngwenya, N.; Gaszynski, C.; Ikumi, D. A review of winery wastewater treatment: A focus on UASB biotechnology optimisation and recovery strategies. *J. Environ. Chem. Eng.* **2022**, *10*, 108172. [CrossRef]
75. Khan, I.G.; Barate, D.L. Effect of various parameters on activity of pectinase enzyme. *Int. J. Adv. Res.* **2016**, *4*, 853–862.

76. Utami, A.P.; Fahrurrozi, F.; Meryandini, A. Production and immobilization pectinase from *Bacillus* sp. 2P11 using alginate beads. *Biodiversitas* **2022**, *23*, 3960–3966. [CrossRef]
77. Malik, W.A.; Javed, S. Biochemical Characterization of Cellulase from *Bacillus subtilis* Strain and its Effect on Digestibility and Structural Modifications of Lignocellulose Rich Biomass. *Front. Bioeng. Biotechnol.* **2021**, *9*, 800265. [CrossRef]
78. Abada, E.A.; Elbaz, R.M.; Sonbol, H.; Korany, S.M. Optimization of cellulase production from *Bacillus albus* (MN755587) and its involvement in bioethanol production. *Pol. J. Environ. Stud.* **2021**, *30*, 2459–2466. [CrossRef] [PubMed]
79. da Silva, R.N.; de Andrade Melo, L.F.; Finkler, C.L.L. Optimization of the cultivation conditions of *Bacillus licheniformis* BCLLNF-01 for cellulase production. *Biotechnol. Rep.* **2021**, *29*, e00599. [CrossRef]
80. Cai, Z.; Wang, Y.; You, Y.; Yang, N.; Lu, S.; Xue, J.; Xing, X.; Sha, S.; Zhao, L. Introduction of Cellulolytic Bacterium *Bacillus velezensis* Z2.6 and Its Cellulase Production Optimization. *Microorganisms* **2024**, *12*, 979. [CrossRef]
81. Widnyana, I.K.; Javandira, C. Activities *Pseudomonas* spp. and *Bacillus* sp. to stimulate germination and seedling growth of tomato plants. *Agric. Agric. Sci. Procedia* **2016**, *9*, 419–423. [CrossRef]
82. Song, P.; Zhao, B.; Sun, X.; Li, L.; Wang, Z.; Ma, C.; Zhang, J. Effects of *Bacillus subtilis* HS5B5 on Maize Seed Germination and Seedling Growth under NaCl Stress Conditions. *Agronomy* **2023**, *13*, 1874. [CrossRef]
83. Li, H.; Yue, H.; Li, L.; Liu, Y.; Zhang, H.; Wang, J.; Jiang, X. Seed biostimulant *Bacillus* sp. MGW9 improves the salt tolerance of maize during seed germination. *AMB Express* **2021**, *11*, 74. [CrossRef]
84. Miljaković, D.; Marinković, J.; Tamindžić, G.; Đorđević, V.; Tintor, B.; Milošević, D.; Ignjatov, M.; Nikolić, Z. Bio-Priming of Soybean with *Bradyrhizobium japonicum* and *Bacillus megaterium*: Strategy to Improve Seed Germination and the Initial Seedling Growth. *Plants* **2022**, *11*, 1927. [CrossRef] [PubMed]
85. Buntić, A.; Stajković-Srbinović, O.S.; Knežević, M.M.; Rasulić, N.V.; Kuzmanović, Đ.Ž.; Dimitrijević-Branković, S.; Delić, D.I. The effect of bacterial isolates from rhizosphere soils on wheat and barley seed germination. *Zemljište I Biljka* **2019**, *68*, 1–11. [CrossRef]
86. Teker Yıldız, M.; Acar, O. Comparison of Two *Bacillus* Strains Isolated from the Coastal Zone in Barley (*Hordeum vulgare* L.) Under Salt Stress. *Plants* **2025**, *14*, 723. [CrossRef]
87. Górski, R.; Rosa, R.; Niewiadomska, A.; Wolna-Maruwka, A.; Płaza, A. Innovative Spring Barley Cultivation Technology Based on the Use of Microbial Products Together with Living Mulch in Organic Farming. *Agronomy* **2023**, *13*, 1914. [CrossRef]
88. Mosse, K.P.; Patti, A.F.; Christen, E.W.; Cavagnaro, T.R. Winery wastewater inhibits seed germination and vegetative growth of common crop species. *J. Hazard. Mater.* **2019**, *180*, 63–70. [CrossRef] [PubMed]

agronomy

MDPI

Article

Biochar Combined with Garbage Enzyme Enhances Nitrogen Conservation during Sewage Sludge Composting: Evidence from Microbial Community and Enzyme Activities Related to Ammoniation

Jishao Jiang [1,*], Huilin Cui [1], Parag Bhople [2], Caspar C. C. Chater [3,4], Fuqiang Yu [5] and Dong Liu [5,*]

1 Henan Key Laboratory for Environmental Pollution Control, Key Laboratory for Yellow River and Huai River Water Environmental Pollution Control, Ministry of Education, School of Environment, Henan Normal University, Xinxiang 453007, China; 2219183018@stu.htu.edu.cn
2 Crops, Environment, and Land Use Department, Environment Research Centre, Teagasc, Johnstown Castle, Y35 TC98 Wexford, Ireland; paragbhoplebt@gmail.com
3 Royal Botanic Gardens, Kew, Richmond TW9 3AE, UK; caspar.chater@gmail.com
4 Plants, Photosynthesis, and Soil, School of Biosciences, University of Sheffield, Sheffield S10 2TN, UK
5 The Germplasm Bank of Wild Species, Yunnan Key Laboratory for Fungal Diversity and Green Development, Kunming Institute of Botany, Chinese Academy of Sciences, Kunming 650201, China; fqyu@mail.kib.ac.cn
* Correspondence: jiangjishao@htu.edu.cn (J.J.); liudongc@mail.kib.ac.cn (D.L.)

Abstract: Nitrogen loss is an unavoidable problem during composting processes, and the ammonia oxidation process significantly affects nitrogen transformation. The objective of this study was to evaluate nitrogen transformation when garbage enzyme (GE), biochar (BC), pelelith (PL) and combinations thereof were added during sewage sludge composting. Meanwhile, the succession of ammonia-oxidizing bacteria (AOB) and archaea (AOA) were also explored via quantitative polymerase chain reaction and high-throughput sequencing. The results showed that GE + BC and GE + PL treatments decreased ammonia (NH_3) formation by 23.8% and 8.3%, and that of nitrous oxide (N_2O) by 25.7% and 26.3% relative to the control, respectively. Simultaneously, the GE, GE + BC, and GE + PL treatments boosted the succession of AOA and AOB, and increased the activities of ammonia monooxygenase (AMO) and hydroxylamine oxidoreductase (HAO) activities and the gene copies of AOA and AOB. The AMO activities, NH_4-N, NO_3-N, and C/N, significantly affect AOA and AOB community structures. The network analysis predicted that the AMO and HAO were secreted mainly by the *unclassified_Archaea* and *norank_Crenarchaeota*, whereas it also showed that the GE + BC improved microbial associations with AOA, enzymatic activity, and environmental factors. Thus, the addition of garbage enzyme and biochar appears to be a promising mitigation strategy to reduce nitrogen losses during the composting process.

Keywords: composting; sewage sludge; biochar; ammonia oxidase; ammoniation

1. Introduction

Composting is an effective method of harmless utilization and recycling of organic solid waste into high-quality environmentally friendly organic fertilizers [1]. As a kind of fertilizer, nitrogen (N) content has been the main standard to evaluate the quality of compost, thus N cycling is of central importance among other aspects such as humification and greenhouse gases [2–4]. During composting, the N loss, in the form of ammonia (NH_3) via the transformation of ammonium-N (NH_4-N) through ammonification, was generated during the composting process [5]. However, NH_4-N could strategically be converted into nitrate (NO_3-N) through nitrification, effectively controlling N loss in the form of NH_3 emissions [6]. This nitrification is directly impacted by ammonia oxidizers through the ammoxidation process, which likely changes the subsequent N cycle and the compost quality as well [7]. Therefore, it becomes highly important to explore the community succession pattern of microorganisms related to ammonia oxidation and their convulsive effects on composting processes.

Bacteria and archaea both are potential ammonia oxidizing microorganisms (ammonia oxidizing bacteria—AOB, and ammonia-oxidizing archaea—AOA) during the process of composting. However, to date, the consensus about AOA or AOB regarding their predominance during composting is inadequate. Zeng et al. [8] reported that gene copies of AOA were more abundant, while AOB were not detected during the thermophilic and cooling stages of the composting of agricultural waste. Yan et al. [9] also reported that AOA outnumber AOB by 2–3 times during cattle manure composting. Contrastingly, Wu et al. [10] recently reported that AOB played the dominant role in comparison to the AOA during the composting of agricultural waste when zeolite and biochar were used as substrates. Previously, Yan et al. [9] had also found similar results during cattle manure composting. These ambiguities demonstrate that community diversity and the quantity of AOA and AOB may be differentially impacted by various factors, such as the raw materials used for composting and the temperature, moisture content, and O_2 concentration, warranting further investigation. In addition, ammonia monooxygenase (AMO) and hydroxylamine oxidoreductase (HAO) were the main enzymes reported during the ammonia oxidation process [11], which directly determined the final NO_3-N content of the compost. The NH_4-N is first converted into hydroxylamine (NH_2OH) using AMO, and then this product may further be oxidated to nitrite by HAO [12]. However, to date, the quantification and reactivity levels of AMO and HAO during organic solid-waste composting has been limited.

The N from the compost can be reserved and the quality of the end compost product could be enhanced by adding appropriate additives. A fermentation solution of waste vegetables and fruits, along with water and brown sugar, called garbage enzyme (GE), has been widely used in wastewater treatment, soil fertilizer, and heavy metal ion absorption research [13], mainly because of its functional capacities to decompose, compose, transform, and catalyze any chemical reaction [13]. However, due to its low pH (~3.3 units) GE could potentially slow down the growth of microorganisms and prolong the composting process [14]. In addition, it is well known that biochar (BC) and pelelith (PL) are widely used additives during organic waste composting that reduce NH_3 emissions and N loss. Their usage is mainly based on their high total-pore volume and adsorption capacities [15], which increase porosity and eliminate anaerobic pocket formation [16,17]. Used together, BC and PL could provide habitat and refuge for bacteria, protect bacteria from low pH levels, and the generate high temperatures via high specific surface area and total pore volume [18]. Furthermore, BC contains alkali metal salts and organic functional groups, and thus has a strong pH-buffering capacity [19]. These advantages are likely to regulate the succession of ammonia oxidizers and improve the ammonia oxidation process, ultimately reducing N loss. Furthermore, BC and PL are inexpensive and convenient, which makes them ideal additives for organic waste composting. However, only a few studies have attempted to evaluate the effects on composting processes and the successions of AOA and AOB of GE combined with BC or PL.

Therefore, the objectives of this research were: (i) to gain insights into the effects of BC and PL combined with GE on N transformation and the activities of AMO and HAO enzymes; (ii) to quantify gene copies and identify the community composition and succession patterns of the AOA and AOB during organic composting; and, finally, (iii) to find the environmental factors driving changes in AOA and AOB community dynamics during the composting of sewage sludge.

2. Materials and Methods

2.1. Raw Materials, Additives, and Design

A laboratory simulation of the composting experiment was conducted at the School of Environment, Henan Normal University (Xinxiang, China; N 35°19′39.522″, E 113°54′55.613″)

for 24 days. The raw materials were dewatered sewage sludge (SS) and saw dust (SD), which were obtained from the Xiaoshangzhuang sewage treatment plant and a local furniture factory in Xinxiang, respectively. The physicochemical characteristics of the sewage sludge have been reported in our previous work [20,21]. The GE was prepared according to the procedure described by Arun and Sivashanmugam [13] using a mixture of brown sugar, waste fruits, and water. The characteristics of the GE were: pH, 3.1; total organic carbon (TOC), 3.1 g L^{-1}; and total nitrogen (TN), 4.1 g L^{-1}, respectively. The BC was purchased form the pingdingshan green source activated carbon Co., Ltd., where the primary material used was wood pellets. Meanwhile, the BC and PL were shredded to a size of 3–5 mm particles, with the total pore volume (TPV) of 0.05 and 0.02 cm^3 g^{-1}, and a BET surface area (SSA) of 14.2 m^2 g^{-1} and 6.5 m^2 g^{-1}, respectively.

The composting experiments consisted of four different treatments. Fresh SS and SD were mixed in a ratio of 4:1 and used as the control treatment. The second treatment, marked as GE, received 2% GE solution (on a dry weight basis). The third treatment (BC + GE), received a 2% GE solution and 10% BC (on a dry weight basis). Finally, the fourth treatment (BC + PL) was composed of a 2% GE solution and 10% PL. The fermentation process refers to a similar procedure to the one described by Jiang et al. [20,21]. During the composting process, sufficient O_2 was supplied by periodical turning and ventilating of the composting pile: 4 instances of ventilating and 2 of turning, during the first 12 days; with only 2 instances of ventilating during the final 13–24 days.

2.2. Sampling and Determination of Physicochemical Properties

The compost was sampled on days 0, 4, 8, 13, and 24 according to temperature changes. Before sampling, the composting mixture needed to be overturned evenly. Three samples from each treatment were collected and stored separately. Each sample was further divided into four parts of which one part was used for the analysis of the total nitrogen (TN) after air-drying. The second part was used to measure physicochemical indexes and enzymatic activities, and was stored at a cool temperature of 4 °C until analysis. While the fourth part was immediately frozen at −80 °C for microbial assays. The NH_3 and N_2O emissions were measured through a portable gas detector [22,23]. A compost extract was obtained from the mixture of compost and deionized water with a ratio of 1:10 (*w*/*v*). After centrifugation and filtration, the pH, EC, NO_3-N, and NH_4-N were measured by following standard laboratory methods [24]. The TN concentration of the pulverized samples was analyzed by an elemental analyzer (Elemental Vario EL, Frankfurt, Germany) [23].

2.3. Enzyme Activity Analysis

Quantification with fluorescein diacetate as a substrate: Fluorescein diacetate hydrolase (FDA) activity was measured at 620 nm using a colorimetric method. The AMO activity of the composting samples was determined using the AMO ELISA kit, and calculated at 420 nm using a colorimetric method [10]. To measure HAO activity, the samples were subjected to a reaction system, which consisted of compost extract, $K_3[Fe(CN)_6]$ (0.01 mol/L), EDTA (0.04 mmol/L), Tris-HCl (10 mmol/L), and $HONH_3Cl$-N (15 mg/L). The system was placed in a water bath at 25 °C for 5 min, and the decrease in $K_3Fe(CN)_6$ per minute was measured to quantify HAO activity [25].

2.4. Microbial Analysis

Composting samples on days 4, 13, and 24 were selected to study the structure and quantity of the AOA and AOB. Total genomic DNA was extracted from a 0.5 g compost sample according to the instructions of the MoBioPowerSoil® DNA Isolation Kit (12888) (MO BIO Laboratories, Inc., Carlsbad, CA, USA). The quality of genomic DNA was determined by 1% agarose gel electrophoresis, followed by NanoDrop2000 (Thermo Scientific Company, Waltham, MA, USA) to determine the DNA concentration and purity. The AOB and AOA gene copies were detected by q-PCR using an ABI 7300 fluorescence quantitative PCR instrument (Applied Biosystems, Waltham, MA, USA) with the primers Arch-amoAF (STAATGGTCTGGCTTAGACG)/Arch-amoAR (GCGGCCATC-CATCTGTATGT) and AmoA-1F (GGGGTTTCTACTGGTGGT)/AmoA-1R (CCCCTCK-GSAAAGCCTTCTTC), respectively [10]. The OD_{260} values of AOA and AOB plasmids were measured with a UV spectrophotometer (NanoDrop2000, Thermo Fisher Scientific, USA) and then converted into a copy number to draw a standard curve. The linearity (R^2) of the standard curves of AOA and AOB genes was 0.9986 and 0.9999, respectively.

Using the previously extracted DNA as a template, PCR amplification of the archaeal and bacterial 16S rRNA gene was performed with the same primers as stated above. The thermal cycling steps for PCR amplification of AOA and AOB were: pre-deformation at 95 °C for 3 min, 1 cycle; deformation at 95 °C for 30 s; annealing at 55 °C for 30 s and extension at 72 °C for 45 s for a total of 27 cycles; and final extension at 72 °C for 10 min (PCR instrument: ABI GeneAmp® 9700, Foster City, CA, USA). The 16S rDNA sequences of the AOA and AOB genes were analyzed using the Illumina MiSeq paired-end 300 bp protocol (Illumina, Inc., San Diego, CA, USA).

2.5. Statistical Analyses

All indicator data were collected in triplicate. IBM SPSS Statistics 20.0 was used to calculate the mean ± standard deviation of the physicochemical indexes and enzymes. The differences in enzymatic activities were evaluated by principal component analysis (PCA) using CANOCO 5.0. While the online platform i-Sanger (http://www.i-sanger.com/, accessed on 1 June 2023) was used to conduct AOA and AOB microbial community diversity analyses. Correlations between the environmental factors, enzyme activities, and microbial community composition of AOA and AOB were assessed by redundancy analysis (RDA) using CANOCO 5.0. However, many environmental factors can affect the distribution of the microbial structure during the process of composting, but many of them have a strong collinearity. Therefore, the driving environmental factor should be screened prior to environmental factor analysis in order to remove the factors that have a strong collinearity. In the present study, the variance inflation factor (VIF) was measured and the environmental factors with a VIF > 20 were removed from the subsequent analysis. Based on the results of the VIF values (Table S1), pH, TOC, TN, and HAO were removed. Therefore, only the EC, MC, NH_4-N, NO_3-N, C/N, FDA, and AMO were used for the redundancy analysis (RDA) using Canoco 5.0 software.

3. Results and Discussion

3.1. Dynamic Changes in the NH_3 and N_2O Emissions

NH_3 and N_2O emissions are the main sources of N loss during a well-controlled composting process [26,27]. As shown in Figure 1a, the NH_3 emitted during the days 3–12 made up approximately 82.0–88.2% of the total gas emissions. With rapid decomposition of the organic matter (OM), the NH_3 emissions increased quickly at the initial stage and reached to the maximums on day 6 for all treatments except the GE + BC treatments on day 8. The maximum emissions of NH_3 occurred at levels of 81.4, 43.9, 70.5, and 71.3 mg/d in the CK, GE, GE + BC, and GE + PL treatments, respectively. After the 8th day of the composting process, the NH_3 emissions were stable and similar to those observed on day 1. During the whole process, the cumulative NH_3 emissions in the treatments of the CK, GE, GE + BC, and GE + PL treatments were 528, 307, 402, and 485 mg, respectively (Figure 1a). Compared to CK, the NH_3 emissions in the GE, GE + BC, and GE + PL treatments reduced by 41.8%, 23.8%, and 8.3%, respectively. These results indicate that additives were effective in compensating the loss of N via NH_3 emissions. These results are in line with previous studies, which suggested that the low-pH and enzyme-rich activities of GE and the adsorption capacity of BC and PL are major sources of reduced N mineralization in composting processes where GE and BC or PL were used as a process catalyst and additives, respectively [17]. However, in our study, the values in the GE + BC and GE + PL treatments were significantly higher than the GE treatments. It can be said that the benefits of adding BC and PL may have been underestimated in previous studies and that such additions are far more advantageous for the growth of microorganisms, strengthening the ammonification process compared to the GE-only treatments with no additives [28]. In addition, the SSA and TPV of BC during different composting processes were all greater than the PL during the entire composting process (Table 1); therefore, the GE + BC treatments may further adsorb additional NH_3 compared to the GE + PL treatments and, ultimately, may result in reduced NH_3 emissions.

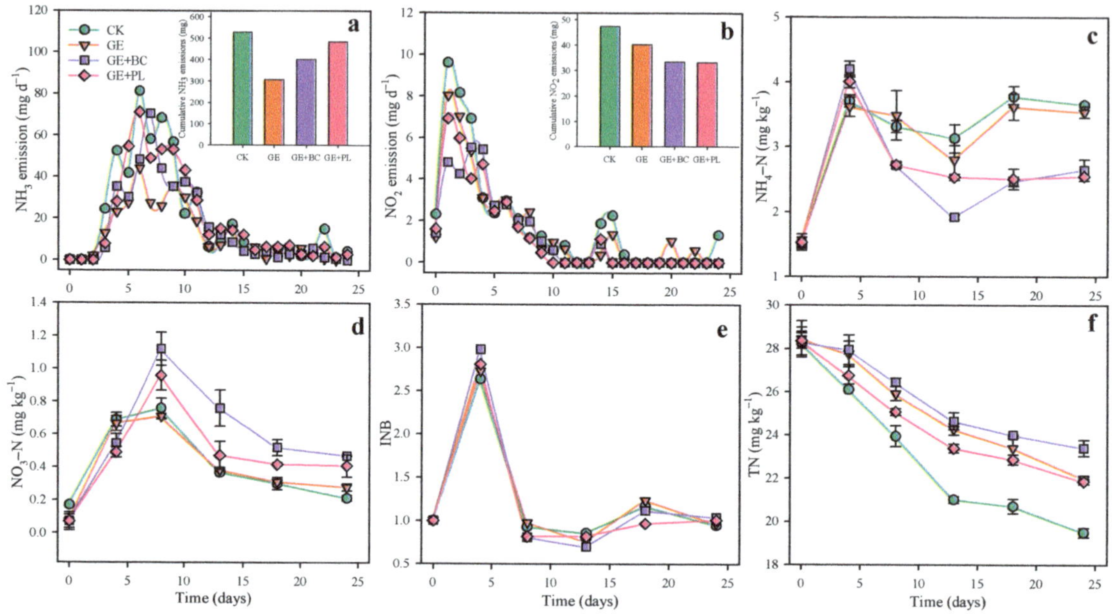

Figure 1. NH_3 (**a**), N_2O (**b**) emissions, NH_4-N (**c**), NO_3-N (**d**), the INB index (**e**), and TN (**f**).

Table 1. Changes in specific surface areas (SSA) and total pore volume (TPV) of PL and BC during the composting process.

Composting Time (Days)	BC		PL	
	SSA (m^2 g^{-1})	TPV (cm^3 g^{-1})	SSA (m^2 g^{-1})	TPV (cm^3 g^{-1})
0	14.19 ± 0.34	0.050 ± 0.001	6.53 ± 0.26	0.019 ± 0.001
4	7.23 ± 0.11	0.039 ± 0.002	3.49 ± 0.11	0.021 ± 0.003
8	6.54 ± 0.38	0.038 ± 0.003	2.35 ± 0.23	0.022 ± 0.002
13	6.13 ± 0.64	0.034 ± 0.001	4.00 ± 0.88	0.023 ± 0.009
18	6.11 ± 0.40	0.035 ± 0.001	2.80 ± 0.30	0.017 ± 0.006
24	5.34 ± 0.35	0.035 ± 0.005	3.16 ± 0.15	0.015 ± 0.001

Similar to the NH_3 emissions, the intensity of N_2O emissions was higher during the initial thermophilic period (days 1–6) (Figure 1b), and similar results were also found that by Awasthi et al. [15] during composting. The peak values of N_2O emissions appeared on day 1 with the values of 9.6, 8.0, and 6.9 mg d^{-1} in the CK, GE, and GE + PL treatments, respectively, while the GE + BC treatments on day 3 had a value of 5.5 mg d^{-1}. The cumulative N_2O emissions for the four treatments were 47.3, 40.3, 33.5, 33.1 mg, respectively. The three additive treatments showed reduced emissions of N_2O by 10.4%, 25.7, and 26.3% relative to the control treatment. Geng et al. [29] reported that biochar could reduce cumulative N_2O releases by 47.7% during kitchen waste composting. The BC and PL not only have a large SSA that can adsorb the emitted N_2O, but also may increase the TPV of the composting pile and impede the formation of an anaerobic environment, thus further decreasing the N_2O emissions caused by the denitrification process and mediated by denitrifying microbial communities.

3.2. Dynamic Changes in NH_4-N, NO_3-N, and TN Concentrations

NH_4-N is the substrate directly acted upon by the ammonia-oxidizing microorganisms and, therefore, has a significant influence on subsequent nitrification processes [30]. The NH_4-N content increased sharply (Figure 1c) because of the intense decomposition of nitrogen from the organic substrate where all treatments showed maximum values on day 4, in a range of 3.71, 3.62, 4.20, and 4.01 mg kg^{-1}. The values for GE + BC and GE + PL were greater than those for the CK and GE treatments, which might be attributed to the low pH of GE, which may slightly inhibit the decomposition of organic matter [23]. Meanwhile, the BC and PL alleviated the inhibition of GE, promoted the degradation of organic matter, and effectively absorbed the emitted NH_3 [17]. Following this, the NH_4-N contents in the GE + BC and GE + PL treatments showed an obvious decrease certainly due to the intense nitrification and, therefore, perhaps not greatly different within the CK and GE treatments. At the end of the composting process, the NH_4-N contents for the GE + BC and GE + PL treatments (2.65 and 2.55 mg kg^{-1}) were less than those for CK and GE (3.65 and 3.54 mg kg^{-1}).

Similarly, the NO_3-N contents firstly increased during days 0–8 and then decreased until the end of the composting period (Figure 1d). This is in line with the previous study conducted on SS composting [17]. The difference was that the maximum NO_3-N contents appeared on day 8, which lagged behing the peak values of the NH_4-N contents. Except for days 0 and 4, the NO_3-N contents for the GE + BC and GE + PL treatments were higher than those for the CK and GE treatments. At the end of composting period, the NO_3-N contents for the four treatments were 0.21, 0.28, 0.47, and 0.41 mg/kg, respectively, and in the order of GE + BC > GE + PL > GE treatment (123.8%, 95.2%, 32.9%) relative to the control treatment. Together with those of NH_4-N, the overall results showed the GE + BC and GE + PL treatments enhanced ammonification and nitrification during the composting process, thus improving the NO_3-N content of the end compost.

The index of the N balance (INB) was used to evaluate the total variation in dissolved inorganic N in the incubation system, such as the soil incubation and wastewater treatment, which was calculated based on the previous study [31]. As shown in Figure 1e, the INB in all treatments sharply increased during days 0–4 in all treatments, and then began to decrease gradually during the latter 4–8 days. The values of INB during days 0–4 ranged from 0.81 to 2.98, suggesting that the N loss was a result of the ammonification of the organic N during this stage [31]. The NH_3 emissions via ammonification during days 0–4 could be explained by this phenomenon (Figure 1a). And then, the IBN became almost near 1.0, which indicates no remarkable amount of DIN production from organic matter and no N loss emitted from NH_3 and N_2O during this stage [32].

Figure 1f shows the TN content for all four treatments with a descending trend during the whole composting process. The rapid decrease during days 0–12 was likely because of the intense emissions of NH_3 and N_2O (Figure 1a,b). The TN contents of the end compost in the CK, GE, GE + BC, and GE + PL treatments therefore reduced by 30.9%, 22.6%, 23.4%, and 21.9% relative to the initial values. This indicated that the BC and PL were correlate with GE and effectively reduced the N loss from NH_3 and N_2O emissions, especially within the GE + BC treatment.

3.3. Dynamic Changes in FDA, AMO, and HAO Activities

Figure 2a shows that the additives (in the GE, GE + BC, and GE + PL treatments) significantly increased the fluorescein diacetate hydrolase (FDA) activities at the initial stage relative to the control treatment, and the activity of FDA for the GE + BC treatments were always highest among the four treatments during the whole composting process. During the whole stage, the FDA activities of the four treatments declined. Although, these results were inconsistent with our previous study [17], showing maximum FDA activities in the similar treatment of additives on day 4. However, an intense decrease in FDA activities in this study in the GE, GE + BC, and GE + PL treatments was detected during days 0–4. This can be attributed to the fact that, during this stage, the OM contents decreased rapidly and all the FDA enzymes in the system were consumed, leaving the system enzyme-deficit in the latter stages. This phenomenon needs further investigation and can be a potential objective in forthcoming studies.

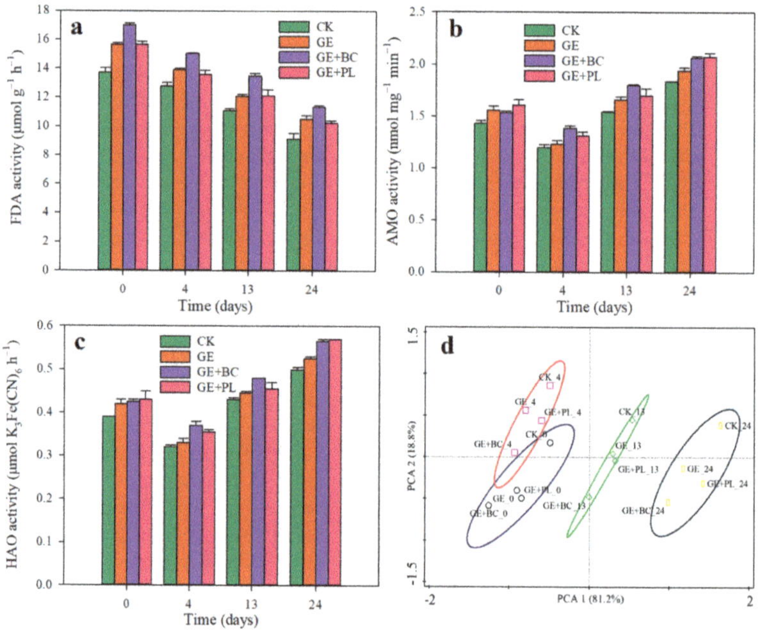

Figure 2. The activities of FDA (**a**), AMO (**b**), HAO (**c**), and the principal component analysis of FDA, AMO, and HAO (**d**).

Few studies have focused on AMO activities during the composting process, while most studies have always evaluated AMO activities by the abundance of AOA or AOB genes [8,9,33]. As shown in Figure 2b, the four treatments decreased sharply during days 0–4, and the AMO activities reached a minimum on day 4, with the values of 1.20, 1.23, 1.39, and 1.32 nmol·mg^{-1}·min^{-1}, respectively. This decrease may be due to the ammoxidation of large quantities of NH_4-N consuming a large amount of AMO enzyme. After that, the AMO activities gradually increased with the decrease in NH_4-N (Figure 1c). The averaged AMO activities during the entire process in the CK, GE, GE + BC, and GE + PL treatments were 1.50, 16.0, 1.70, and 1.67 nmol·mg^{-1}·min^{-1}, respectively.

The HAO is the main enzyme for converting hydroxylamine into nitrite nitrogen, and is directly related to the NO_3-N of the end product compost [8]. Similar to AMO activities, HAO activities in the four treatments reduced drastically during days 0–4, and then gradually increased until the end of the composting period (Figure 2c). The rapid decrease in HAO activity also corresponded to the rapid increase in NO_3-N content (Figure 1d). The average values of the HAO in the CK, GE, GE + BC, and GE + PL treatments' activities during the entire process were 0.41, 0.43, 0.43, and 0.45 μmol $K_3Fe(CN)_6$ h^{-1}, respectively. These results showed that GE combined with BC and PL increased the AMO and HAO activities and elevated the process of ammoxidation, thus improving the subsequent nitrification process and augmenting the final NO_3-N content of the compost.

Furthermore, the PCA analysis showed that the three treatments (the GE, GE + BC, and GE + PL treatments) affected the FDA, AMO, and HAO activities relative to the control treatment (Figure 2d) during the whole composting period. Among the three treatments, samples for the GE + BC treatments on days 4, 13, and 24 were very different to the samples for the GE and GE + PL treatments, which indicated that BC connected GE could effectively change enzymatic activity and has a greater impact on composting process compared to the GE and GE + PL treatments.

3.4. Dynamic Changes in the Abundances of Archaea and Bacteria of the amoA Gene

AOA and AOB participate in the oxidation process of ammonia during the composting process [8]. Figure 3 shows the abundance of AOA and AOB genes ranging from 1.12×10^5–4.71×10^5 to 3.28×10^4–9.13×10^4 copies g^{-1}. During the whole composting process, the gene copies of AOA were higher than those of AOB, which shows that AOA has a greater contribution in accelerated nitrification in this study. This result was in line with the results of Yan et al. [34] on the composting of cattle manure and those of Zeng et al. [8] on the composting of agricultural waste.

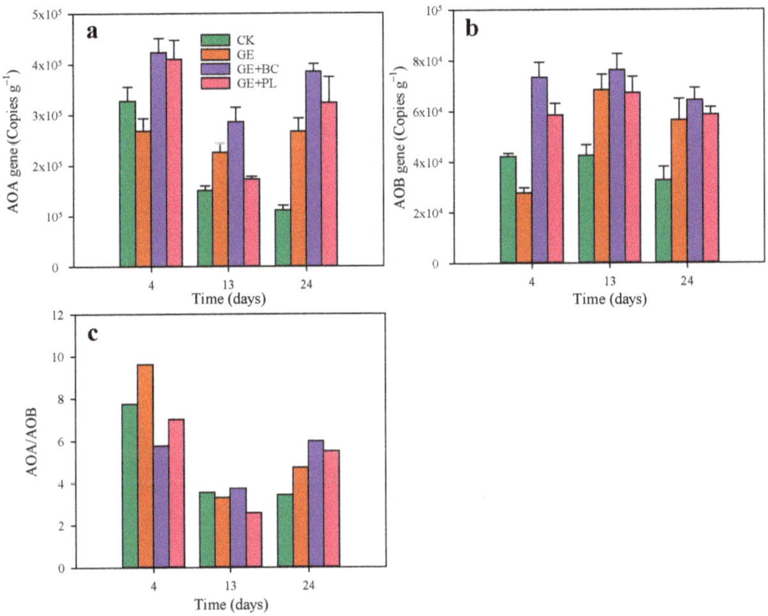

Figure 3. The abundances of AOA (**a**) and AOB (**b**) genes and the ratio of AOA to AOB (**c**).

As shown in Figure 3a, the AOA gene in the CK, GE, GE + BC, and GE + PL treatments during the thermophilic phase (4 d) had the highest values, with values of 3.27×10^5, 2.68×10^5, 4.23×10^5, and 4.10×10^5 copies g^{-1}, respectively. The abundance of the AOA gene in GE was lower than that in the control treatment because of the inhibition of GE by microorganisms. Whereas those in GE + BC and GE + PL were higher than the control mainly due to BC and PL alleviating the inhibition of GE. At the cool stage (13 d), the abundance of the AOA gene decreased, which can be attributed to probable relatively higher temperature during the thermophilic stage slowing down the growth of AOA. After this stage, the temperature dropped and the abundance of the AOA gene simultaneously increased until compost maturity (24 d).

Relative to AOA, the AOB gene had a similar change during the thermophilic phase (Figure 3b). The abundance of AOB genes in the GE + BC and GE + PL treatments on day 4 increased by 73.9% and 38.7%, while in the GE treatment it reduced by 33.8% relative to the control. Wu et al. [10] also reported that BC promoted the abundance of AOB during the composting of agricultural waste. Lin et al. [35] reported that biochar significantly increased the abundance of the AOA gene, and found that N_2O emissions had an extreme relationship with the abundance of the AOA gene. As the composting process proceeded, the inhibition of GE weakened or was eliminated after the thermophilic phase. This change gradually supported the growth of AOA resulting in an increased abundance of AOA genes during cooling stage. During the cooling and maturity stages, the abundance of the AOA genes in the three additive treatments raged from 5 to 7 $\times$ 10^5 copies g^{-1}, while this value was 3–4 $\times$ 10^5 copies g^{-1} in the control.

The ratio of AOA to AOB copies initially decreased during days 4–13, and then increased during days 13–24 (Figure 3c). The results also showed that the AOA copies increased at the cooling and maturity stages relative to AOB, which indicated that AOA could grow well at lower temperatures during the composting process. Sims et al. [5] also reported the better adaptation of AOA than AOB to the low-temperature environment of natural wetlands. Meanwhile, AOB increased during the thermophilic phase, meaning that the high temperature suited the growth of AOB, which also corresponded with the high NH_4-N content (Figure 1c), providing further optimal conditions for AOB [34].

As the results of the AOA and AOB genes shown, the addition of GE slowed down the growth of AOA and AOB, while the addition of BC and PL could eliminate this inhibition during the thermophilic stage. In addition, the gene copies of AOA and AOB in GE + BC and GE + PL were also greater than in the GE and CK treatments during the whole process. This shows that BC and PL combined with GE could improve the growth environment for AOA and AOB, and strengthen ammoxidation during the composting process, which significantly reduces N_2O emissions (Figure 1b) and increases NO_3-N content (Figure 1d). Meanwhile, other nitrogen functional genes, such as the ammonia monooxygenase (*amoA*), nitrite reductase (*nirK* and *nirS*), and nitrous oxide reductase (*nosZ*) genes, also had close relationships with NH_3 and N_2O emissions [33]. Thus, further research will be necessary to evaluate all nitrogen-functional genes after the addition of additives, which is likely to reveal the mechanism of the nitrogen loss reduction. In addition, this also has important guiding significance for selecting suitable additives to reduce nitrogen loss.

3.5. Community Composition of AOA and AOB

Based on the 16S rDNA sequence data of AOA, a total of 171, 053 high-quality sequences were obtained from 12 samples, and clustered into 51 OTUs based on a $\geq$3% dissimilarity cutoff. Meanwhile, 252,871 sequences and only 14 OTUs were obtained for AOB from the same samples. The results were in line with the relative abundances (Ras) of AOA and AOB genes, and once more affirmed that AOA has a greater contribution to nitrification in this study, relative to AOB.

Principal coordinates analyses (PCoAs), based on unweighted UniFrac distances and the OUT level, were also conducted to explore the difference in species composition among the different treatments. The results showed that the grouping ellipse sizes in GE + BC and GE + BC were bigger than those in the CK and GE treatments, which indicated that BC and PL correlated with GE significantly, which altered the community composition of the AOA (Figure S1).

As shown in Figure 4a, only Crenarchaeota and Thaumarchaeota could be identified as AOA during the composting processes, which is also in line with our previous study when a urease inhibitor and a nitrification inhibitor were added during SS composting [36]. The RAs of theses four phyla were 27.55–64.21%, 34.59–63.33%, 0–24.33%, and 0–1.24%, respectively. Significant differences in AOA between treatments (Figure S2) showed that the RA of an unclassified phylum for the GE treatments was lower than for CK, while the RAs for the GE + BC, and GE + PL treatments were greater than the control treatment during thermophilic phase. In the cooling and maturity phases, the RAs of the unclassified phylum in the GE, GE + BC, and GE + PL treatments were all higher than that in the control. The RAs of Crenarchaeota in the GE + BC and GE + PL treatments were lower than in CK during the entire composting process. However, the RA in the GE treatment during thermophilic stage was higher than in the control, while the RA in the GE treatment during the cooling stage was lower than in the CK treatment. Crenarchaeota widely exist in the ocean, hot springs, and soil, and are the most abundant and most widely distributed AOA in nature [37]. These results not only showed that BC and PL together with GE resulted in the transition from Crenarchaeota to an unclassified phylum, but also indicated the complexity of the community structure of AOA. Thaumarchaeota, as the newly discovered AOA phylum, are highly diverse and widely distributed in various environments, but their physiological metabolism, genetic characteristics, and ecological functions need further study [38]. Thaumarchaeota were only found in the samples of GE + BC_13, CK_24, and GE_24. At the genus level, *unclassified_Archaea, norank_Crenarchaeota, Nitrososphera,* and *norank_p_environmental_samples* were the dominant genera of AOA (Figure 4b). These three genera, *unclassified_Archaea, norank_Crenarchaeota,* and *norank_p_environmental_samples,* accounted for 75.7–100% of the total sequencing reads. Lin et al. [39] also reported that all of the detected genera belonged to no rank or unclassified genus when treating saline wastewater in biofilm reactors.

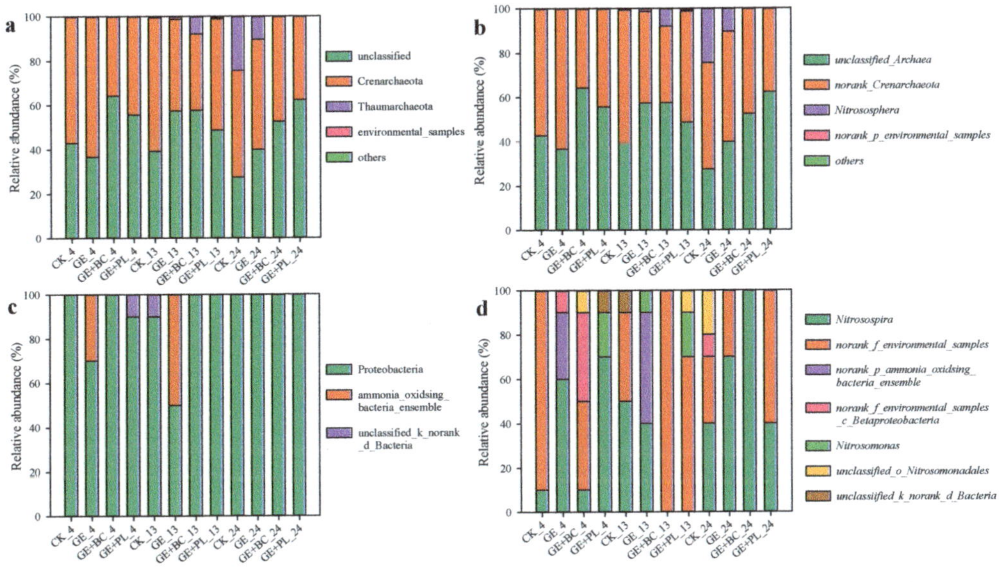

Figure 4. The community composition AOA at the phylum level (**a**) and at the genus level (**b**); the community composition AOB at the phylum level (**c**) and at the genus level (**d**).

As shown in Figure 4c, only three phyla of AOB, Proteobacteria, ammonia_oxidsing_bacteria_ensemble, and unclassified_k_norank_d_Bacteria, were detected in all of the samples during the composting process. The Proteobacteria was the dominant phylum in all samples, while ammonia_oxidsing_bacteria_ensemble was only found in the GE_4 and GE_13 samples, with the RAs of 30% and 50%, respectively. At the genus level, there were significant differences in the AOA composition between different periods and treatments (Figure 4d). *Nitrosospira* (0–100%), *norank_f_environmental_samples* (0–100%), *norank_p_ammonia_oxidsing_bacteria_ensemble* (0–50%), *norank_f_environmental_samples_c_Betaproteobacteria* (0–40%), *Nitrosomonas* (0–20%), *unclassified_o_Nitrosomonadales* (0–20%), and *unclassiified_k_norank_d_Bacteria* (0–10%) were the main genera during the composting process. The no-rank and unclassified genera accounted for 75.7–100% of the total sequencing reads. Related to the no-rank and unclassified genera for AOA, these results indicate the complexity of the community structure of AOA and AOB during composting, which urgently requires advanced isolation and identification techniques and the updating of gene databases. Studies have found that Nitrosospira and Nitrosomonas were the dominant genus for nitrification in many ecosystems, such as sea [30], soil [36], wastewater [39].

On day 4, the RAs of Nitrosospira in the GE and GE + PL treatments were significantly greater than in the CK and GE + BC treatments. This can be mainly attributed to the high temperature and NH_4-N content in the GE + BC treatment, which ultimately inhibited the Nitrosospira genus [40]. Going through the high temperature of the thermophilic stage, the Nitrosospira in GE + PL and GE + BC could not be detected on day 13, meaning that the composting conditions were accrual for this specific bacterial genus. With the decrease in temperature during the maturity stages, Nitrosospira became the dominant genus with RAs of 40%, 70%, 100%, and 40% in the CK, GE, GE + BC, and GE + PL treatments, and its RAs in the GE + BC treatment reached up to 100%. The high NO_3-N content could support this inference. Nitrosomonas was the dominant species among the AOB, and Lu et al. [31] found that the RAs of Nitrosomonas were up to 79.8% in an anammox-inoculated wastewater treatment system. However, Nitrosomonas was only detected in the samples GE + PL_4, GE_13, and GE + PL_13 with RAs of 20%, 10%, and 20%, respectively.

3.6. Relationships between Physicochemical Indexes, Enzymatic Activity, and Ammonia-Oxidizing Microorganisms

The RDA shows that the selected environmental variables (EC, MC, NH_4-N, NO_3-N, and C/N) and AMO enzymatic activities could explain 70.8% and 60.1% of the variations in the AOA and AOB community, respectively (Figure 5a,b). Among the seven selected tactors, FDA and AMO activities, NH_4-N, NO_3-N, and C/N significantly affect the AOA and AOB community structures. FDA represented the overall level of all enzymes and was closely related to the N transformation during the composting process [23]. The AMO enzyme is secreted directly by ammonia-oxidizing microorganisms, consisting of AOA and AOB. Wu et al. [10] also reported that AMO activity had markedly positive relationships with AOA and AOB numbers. As the substrate of ammonia-oxidizing microorganisms, NH_4-N content directly affected their structure and abundance. NO_3-N, as an indirect product of ammonia oxidation, could reflect the evolution of ammonia-oxidizing microorganisms to a certain extent. Shi et al. [6] found that NO_3-N solely explained 27.3% of the variation in the AOB species during food waste composting. The initial C/N ratio of composting substrates affects the reproduction of all microorganisms and plays a vital role during the composting process. Lin et al. [35] reported that AOB were positively correlated with NH_4-N and C/N, while having an opposite correlation with NO_3-N.

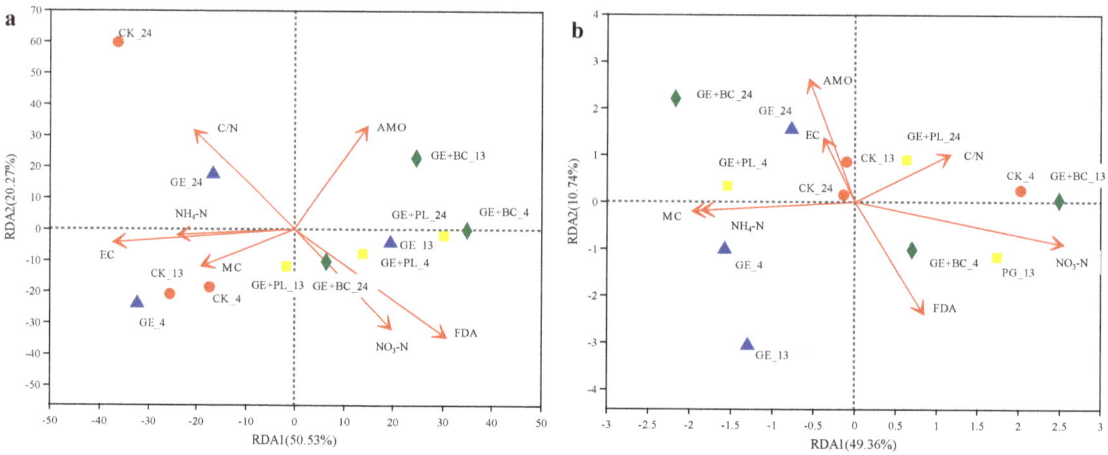

Figure 5. The RDA analysis of enzyme activities, physicochemical indexes, and ammonia-oxidizing microorganisms at the genus level: the RDA of AOA (**a**) and AOB (**b**).

In order to analyze the relationships between the three enzymes, eight physiochemical factors, and the detected OTUs (OTU 51 for AOA and OUT 14 for AOB), network analysis was conducted. The yielded nodes for the AOB community in the CK, GE, GE + BC, and GE + PL treatments were only 13, 4, 3, and 3 (Figure S3), while the corresponding nodes for the AOA community were 20, 25, 26, and 22, respectively (Figure 6). Therefore, only the network analysis related to AOA is presented in the current study.

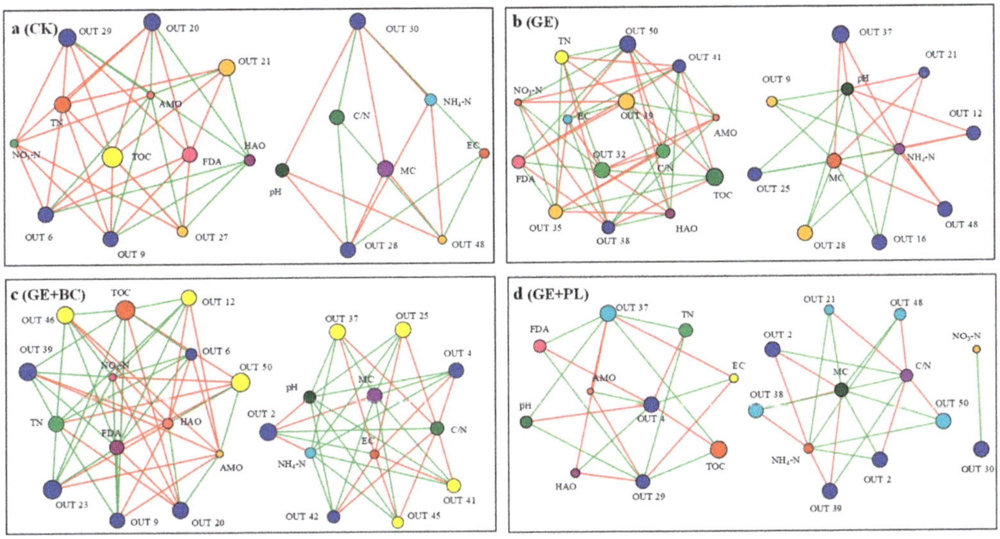

Figure 6. The network analysis of the enzymes, physicochemical characteristics, and the ammonia-oxidizing archaea (AOA) at the OTU level for CK (**a**), GE (**b**), GE + BC (**c**), and GE + PL (**d**) treatments, respectively. A connection represents a significant correlation ($p < 0.05$) according to Spearman's rank analysis (rho > 0.6).

For the AOA, the analysis revealed 20 nodes and 102 edges, with 11 physicochemical indexes and 9 OTUs in CK. The AMO and HAO activities had negative relationships with OTU 9, OTU 6, OTU 20, OTU 29, OTU 27, and OTU 21 (Figure 6). In the GE treatment, there were 25 nodes and 144 edges, with 11 physicochemical indexes and 14 OTUs. The AMO and HAO activities had positive relationships with OTU 32, OTU 35, and OTU 50, while they negatively correlated with OTU 38, and OTU 41 OTU. For the GE + BC treatment, the analysis revealed 26 nodes and 166 edges, with 11 physicochemical indexes and 15 OTUs. The AMO and HAO activities had positive relationships with OTU 6, OTU 9, OTU 12, OTU 23, OTU 39, and OTU 46, while they had negative relationships with OTU 20 and OUT 50. For the GE + PL treatments, there were 22 nodes and 86 edges, with 11 physicochemical indexes and 11 OTUs. The AMO and HAO activities had positive relationships with OTU 29 and OTU 37, while they had negative relationship with OTU 4.

The OTUs had positive relationships with all AMO and HAO activities belonging to the *unclassified_Archaea* and *norank_Crenarchaeota* genera, indicating that these two genera most probable secreted ammoxidation-related enzymes. Among the three treatments, the close relationships between environmental factors, enzymatic activities, and AOA, were significantly promoted as revealed by the increased numbers of nodes and edges, and the OTUs that had a positive relationship with AMO and HAO activities were also increased in the GE + BC treatment, which showed that the combined usage of GE and BC has the greatest potential to enhance the relationship between enzymes and microorganisms related to ammonia oxidation processes compared to other treatments. Accordingly, this treatment increased the rate of nitrification and reduced NH_3 emissions and N loss during the composting process. Similar results were also reported by Jiang et al. [17] during SS composting.

4. Conclusions

The GE + BC and GE + PL treatments clearly promoted the activities of AMO and HAO and the gene copy numbers of AOA and AOB, thus improving NH_4-N oxidation and NO_3-N production. Meanwhile, the GE + BC and GE + PL treatments decreased NH_3 emissions by 23.8% and 8.3%, respectively. Network analysis showed that the co-application of garbage enzyme with biochar enhanced the positive connections between environmental factors, enzymatic activities, and the dominant genus within AOA. In addition, RDA indicated that FDA and AMO activities, NH_4-N, NO_3-N, and C/N significantly affect the community structures of AOA and AOB. Therefore, GE combined with BC might be a useful strategy to reduce NH_3 emissions and accelerate ammoxidation during SS composting. Due to the complexity of the community structures of AOA and AOB during composting, further research will be necessary to exploit advanced isolation and identification techniques and the updating of the gene database.

Supplementary Materials: The following supporting information can be downloaded at: https://www.mdpi.com/article/10.3390/agronomy14061162/s1, Table S1: The analyze of variance inflation factor; Figure S1: The principal co-ordinates analysis (PCoA) analysis based on unweighted UniFrac distances for the ammonia-oxidizing microorganisms on OUT level: (a) the PCoA nanlysis of AOA, (b) the PCoA nanlysis of AOB; Figure S2: Significant differences in ammonia-oxidizing archaea (AOA) at the phylum level between treatments; Figure S3: The network analysis among the enzymatic activities, physicochemical characteristics, and the ammonia oxidizing bacteria (AOB) at the OTU level. A connection represents a significant correlation ($p < 0.05$) according to Spearman's rank analysis (rho > 0.6).

Author Contributions: Funding acquisition, Supervision, Writing—review &editing, J.J. Writing—original draft, Investigation, H.C. Data curation, Formal analysis, Investigation, P.B. Writing—review & editing, C.C.C.C. Funding acquisition, Supervision, F.Y. Methodology, Resources, Writing—review & editing, Supervision, Validation, D.L. All authors have read and agreed to the published version of the manuscript.

Funding: This research was funded by the National Natural Science Foundation of China (41805123), the Natural Science Foundation of Henan Province (232300421247), and the Yunnan Revitalization Talent Support Program 'Young Talent' Project (YNQR-QNRC-2019-025).

Data Availability Statement: The original contributions presented in the study are included in the article, further inquiries can be directed to the corresponding author.

Acknowledgments: The bacterial community analyses were conducted on the online platform of Majorbio Cloud Platform.

Conflicts of Interest: The authors declare no conflicts of interest.

References

1. Zhao, Y.; Li, W.; Chen, L.; Meng, L.; Zheng, Z. Effect of enriched thermotolerant nitrifying bacteria inoculation on reducing nitrogen loss during sewage sludge composting. *Bioresour. Technol.* **2020**, *311*, 123461. [CrossRef] [PubMed]
2. Ruiz-Barrera, O.; Ontiveros-Magadan, M.; Anderson, R.C.; Byrd, A.; Hume, M.E.; Latham, E.A.; Nisbet, D.J.; Arzola-Alvarez, C.; Salinas-Chavira, J.; Castillo-Castillo, Y. Nitro-treatment of composted poultry litter; effects on Salmonella, *E. coli* and nitrogen metabolism. *Bioresour. Technol.* **2020**, *310*, 123459. [CrossRef] [PubMed]
3. Lei, L.; Gu, J.; Wang, X.; Song, Z.; Yu, J.; Wang, J.; Dai, X.; Zhao, W. Effects of phosphogypsum and medical stone on nitrogen transformation, nitrogen functional genes, and bacterial community during aerobic composting. *Sci. Total Environ.* **2021**, *753*, 141746. [CrossRef] [PubMed]
4. Huang, D.; Gao, L.; Cheng, M.; Yan, M.; Zhang, G.; Chen, S.; Du, L.; Wang, G.; Li, R.; Tao, J.; et al. Carbon and N conservation during composting: A review. *Sci. Total Environ.* **2022**, *840*, 156355. [CrossRef]
5. Oishi, R.; Tada, C.; Asano, R.; Yamamoto, N.; Suyama, Y.; Nakai, Y. Growth of Ammonia-Oxidizing Archaea and Bacteria in Cattle Manure Compost under Various Temperatures and Ammonia Concentrations. *Microb. Ecol.* **2012**, *64*, 787–793. [CrossRef]
6. Shi, S.; Zou, D.; Wang, Q.; Xia, X.; Zheng, T.; Wu, C.; Gao, M. Responses of ammonia-oxidizing bacteria community composition to temporal changes in physicochemical parameters during food waste composting. *RSC Adv.* **2016**, *6*, 9541–9548. [CrossRef]

7. Zhou, S.; Wen, X.; Cao, Z.; Cheng, R.; Qian, Y.; Mi, J.; Wang, Y.; Liao, X.; Ma, B.; Zou, Y.; et al. Modified cornstalk biochar can reduce ammonia emissions from compost by increasing the number of ammonia-oxidizing bacteria and decreasing urease activity. *Bioresour. Technol.* **2021**, *319*, 124120. [CrossRef] [PubMed]
8. Zeng, G.; Zhang, J.; Chen, Y.; Yu, Z.; Yu, M.; Li, H.; Liu, Z.; Chen, M.; Lu, L.; Hu, C. Relative contributions of archaea and bacteria to microbial ammonia oxidation differ under different conditions during agricultural waste composting. *Bioresour. Technol.* **2011**, *102*, 9026–9032. [CrossRef] [PubMed]
9. Yan, L.; Li, Z.; Wang, G.; Gao, Y.; Wang, Y.; Gu, J.; Wang, W. Diversity of ammonia-oxidizing bacteria and archaea in response to different aeration rates during cattle manure composting. *Ecol. Eng.* **2016**, *93*, 46–54. [CrossRef]
10. Wu, X.; Ren, L.; Zhang, J.; Peng, H. Effects of zeolite and biochar addition on ammonia-oxidizing bacteria and ammonia-oxidizing archaea communities during agricultural waste composting. *Sustainability* **2020**, *12*, 6336. [CrossRef]
11. Kuypers, M.M.M.; Marchant, H.K.; Kartal, B. The microbial nitrogen-cycling network. *Nat. Rev. Microbiol.* **2018**, *16*, 263–276. [CrossRef] [PubMed]
12. Zhang, L.; Dong, H.; Zhang, J.; Chen, Y.; Zeng, G.; Yuan, Y.; Cao, W.; Fang, W.; Hou, K.; Wang, B.; et al. Influence of FeONPs amendment on nitrogen conservation and microbial community succession during composting of agricultural waste: Relative contributions of ammonia-oxidizing bacteria and archaea to nitrogen conservation. *Bioresour. Technol.* **2019**, *287*, 121463. [CrossRef] [PubMed]
13. Arun, C.; Sivashanmugam, P. Identification and optimization of parameters for the semi-continuous production of garbage enzyme from pre-consumer organic waste by green RP-HPLC method. *Waste Manag.* **2015**, *44*, 28–33. [CrossRef] [PubMed]
14. Negi, S.; Mandpe, A.; Hussain, A.; Kumar, S. Collegial effect of maggots larvae and garbage enzyme in rapid composting of food waste with wheat straw or biomass waste. *J. Clean. Prod.* **2020**, *258*, 120854. [CrossRef]
15. Awasthi, M.K.; Duan, Y.; Awasthi, S.K.; Liu, T.; Zhang, Z. Influence of bamboo biochar on mitigating greenhouse gas emissions and nitrogen loss during poultry manure composting. *Bioresour. Technol.* **2020**, *303*, 122952. [CrossRef] [PubMed]
16. Ravindran, B.; Nguyen, D.D.; Chaudhary, D.K.; Chang, S.W.; Kim, J.; Lee, S.R.; Shin, J.; Jeon, B.-H.; Chung, S.; Lee, J. Influence of biochar on physico-chemical and microbial community during swine manure composting process. *J. Environ. Manag.* **2019**, *232*, 592–599. [CrossRef] [PubMed]
17. Jiang, J.; Wang, Y.; Yu, D.; Zhu, G.; Cao, Z.; Yan, G.; Li, Y. Comparative evaluation of biochar, pelelith, and garbage enzyme on nitrogenase and nitrogen-fixing bacteria during the composting of sewage sludge. *Bioresour. Technol.* **2021**, *333*, 125165. [CrossRef] [PubMed]
18. Ding, S.; Zhou, D.; Wei, H.; Wu, S.; Xie, B. Alleviating soil degradation caused by watermelon continuous cropping obstacle: Application of urban waste compost. *Chemosphere* **2021**, *262*, 128387. [CrossRef] [PubMed]
19. Wang, G.; Li, Y.; Sheng, L.; Xing, Y.; Liu, G.; Yao, G.; Ngo, H.; Li, Q.; Wang, X.C.; Li, Y.; et al. A review on facilitating bio-wastes degradation and energy recovery efficiencies in anaerobic digestion systems with biochar amendment. *Bioresour. Technol.* **2020**, *314*, 123777. [CrossRef] [PubMed]
20. Jiang, J.; Pan, Y.; Yang, X.; Liu, J.; Miao, H.; Ren, Y.; Zhang, C.; Yan, G.; Lv, J.; Li, Y. Beneficial influences of pelelith and dicyandiamide on gaseous emissions and the fungal community during sewage sludge composting. *Environ. Sci. Pollut. Res.* **2019**, *26*, 8928–8938. [CrossRef]
21. Jiang, J.; Wang, Y.; Liu, J.; Yang, X.; Ren, Y.; Miao, H.; Pan, Y.; Lv, J.; Yan, G.; Ding, L.; et al. Exploring the mechanisms of organic matter degradation and methane emission during sewage sludge composting with added vesuvianite: Insights into the prediction of microbial metabolic function and enzymatic activity. *Bioresour. Technol.* **2019**, *286*, 121397. [CrossRef] [PubMed]
22. Han, Z.; Sun, D.; Wang, H.; Li, R.; Bao, Z.; Qi, F. Effects of ambient temperature and aeration frequency on emissions of ammonia and greenhouse gases from a sewage sludge aerobic composting plant. *Bioresour. Technol.* **2018**, *270*, 457–466. [CrossRef] [PubMed]
23. Jiang, J.; Wang, Y.; Yu, D.; Yao, X.; Han, J.; Cheng, R.; Cui, H.; Yan, G.; Zhang, X.; Zhu, G. Garbage enzymes effectively regulated the succession of enzymatic activities and the bacterial community during sewage sludge composting. *Bioresour. Technol.* **2021**, *327*, 124792. [CrossRef] [PubMed]
24. Jiang, J.; Kang, K.; Chen, D.; Liu, N. Impacts of delayed addition of N-rich and acidic substrates on nitrogen loss and compost quality during pig manure composting. *Waste Manag.* **2018**, *72*, 161–167. [CrossRef] [PubMed]
25. Zhang, M.; He, T.; Chen, M.; Wu, Q. Ammonium and hydroxylamine can be preferentially removed during simultaneous nitrification and denitrification by Pseudomonas taiwanensis EN-F2. *Bioresour. Technol.* **2022**, *350*, 126912. [CrossRef] [PubMed]
26. Guo, H.; Gu, J.; Wang, X.; Yu, J.; Nasir, M.; Zhang, K.; Sun, W. Microbial driven reduction of N_2O and NH_3 emissions during composting: Effects of bamboo charcoal and bamboo vinegar. *J. Hazard. Mater.* **2020**, *390*, 121292. [CrossRef] [PubMed]
27. Liu, C.; Zhang, X.; Zhang, W.; Wang, S.; Fan, Y.; Xie, J.; Liao, W.; Gao, Z. Mitigating gas emissions from poultry litter composting with waste vinegar residue. *Sci. Total Environ.* **2022**, *842*, 156957. [CrossRef] [PubMed]
28. Li, X.; Zhao, Y.; Xu, A.; Chang, H.; Lin, G.; Li, R. Conductive biochar promotes oxygen utilization to inhibit greenhouse gas emissions during electric field-assisted aerobic composting. *Sci. Total Environ.* **2022**, *842*, 156929. [CrossRef] [PubMed]
29. Geng, X.; Yang, H.; Gao, W.; Yue, J.; Mu, D.; Wei, Z. Greenhouse gas emission characteristics during kitchen waste composting with biochar and zeolite addition. *Bioresour. Technol.* **2024**, *399*, 130575. [CrossRef] [PubMed]
30. Li, M.; He, H.; Mi, T.; Zhen, Y. Spatiotemporal dynamics of ammonia-oxidizing archaea and bacteria contributing to nitrification in sediments from Bohai Sea and South Yellow Sea, China. *Sci. Total Environ.* **2022**, *825*, 153972. [CrossRef] [PubMed]

31. Lu, J.; Hong, Y.; Wei, Y.; Gu, J.; Wu, J.; Wang, Y.; Ye, F.; Lin, J.-G. Nitrification mainly driven by ammonia-oxidizing bacteria and nitrite-oxidizing bacteria in an anammox-inoculated wastewater treatment system. *AMB Express* **2021**, *11*, 158. [CrossRef] [PubMed]
32. Hong, Y.; Wang, Y.; Wu, J.; Jiao, L.; He, X.; Wen, X.; Zhang, H.; Chang, X. Developing a mathematical modeling method for determining the potential rates of microbial ammonia oxidation and nitrite oxidation in environmental samples. *Int. Biodeterior. Biodegrad.* **2018**, *133*, 116–123. [CrossRef]
33. Liu, B.; Chen, W.; Wang, Z.; Guo, Z.; Li, Y.; Xu, L.; Wu, M.; Yin, H. The Impact of Bacillus coagulans X3 on Available Nitrogen Content, Bacterial Community Composition, and Nitrogen Functional Gene Levels When Composting Cattle Manure. *Agronomy* **2024**, *14*, 587. [CrossRef]
34. Yan, L.; Wang, G.; Ai, S.; Huo, Z.; Wang, Y.; Gu, J.; Wang, W. Abundance of ammonia-oxidizing bacteria and archaea under different ventilation strategies during cattle manure composting. *J. Environ. Manag.* **2018**, *212*, 375–383. [CrossRef] [PubMed]
35. Lin, X.; Al-Dhabi, N.A.; Li, F.; Wang, N.; Peng, H.; Chen, A.; Wu, G.; Zhang, J.; Zhang, L.; Huang, H.; et al. Relative contribution of ammonia-oxidizing bacteria and denitrifying fungi to N_2O production during rice straw composting with biochar and biogas residue amendments. *Bioresour. Technol.* **2023**, *390*, 129891. [CrossRef] [PubMed]
36. Hu, J.; Richwine, J.D.; Keyser, P.D.; Li, L.; Yao, F.; Jagadamma, S.; DeBruyn, J.M. Ammonia-oxidizing bacterial communities are affected by nitrogen fertilization and grass species in native C4 grassland soils. *PeerJ* **2021**, *9*, e12592. [CrossRef] [PubMed]
37. Brochier-Armanet, C.; Boussau, B.; Gribaldo, S.; Forterre, P. Mesophilic crenarchaeota: Proposal for a third archaeal phylum, the Thaumarchaeota. *Nat. Rev. Microbiol.* **2008**, *6*, 245–252. [CrossRef]
38. Huang, M.; Chai, L.; Jiang, D.; Zhang, M.; Zhao, Y.; Huang, Y. Increasing aridity affects soil archaeal communities by mediating soil niches in semi-arid regions. *Sci. Total Environ.* **2019**, *647*, 699–707. [CrossRef] [PubMed]
39. Lin, Z.; Huang, W.; Zhou, J.; He, X.; Wang, J.; Wang, X.; Zhou, J. The variation on nitrogen removal mechanisms and the succession of ammonia oxidizing archaea and ammonia oxidizing bacteria with temperature in biofilm reactors treating saline wastewater *Bioresour. Technol.* **2020**, *314*, 123760. [CrossRef] [PubMed]
40. Zhang, J.; Luo, L.; Gao, J.; Peng, Q.; Huang, H.; Chen, A.; Lu, L.; Yan, B.; Wong, J.W.C. Ammonia-oxidizing bacterial communities and shaping factors with different Phanerochaete chrysosporium inoculation regimes during agricultural waste composting. *RSC Adv.* **2016**, *6*, 61473–61481. [CrossRef]

agronomy

MDPI

Article

Closing the Loop in Opuntia Cultivation: Opportunities and Challenges in Residue Valorization

Alan Jesús Torres-Sandoval [1], Yolanda Donají Ortiz-Hernández [2,*], María Elena Tavera-Cortés [3], Marco Aurelio Acevedo-Ortiz [1,*] and Gema Lugo-Espinosa [1]

1 Secretaría de Ciencias, Humanidades, Tecnología e Innovación—Instituto Politécnico Nacional, Centro Interdisciplinario de Investigación para el Desarrollo Integral Regional Oaxaca, Santa Cruz Xoxocotlán 71230, Mexico; atorress1404@alumno.ipn.mx (A.J.T.-S.); glugoe@ipn.mx (G.L.-E.)

2 Instituto Politécnico Nacional, Centro Interdisciplinario de Investigación para el Desarrollo Integral Regional Oaxaca, Santa Cruz Xoxocotlán 71230, Mexico

3 Instituto Politécnico Nacional, Unidad Profesional Interdisciplinaria de Ingeniería y Ciencias Sociales y Administrativas, Ciudad de México 08400, Mexico; mtavera@ipn.mx

* Correspondence: yortiz@ipn.mx (Y.D.O.-H.); macevedoo@ipn.mx (M.A.A.-O.)

Abstract: Global food systems face growing pressure from population expansion and climate change, making the identification of resilient crops a priority. The nopal cactus (*Opuntia* spp.) stands out for its capacity to thrive in arid environments and for its cultural and economic importance in Mexico. This study analyzes worldwide research trends and evaluates evidence from Mexico to identify opportunities and strategies for closing production cycles through residue valorization. Scientific output over the past decade shows steady growth and a thematic transition from basic agronomic and compositional studies toward sustainability, bioactive compounds, and circular economy approaches. In the Mexican context, applied studies demonstrate that *Opuntia* spp. cladodes residues can be transformed into composts with C/N ratios between 12 and 26, improving soil organic matter and nutrient availability. Biofertilizers produced through anaerobic fermentation enhanced phosphorus solubility in alkaline soils, while direct residue incorporation increased carrot and tomato yields up to threefold. Farmers recognize these practices as low-cost and compatible with local systems. Nevertheless, the lack of standardized protocols and scalable models limits widespread adoption. Strengthening research collaboration, policy incentives, and technology transfer could position Mexico as a leader in sustainable Opuntia value chains and advance circular economy practices in smallholder farming systems.

Keywords: Opuntia; waste valorization; composting; bibliometrics; systematic review; circular economy; sustainable agriculture; food security

1. Introduction

Faced with increasing pressure on global food systems, driven by rapid population growth and the challenges posed by climate change, identifying resilient crops capable of withstanding harsh environmental conditions has become a pressing priority. In this context, the nopal (*Opuntia* spp.) stands out as an exceptional resource, uniquely adapted to thrive in arid and semi-arid regions [1]. Its stems, known as cladodes are photosynthetic and function as natural water reservoirs, a physiological feature that underpins its remarkable drought tolerance and productivity under adverse conditions. The Food and Agriculture Organization of the United Nations (FAO) has highlighted its potential to strengthen food security, recognizing its remarkable adaptation to drought, efficient use of water, and versatility as food for humans and livestock [1,2].

Mexico is the center of *Opuntia* spp. biological diversity and domestication [3]. National production of nopalitos (young pads) surpasses 860,000 tons annually, cultivated on more than 12,600 hectares [4]. Leading production areas include Morelos, Estado de México, and Mexico City, with Milpa Alta alone accounting for 22% of national output, supporting over 7000 families and preserving strong cultural and ecological traditions [5–7]. Thus, Opuntia plays a dual role in Mexico: it is both a staple food and a symbol of national identity, while also representing an expanding agro-industrial sector with significant economic potential. However, this growing industry faces important challenges, such as the lack of standardized cultivation and processing practices and the environmental impacts of intensive production. Addressing these requires research that validates the crop's benefits while also developing strategies to ensure sustainable production and safe commercialization.

A central challenge is how to expand the economic use of *Opuntia* spp. without compromising its cultural and ecological foundations. The objective of this study is to evaluate global research trends on *Opuntia* spp. and to examine practical evidence from Mexico in order to identify opportunities, limitations, and strategies for closing production cycles through residue valorization in nopal cultivation. To achieve this objective, the study combines an analysis of international scientific production between 2014 and 2024 with a targeted assessment of applied cases of residue utilization in Mexico, with particular attention to composting and related processes. This dual perspective makes it possible to identify global research trajectories while also evaluating practical experiences that can inform the sustainable management of cladodes residues.

To guide this effort, the study addresses three research questions: (a) What are the prevailing scientific trends and thematic priorities in *Opuntia* spp. research worldwide? (b) What evidence exists in Mexico regarding residue valorization, and what are the key agronomic and product-quality outcomes of these interventions? (c) How do these local practices align with or diverge from international innovation pathways, and what opportunities and barriers influence their adoption at scale? By answering these questions, this work seeks to generate actionable insights for sustainable residue management and the development of integrated value chains. In doing so, it fills a critical knowledge gap, linking international innovation with local practices to promote low-cost recycling technologies, improve soil health, and strengthen the resilience of smallholder production systems in Mexico.

2. Materials and Methods

This study employed a mixed methodological design to analyze global scientific production on *Opuntia* spp. and to evaluate evidence of residue valorization practices in Mexico. The approach combined a bibliometric mapping of research outputs with a structured screening of empirical studies, enabling both an international overview and a context-specific assessment. Together, these methods provided complementary insights into how global trends connect with local practices in composting and biofertilizer development.

2.1. Global Research Analysis

The global dimension of Opuntia research was examined using bibliometric techniques to characterize publication dynamics, citation patterns, and thematic clusters [8–10]. Indicators such as publication growth, citation impact, and keyword co-occurrence were assessed to identify influential authors, institutions, and research areas. Scopus served as the primary database, with queries designed to retrieve records from titles, abstracts, and keywords. The search combined Opuntia-related terms ("nopal", "prickly pear", "cactus pear") with thematic terms such as "sustainability", "food security", "climate change", "soil", "fodder", "animal feed", "economic", and "application".

The dataset covered publications from 1 January 2014, to 31 December 2024, excluding 2025 due to incomplete data. Only articles and book chapters in English and Spanish were included to allow international comparison. Data processing and visualization were carried out with Bibliometrix (version 4.4.3) for descriptive statistics [11], and VOSviewer (version 1.6.20) was used to generate keyword co-occurrence and collaboration network visualizations [12]. A minimum threshold of five keyword occurrences was set for mapping, and institutional networks were analyzed to highlight research communities.

2.2. Screening of Empirical Evidence on Residue Valorization

To complement the global analysis, a structured screening of empirical evidence was conducted to evaluate cladodes residues valorization through composting in Mexico. The screening procedure was informed by PRISMA guidelines [13] but applied in the context of this study as a tool to identify relevant primary research. Records were retrieved from Scopus, PubMed, and Lens.org without publication date restrictions, using the terms: (nopal OR "Opuntia ficus-indica" OR "prickly pear") AND (compost OR "composta" OR composting) AND (Mexico OR México).

Inclusion criteria focused on original empirical research providing primary data, including peer-reviewed articles and thesis chapters. Eligible studies assessed aerobic composting processes using cladodes residues as primary substrates, co-substrates, or additives, and reported process variables (e.g., temperature, pH, C/N ratio) or final product quality (e.g., nutrient content, soil or crop effects). All studies had to be conducted in Mexico and published in Spanish or English.

Exclusion criteria ruled out reviews, opinion papers, editorials, and conference abstracts lacking complete methodology or results. Studies on non-composting uses of Opuntia (e.g., fodder, food, medicinal, cosmetic) were excluded unless composting was a central or secondary focus.

The selection process was carried out in CADIMA using a two-phase protocol: (i) screening of titles and abstracts to remove duplicates and apply inclusion/exclusion criteria, followed by (ii) full-text assessment to confirm eligibility. A PRISMA flow diagram was generated to document each step. To ensure methodological rigor, two researchers independently conducted the screening within the CADIMA platform. Discrepancies were resolved through discussion, and when consensus could not be reached, a third reviewer provided the final decision. Although Cohen's kappa was not formally calculated due to the small number of included studies ($n = 5$), this double-screening and consensus approach strengthened reproducibility and minimized selection bias.

For data extraction, a structured form captured identification details (authors, year, country), study context (objectives, design), intervention characteristics (residues, composting method, duration), and outcomes (effects on soil, conclusions). Due to heterogeneity in design, substrates, and parameters, results were synthesized narratively and thematically, comparing composting methods, product characteristics, and reported effects.

Therefore, the objective of this study is to evaluate global research trends on *Opuntia* spp. and to examine practical evidence from Mexico in order to identify opportunities, limitations, and strategies for closing production cycles through residue valorization in nopal cultivation. The complete datasets supporting the global research analysis (Table S1) and the empirical evidence screening (Table S2) are provided as Supplementary Materials to ensure transparency and reproducibility.

3. Results

3.1. Description and Evolution of Scientific Production on Opuntia spp.

3.1.1. Main Information

The Scopus search yielded a corpus of 789 documents published between 2014 and 2024, reflecting a robust and consistent growth in scientific attention to *Opuntia* spp., with an annual growth rate of 16.79%. This growth underscores the global recognition of Opuntia as a relevant species for food security, climate resilience, and sustainable agricultural practices (Figure 1).

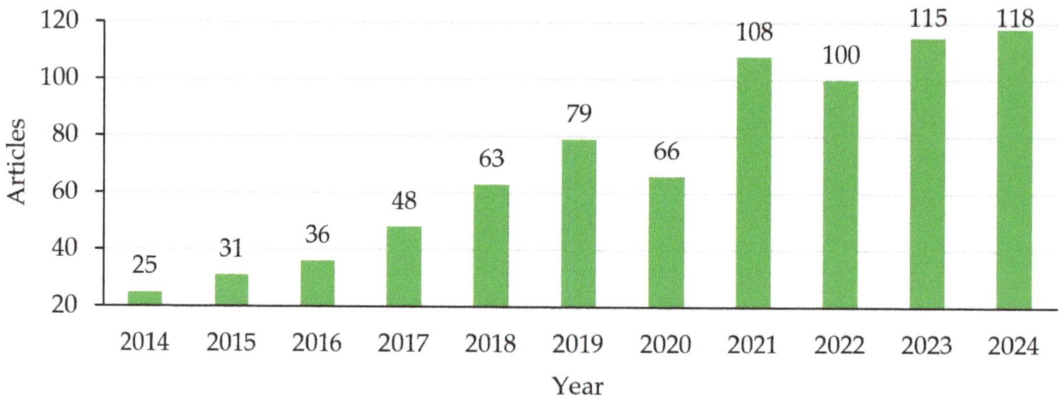

Figure 1. Annual Evolution of Research Publications Related to *Opuntia* spp. (2014–2024).

The trajectory shows steady growth, from 25 publications in 2014 to 118 in 2024, despite a slight dip in 2020 likely attributable to the COVID-19 pandemic. After 2021, output recovered and accelerated, reflecting renewed global attention to climate resilience and sustainable crops. This expansion frames our empirical focus on waste valorization as timely and aligned with the field's pivot toward sustainability. The fourfold increase also signals a consolidated trend toward integrating Opuntia research into discussions on sustainability, waste management, and resource efficiency.

Geographically, scientific production is concentrated in specific regions (Table 1). Mexico leads with 155 documents, followed by Italy (91) and Brazil (72). Other countries such as India, Morocco, and South Africa also contribute significantly, while Switzerland's single publication reached 252 citations, reflecting its global impact [14].

Table 1. Scientific production by country on *Opuntia* spp.

Country	Total Citations	Average Citations	Documents
Mexico	1976	12.70	155
Italy	1189	16.50	91
Brazil	922	10.10	72
India	893	28.80	31
Morocco	668	15.90	42
South Africa	620	23.00	27
Egypt	520	19.30	27
Tunisia	376	13.40	28
Algeria	373	12.00	31
Switzerland	252	252.00	1

The data show two important tendencies. First, countries like Mexico and Brazil have made significant investments in addressing local agronomic challenges, often linked to food security and resource management. Second, Italy's high average citation rate indicates that its research has become a reference point for connecting Mediterranean agricultural systems to broader debates on resilience and waste valorization. Mexico's average citation rate is lower than Italy's, suggesting differences in international visibility and research impact. This gap reflects not only reduced research visibility but also differences in how studies are positioned and disseminated internationally. Italian teams frequently participate in long-standing European consortia, benefiting from strong funding frameworks, access to high-impact journals, and multidisciplinary collaborations that integrate agronomy, chemistry, and bio industrial applications. By contrast, Mexican research is often carried out through smaller, project-based efforts that focus on local agronomic feasibility and sustainability but show limited integration into global innovation networks. These structural differences, particularly access to stable funding, international partnerships, and English-language dissemination channels, help explain why Mexico's substantial scientific productivity has not translated into equally high citation impact.

Meanwhile, India and South Africa, though publishing fewer documents, have positioned themselves in specialized, high-impact niches, often focusing on food, nutritional chemistry, and drought resilience, which gives them disproportionate influence relative to their output. The case of Switzerland illustrates another dynamic: although contributing only one document, its review on invasive alien species by Shackleton et al. (2019) [14] generated substantial influence, demonstrating how comprehensive syntheses can shape global conversations and set the agenda for subsequent research.

3.1.2. Publication Sources and Leading Authors

The analysis of scientific production reveals key publication sources and the leading researchers in the field of Opuntia research. Table 2 lists the top sources, with the Journal of the Professional Association for Cactus Development as the leading outlet, publishing 19 documents. This is followed by journals such as Acta Horticulturae (17 documents), *Opuntia* spp.: Chemistry, Bioactivity and Industrial Applications (16 documents), and Journal of Food Processing and Preservation (13 documents), while Agronomy has published 10 documents reflecting the topic's relevance in specialized research.

Table 2. Top 10 sources in research on Nopal.

Sources	Documents
Journal of the Professional Association for Cactus Development	19
Acta Horticulturae	17
Opuntia spp.: Chemistry, Bioactivity and Industrial Applications	16
Journal of Food Processing and Preservation	13
Tropical Animal Health and Production	13
Polymers	11
Agronomy	10
Foods	10
Revista Brasileira de Engenharia Agricola e Ambiental	9
Revista Caatinga	9

The most productive authors include De Wit, M. (17 publications), who has extensively investigated chemical composition and industrial applications in South Africa; Donato, S. (13), focused on agronomic optimization in Brazil; and Welti-Chanes, J. (10), based in Mexico, who has advanced studies on food processing and value chains (Table 3).

Table 3. Top 10 Authors in Research on Nopal.

Authors	Documents
De Wit, M.	17
Donato, S.	13
Hugo, A.	13
Louhaichi, M.	13
Hassan, S.	12
Welti-Chanes, J.	10
Edvan, R.	8
Fouché, H.	8
Venter, S.	8
Da Silva, J.	7

Table 3 indicates that a small set of authors concentrates a sizable share of output in the Opuntia field. This concentration suggests both thematic leadership and path dependence in collaboration networks. A South African group (De Wit, Hugo, Fouché, Venter) accounts for approximately 42% of top-author documents and is closely associated with industrial and biochemical lines of work (seed oils, mucilage/viscosity, composition and quality assessment), which tend to publish in higher visibility journals and connect to food, cosmetic, and pharmaceutical applications. Brazilian leaders (Donato, Da Silva, Edvan; approximately 26% of the total) are predominantly agronomic, emphasizing plant spacing, fertilization, irrigation, and forage performance. Mexico appears in the top 10 through Welti-Chanes, whose work connects Opuntia to value-added processing and food systems, yet Mexican leadership is otherwise underrepresented among the most prolific authors, reflecting the broader finding that Mexico's contribution is strong in volume but less embedded in global innovation stream.

Mexico and Brazil concentrate on the plant's agronomic and food security applications. The Sankey diagram (Figure 2) visually links these authors to the specific topics they investigate and their countries of affiliation, showcasing the structure of knowledge networks in the field. This structure underpins our subsequent analysis by indicating where expertise and potential partners for residue valorization are concentrated.

The diagram highlights the geographical and thematic specialization within the research. South African research, led by De Wit, M., is heavily linked to keywords like "chemical composition," emphasizing Opuntia's role in nutritional applications. In contrast, Italian research, led by authors like Inglese, P. and Sortino, G., is more focused on value-added topics, such as the study of "phenolic compounds" and "antioxidant activity."

Brazilian authors like Donato, S. share thematic links with South Africa, yet focus on the agronomic aspects within the South American production context. Mexican authors, on the other hand, explore both the resilience of Opuntia crops and their potential for sustainable development and valorization, reflecting the broader cultural and economic importance of the species in Mexico.

The temporal analysis of research activity (Figure 3) reveals different productivity profiles among authors. De Wit, M. and Donato, S., for example, have maintained consistent publication output throughout the decade, with notable productivity peaks, such as De Wit's in 2018–2019. This reflects the sustained research efforts by established research groups [15–18]. De Wit has focused on Opuntia's industrial valorization, particularly in the food, cosmetic, and pharmaceutical industries, investigating aspects like prickly pear seed oil and the hydrocolloid properties of cladodes [18].

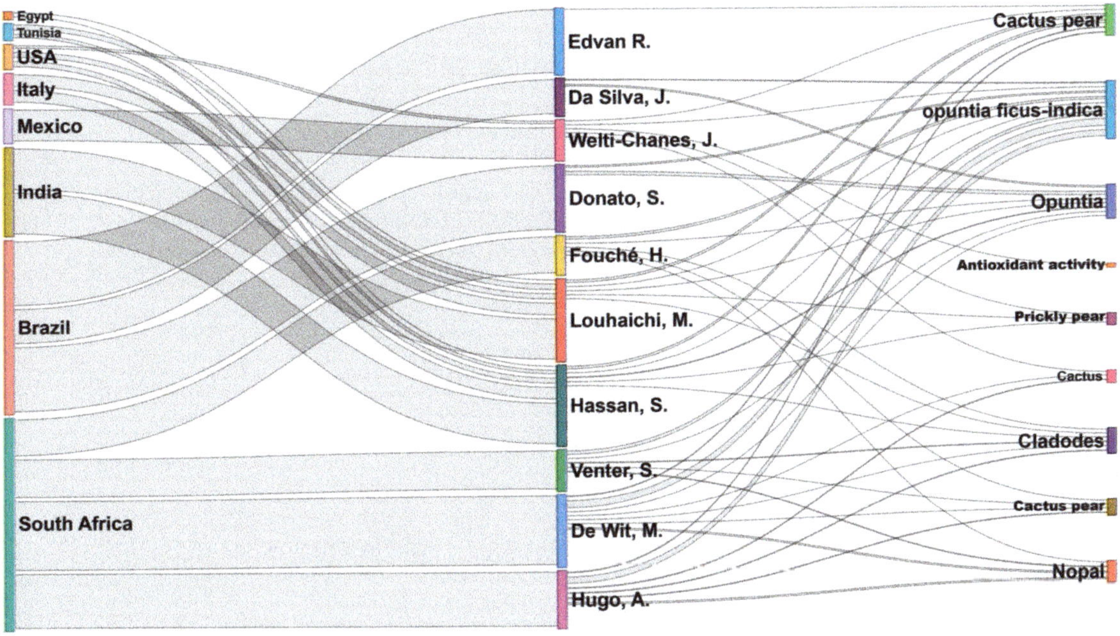

Figure 2. Sankey diagram showing the connections between countries (**left**), leading authors (**center**), and main research keywords (**right**) in *Opuntia* spp. studies. The width of the links reflects the intensity of the association, illustrating both collaboration patterns and thematic focus.

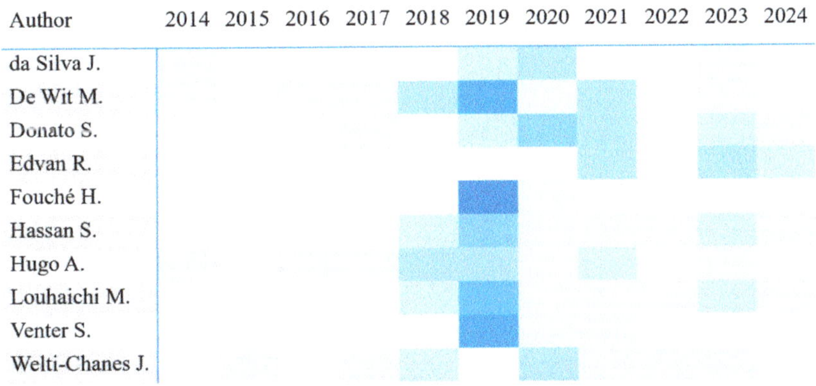

Figure 3. Scientific production of the main authors on *Opuntia* spp. studies over time. Each cell represents the number of publications by an author each year. The color intensity indicates productivity: light blue corresponds to lower frequencies (1 publication), while dark blue indicates higher frequencies (up to 7 publications).

In contrast, Donato, S.'s work centers on plant nutrition and agronomy, with a focus on optimizing nopal as a forage crop. His research evaluates how different agricultural practices, such as plant spacing and organic fertilization, impact plant production and nutrient extraction, aiming to ensure the sustainability of the production system [19,20]. In addition, authors like Fouché, H. and Venter, S. have shown bursts of productivity around 2019. This suggests concentrated research efforts during specific projects, such as Fouché's long-term studies on genotype-environment interactions in nopal cultivars, and Venter's work on promoting *Opuntia* spp. studies as a multipurpose crop to combat food insecurity [21].

The coexistence of these clusters points to a fragmented but complementary research landscape. On one hand, long-standing collaborations contribute stability and continuity to the field, while on the other, the emergence of newer networks suggests growing interest in innovative applications and diversification of research themes. However, the limited connections between clusters highlight a challenge: knowledge produced in one research community is not always fully integrated with that of others. Strengthening cross-cluster collaboration could help bridge gaps between agronomic, biochemical, and sustainability-focused approaches, leading to more comprehensive strategies for the valorization of Opuntia and its byproducts (Figure 4).

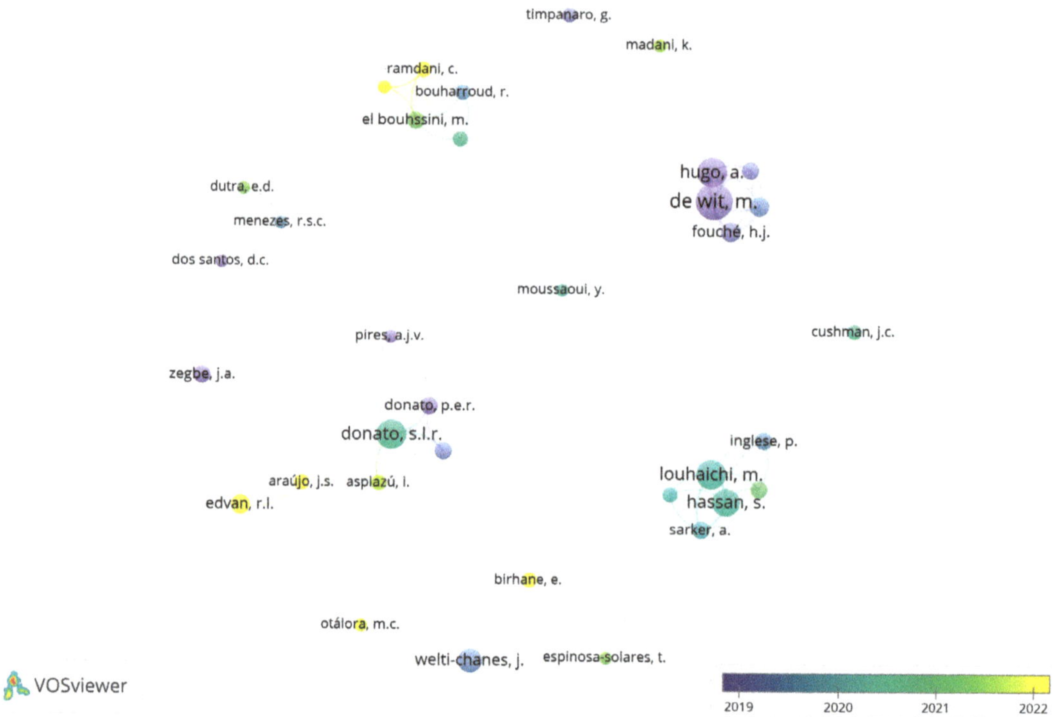

Figure 4. Collaboration networks between authors in *Opuntia* spp. studies research. Each node represents an author, with node size proportional to the number of publications. Links indicate co-authorship connections, while clusters of nodes show groups of researchers who collaborate frequently. Colors correspond to the average year of publication, with darker tones (blue–purple) representing earlier collaborations (2019) and lighter tones (green–yellow) representing more recent ones (2022).

The co-authorship network among the most productive researchers in *Opuntia* spp. studies. The visualization reveals several well-defined clusters rather than a single cohesive global network, indicating a fragmented but potentially complementary research landscape. The largest and densest cluster is anchored by De Wit, Hugo, and Fouché in South Africa [15–18,21]. This group shows long-standing internal connectivity and continuity, largely focused on the biochemical and nutritional dimensions of Opuntia, including seed oil quality, mucilage properties, and industrial applications. The Brazilian cluster, built around Donato, Da Silva, Edvan, and Araujo, is moderately connected and focuses on agronomy and forage systems, often with practical implications for animal feed and crop management [19,20]. The Mediterranean cluster, including Louhaichi, Hassan, Sarker, and Inglese, integrates agronomic and sustainability topics, with strong links between Italy and North Africa to address shared climatic and cropping challenges [2].

Mexico appears primarily in a smaller, emerging cluster around Welti-Chanes and Espinosa-Solares, oriented toward processing, food systems, and early sustainability narratives. However, this Mexican presence is peripheral and less connected to the larger, high-impact clusters in Europe and Africa.

The coexistence of these clusters suggests a field that is dynamic but loosely integrated. Long-standing collaborations provide stability and expertise within specific niches, while the emergence of new networks signals diversification and growing interest in innovative applications such as circular economy and residue valorization. Nevertheless, the limited bridges between clusters imply that knowledge transfer remains incomplete, breakthroughs in biochemical or processing domains may not reach agronomic researchers, and vice versa.

The geographical collaboration map (Figure 5) confirms that research on *Opuntia* spp. is highly international, yet it remains regionally segmented. These regional networks demonstrate strong intra-continental cooperation but also reveal limitations: while intensive within specific areas, cross-continental partnerships are comparatively scarce. The relative absence of connections between regions such as Africa, Latin America, and Asia suggests untapped opportunities for broader integration. Expanding these global linkages could strengthen comparative studies, promote knowledge transfer across different production systems, and foster more comprehensive approaches to waste valorization and sustainable management of Opuntia resources.

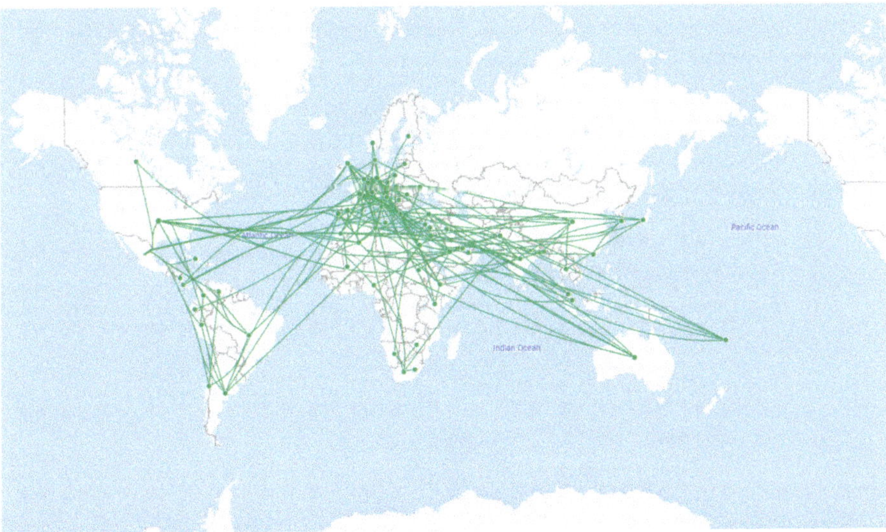

Figure 5. Geographical map of international collaboration networks in *Opuntia* spp. research (2014–2024). Each node represents a country, and connecting lines indicate co-authorship links between institutions from different countries. The density of connections illustrates the extent and intensity of cross-border research collaborations, highlighting regional hubs and transcontinental partnerships.

Opuntia spp. research is highly international but remains regionally segmented. In the Mediterranean region, Italy acts as the primary hub, driving collaborations that radiate into North Africa, notably Tunisia and Algeria, where the crop is important for dryland adaptation and agronomic innovation. These links reflect historical academic exchange and shared climate challenges, allowing joint advances in water-efficient cultivation, phytochemical characterization, and value-added processing.

In the Americas, a clear North American axis is visible between Mexico and the United States, largely oriented toward agronomic improvement, drought tolerance, and sustainable production. Brazil, meanwhile, consolidates its position as South America's anchor, generating dense connections with regional partners to support forage systems and low input agronomy.

Despite this robust collaboration, the map exposes significant cross continental gaps. Networks linking Africa and the Mediterranean to Latin America are minimal, and ties between Latin America and Asia or Oceania are nearly absent. As a result, thematic innovations, such as Europe's nutraceutical and bio industrial focus or South Africa's biochemical characterization do not fully transfer to Mexico and Brazil, where studies remain mainly agronomic.

For Mexico in particular, the map reinforces earlier findings: although well connected to the U.S. and regionally active, it is weakly integrated into the European and African clusters that lead high-impact industrial and biochemical research. Expanding transcontinental partnerships could enhance methodology, increase international visibility, and support policy and funding alignment for circular economy practices. Promoting joint projects with Mediterranean and African groups would also enable comparative studies under similar semi-arid conditions and broaden the translational reach of Mexican residue valorization efforts.

3.1.3. Thematic Analysis of Keywords

The keyword co-occurrence network (Figure 6) highlights an interconnected research landscape structured around five primary clusters.

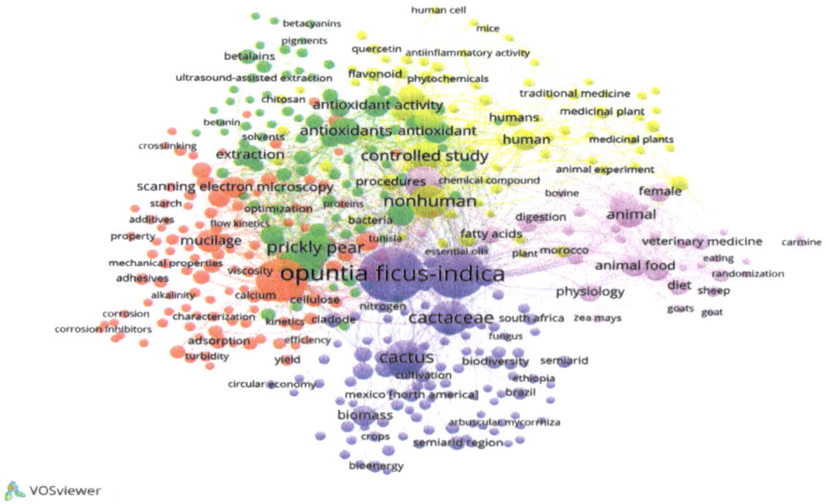

Figure 6. Keyword co-occurrence network of *Opuntia* spp. research (2014–2024). Node size reflects keyword frequency, while colors represent thematic clusters.

This thematic co-occurrence map illustrates the evolution of *Opuntia* spp. research, which has transitioned from agronomic studies and basic composition analyses toward specialized and value-added areas such as human health and sustainability.

Red Cluster: This cluster is one of the densest and centers around core terms like "Opuntia ficus-indica", "prickly pear", and "fruit". Associated keywords such as "phytochemicals", "polyphenols", "polysaccharides", and "betalains" indicate a strong focus on identifying and quantifying the chemical compounds of the plant.

Green Cluster: This cluster emphasizes "antioxidant activity", "phenolic compounds", "bioactive compounds", and "anti-inflammatory" properties. It reflects research into the functional health benefits of Opuntia, with terms like "anti-diabetic properties" and "food additives" highlighting its potential applications in the food and pharmaceutical industries.

Blue Cluster: Linked to the previous clusters through the term "Opuntia", this group centers on the agricultural and environmental aspects of the crop. Keywords like "cladode", "drought", "sustainability", and "water use efficiency" indicate research focused on agronomic management and the crop's role in sustainable agriculture, especially in the context of climate change resilience.

Purple Cluster: This cluster focuses on animal husbandry, with dominant terms such as "animal feed", "fodder", "cattle", and "ruminants". Research in this area is crucial for food security in arid regions, where Opuntia is used as a key nutrient source for livestock, particularly during droughts.

Yellow Cluster: Serving as a bridge between the other topics, this cluster includes terms like "chemistry", "human", "controlled study", and "metabolism". It encompasses research on the effects of young cladode consumption and the basic chemical studies supporting the other areas.

Beyond thematic structure, analyzing the temporal evolution of keywords provides further insight into how the research field has developed and which scientific trends are emerging. Figure 7 presents this dynamic trend, showing the period of activity of the most relevant terms from 2014 to 2024.

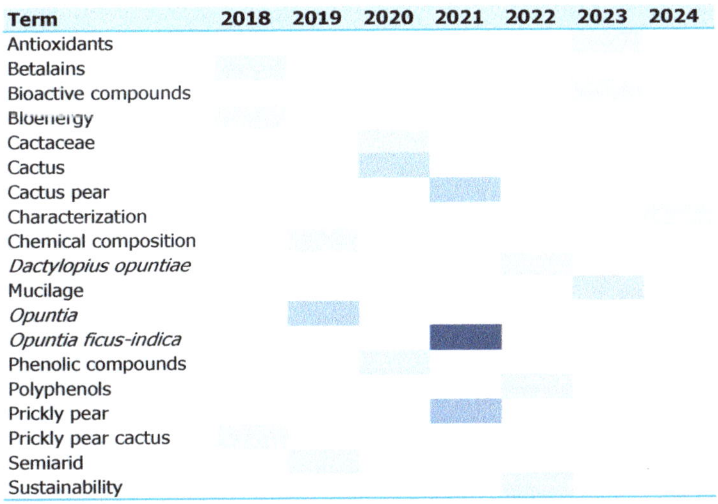

Figure 7. Temporal evolution of the main research terms in *Opuntia* spp. studies (2018–2024). Each row represents a keyword, and columns indicate the years of occurrence. The color intensity reflects frequency, with darker shades (blue) corresponding to higher values (over 100 occurrences) and lighter shades to lower values (as few as 6 occurrences). This visualization highlights shifts in thematic focus over time, showing the transition from foundational terms to more specialized topics in recent years.

A shift from foundational to more specialized research is evident over the years. In the early years of the period, terms like "betalains" and "prickly pear cactus" dominated, reflecting an initial focus on characterization and general studies of the plant. Throughout the decade, core terms like "Opuntia ficus-indica", "prickly pear", and "cactus pear" remained consistently active, serving as the foundation for research development.

The most notable change occurred from 2021 to 2024. During this period, topics such as "bioactive compounds", "antioxidants", "mucilage", and especially "sustainability" gained prominence.

3.1.4. Analysis of the Most Cited Documents

The impact of a research field is often shaped by a set of key publications that become references for future studies. An examination of the most highly cited documents (Table 4) highlights the works that have had the greatest influence on the scientific community.

Table 4. Main Published Documents According to Citation Count.

Title	Reference	Source	Citations
Activated biochar derived from *Opuntia ficus-indica* for the efficient adsorption of malachite green dye, Cu + 2 and Ni + 2 from water.	Choudhary, M., 2020 [22]	Journal of Hazardous Materials	348
The role of invasive alien species in shaping local livelihoods and human well-being: A review.	Shackleton, R., 2019 [14]	Journal of Environmental Management	252
Production and Characterization of Cosmetic Nanoemulsions Containing *Opuntia ficus-indica* (L.) Mill Extract as Moisturizing Agent.	Ribeiro, R., 2015 [23]	Molecules	178
A Contemporary Review on Plant-Based Coagulants for Applications in Water Treatment.	Saleem, M., 2019 [24]	Journal of Industrial and Engineering Chemistry	143
Electrocoagulation-flocculation as a low-cost process for pollutants removal from urban wastewater.	Elazzouzi, M., 2017 [25]	Chemical Engineering Research and Design	117
Layer-by-layer Edible coatings based on mucilages, pullulan and chitosan and its effect on quality and preservation of fresh-cut pineapple (*Ananas comosus*)	Treviño-Garza, M., 2017 [26]	Postharvest Biology and Technology	111
Biomass characterization of Agave and Opuntia as potential biofuel feedstocks.	Yang, L., 2015 [27]	Biomass & Bioenergy	110
Quality enhancement of prickly pears (*Opuntia* sp.) juice through probiotic fermentation using *Lactobacillus fermentum*	Panda, S., 2017 [28]	LWT—Food Science and Technology	104
Green synthesis of surface-passivated carbon dots from the prickly pear cactus as a fluorescent probe for the dual detection of arsenic (III) and hypochlorite ions from drinking water.	Radhakrishnan, K., 2018 [29]	Royal Society of Chemistry Advances	97
Characterization of prickly pear short fiber and red onion peel biocarbon nanosheets toughened epoxy composites.	Ramaswamy, S., 2022 [30]	Polymer Composites	95

The most cited work is that of Choudhary M. et al. (2020), published in Journal of Hazardous Materials, with 348 citations [22]. This study addresses water pollution and demonstrates the synthesis of activated biochar from *Opuntia ficus-indica* cladodes, which effectively adsorb, remove organic dyes, and heavy metals from wastewater. This research is significant for offering a low-cost, biomass-based solution to a pressing environmental issue, positioning *Opuntia* spp. not just as an agricultural crop but as a critical raw material for bioremediation.

The second most cited work, by Shackleton R. et al. (2019) in Journal of Environmental Management, with 252 citations [14], discusses the role of invasive alien species in shaping local livelihoods. While not solely focused on Opuntia, the paper highlights the species as a key case study. It emphasizes *Opuntia ficus-indica's* dual nature: some varieties, like *Opuntia stricta*, are problematic, while others, such as *Opuntia ficus-indica*, provide significant benefits as both food and forage, particularly in arid regions.

In third place is the study by Ribeiro R. et al. (2015), published in Molecules, with 178 citations [23]. This work explores the production of emulsions containing *Opuntia ficus-indica* extract as a moisturizing agent for skin. It is highly cited due to its contribution to linking the plant's chemical composition—rich in carbohydrates and polyphenols—to value-added applications in cosmetics, paving the way for the sustainable valorization of Opuntia biomass.

Other significant publications reinforce the versatility of *Opuntia* spp. Studies by Saleem M. (2019) [24] and Elazzouzi M. (2017) [25] explore its use in water treatment, offering a sustainable alternative to conventional chemicals. Yang L. et al. (2015) [27] highlight Opuntia's potential as a biofuel feedstock. Furthermore, Treviño-Garza M. et al. (2017) [26] demonstrate the use of its mucilage for edible coatings that extend the shelf life of fresh fruits, contributing to the reduction of food waste.

3.2. *Empirical Evidence on Residue Valorization in Mexico*

Global trends in *Opuntia* spp. research reveal a thematic transition from basic characterization toward sustainability-oriented approaches. In the Mexican context, however, residue valorization remains comparatively incipient despite its strategic potential for closing production cycles. Composting stands out as a promising avenue: it addresses seasonal waste streams, reduces reliance on mineral fertilizers associated with greenhouse gas emissions, and provides low-cost bio-inputs that can benefit small and medium producers in production centers such as Milpa Alta.

To ground these insights in practice, this study systematically examined the body of empirical evidence generated in Mexico. The objective was to determine what types of interventions have been tested, under which conditions, and with what agronomic and product-quality outcomes. This evidence-based assessment makes it possible to highlight feasible pathways and identify gaps for scaling residue valorization in real production systems.

3.2.1. Selection of Relevant Studies

A comprehensive literature screening identified 721 initial records across Scopus ($n = 544$), PubMed ($n = 19$), and Lens.org ($n = 158$). After removing 129 duplicates, 592 records were screened by title and abstract. Of these, 548 were excluded for not meeting the predefined inclusion criteria. The full texts of 44 articles were retrieved, although 5 could not be accessed. After full-text evaluation, 34 studies were excluded: 12 did not specifically address Opuntia cladodes residues, 18 did not involve compost preparation or application, and 3 were not original research articles. Ultimately, only 5 studies fulfilled all eligibility requirements and were included for detailed analysis (Figure 8).

This narrowing from hundreds of records to five eligible studies indicates a thin but actionable evidence base. For the purposes of this article, this compact corpus functions as a practical sample from which to infer (i) operational conditions that consistently yield agronomically useful products and (ii) design elements that merit standardization to enable replication and scale in Mexican Opuntia systems.

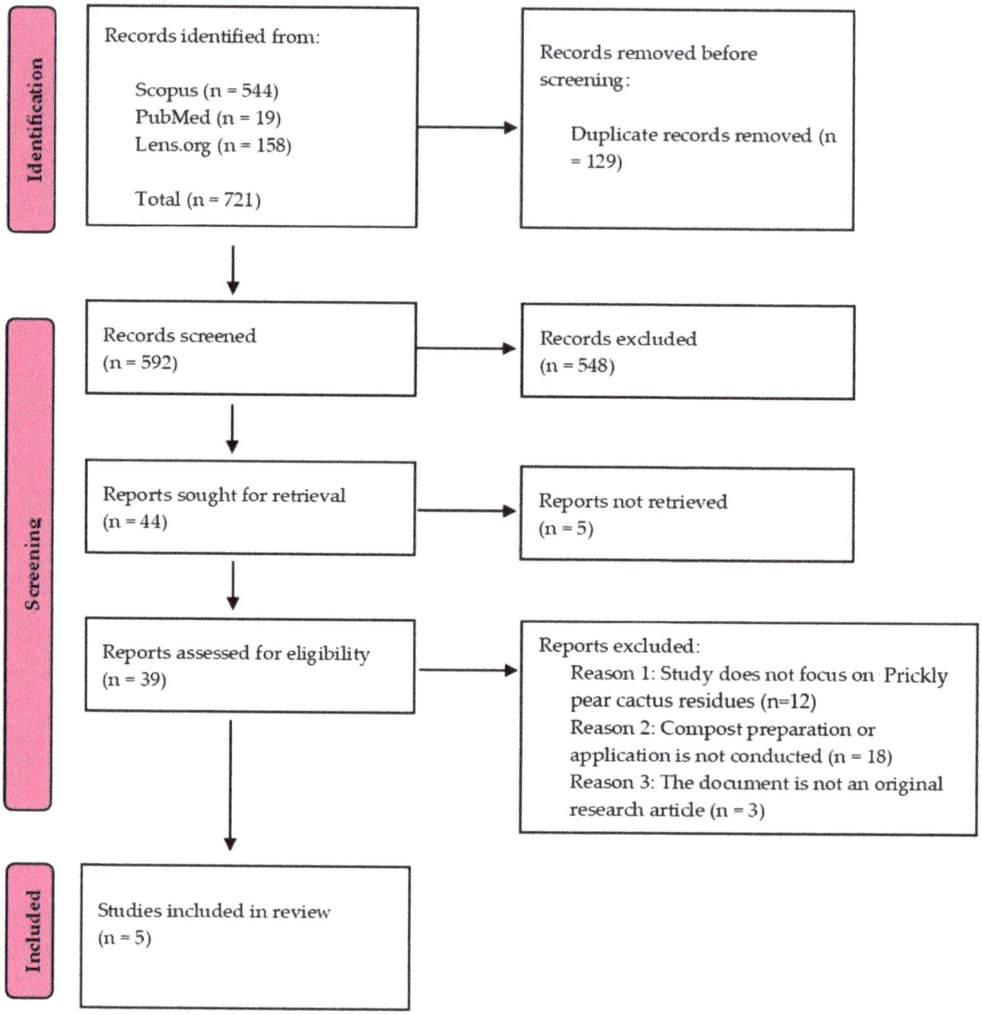

Figure 8. PRISMA Flow Diagram.

3.2.2. Characterization of the Mexican Studies

A total of five Mexican studies were analyzed, published between 2016 and 2025, covering a wide range of scopes and methodologies. Four were field trials with comparative experimental designs, while one presented a technical and economic assessment aimed at standardizing composting practices. Their objectives ranged from evaluating and optimizing composting processes to testing the agronomic performance of biofertilizers and assessing crop-level outcomes (Table 5). This diversity illustrates both the versatility of cladodes residues and the lack of a consolidated research agenda integrating agronomic, technical, and economic dimensions.

Table 5. Extracted Information Matrix.

Author and Year	Objective	Study Design	Intervention (Residues, Method, and Duration)	Application and Comparison	Quantitative Results and Observations	Key Conclusions
Escamilla-García et al. (2025) [31]	Standardize the composting process of cladodes residues for agricultural use.	Technical and economic analysis.	Windrow composting with manual turning using cladodes residues, cow manure, and grass for 120 days.	Not applicable. The study focused on compost production.	Final compost with a C:N ratio of 26.5 and 32.9% organic matter.	The standardized process is a technically and economically viable solution for the valorization of cladodes.
Cruz-Méndez et al. (2021) [32]	Evaluate cladodes residues and biochar-based biofertilizer to improve alkaline soils.	Greenhouse experiment with randomized block design.	Anaerobic fermentation of cladodes residues and rabbit manure for 50 days to produce a biofertilizer.	Applied to tomatillo crops, compared with chemical fertilization and a control.	Improved water retention capacity and available phosphorus in the soil.	The Opuntia biofertilizer is a suitable amendment for improving low-buffering-capacity soils.
Rodas-Gaitán et al. (2019) [33]	Evaluate the effect of biodynamic preparations (BD) in the composting of cladodes and moringa.	Field experiment with a 2 × 2 factorial design.	Static pile composting with passive aeration using cladodes/moringa and manure for 1 year.	Not applicable. Focused on the composting process and product stability.	Final compost C:N ratio ranged from 12–14.	Using cladodes residues as a substrate produces high-quality compost.
González-Torres et al. (2024) [34]	Investigate the effect of raw cladodes and goat manure on soil and crops.	Field experiment with randomized block design.	Direct application to the soil of raw cladodes residues and goat manure. Duration: N/A.	Applied to tomato and carrot crops, compared with control and residue combinations.	Increased carrot yields up to 3.3 times and tomato yields up to 3.9 kg/plant.	Adding Opuntia components to the soil is a viable alternative that enhances horticultural crop production.
Alvarado-Raya et al. (2016) [35]	Compare compost, fresh manure, and synthetic fertilizer in Opuntia cultivation.	Comparative field trial.	Composting on concrete beds with manual turning using manure, cladodes residues, and garden waste for 3 months.	Applied to Opuntia crop, compared with fresh manure and synthetic fertilizer.	No significant differences in yield among treatments. Compost influenced the firmness and shelf life of the cladodes. Final compost had a C:N ratio of 25.5.	Manure compost is an adequate substitute in vegetable Opuntia fertilization without negatively affecting yield.

Collectively, these designs cover the spectrum from process standardization to field performance, offering enough variation to distill operational levers (C/N balancing, aeration regimes, and co-substrate choices) with direct relevance for producers.

Table 6 summarizes the key physicochemical properties of the different residue valorization products identified in the Mexican case studies. It highlights how compost and biofertilizers vary according to the process used. This comparative overview provides a baseline to evaluate the agronomic potential and quality standards of products derived from cladodes residues.

Table 6. Physicochemical Characteristics of Valorization Products.

Product	C/N Ratio	pH	Moisture	Organic Matter	Electrical Conductivity (EC)	Main Nutrients
Aerobic Composting with Manual Turning	25.5:1	7.8–9.0 (slightly alkaline)	N/A	High (exact% not provided)	N/A	Assumed N, P, K, Ca, Mg from cattle manure [35,36]
Standardized Compost (120 days)	26.5:1	8.6	40%	32.9%	2.89 mS/cm	N: 1120 mg/L, P: 355 mg/L, K: 990 mg/L, Ca: 335 mg/L, Mg: 187 mg/L, Zn: 8.6 mg/L, Fe: 15.1 mg/L [31,37–40]
Biodynamic Static Compost (1 year)	12:1–14:1	7.8–9.0 (alkaline)	N/A	Higher in prickly pear (exact% not reported)	N/A	Increased P, Ca, Mg, Fe, Zn, Mn (exact values not provided) [33,40,41]
Liquid Biofertilizer (Anaerobic Fermentation)	N/A	5.9 (acidic)	N/A	1.8% (total organic carbon)	3.83 mS/cm	K: 5568 mg/L, N: 1120 mg/L, P: 355 mg/L, Fe: 15.1 mg/L, Ca, Mg, Zn [32,37,39]

This data reveals important contrasts between valorization strategies and their resulting products. Aerobic composting, even under non standardized manual turning, generates alkaline materials with moderate to high organic matter and balanced macronutrients. Long-duration biodynamic composting produces more mature material (lower C/N ratios between 12:1 and 14:1) and enriched mineral profiles, supporting its potential for improving soil fertility and structure. In contrast, anaerobic fermentation yields liquid biofertilizers with acidic pH and high electrical conductivity, indicating concentrated nutrients readily available to plants, particularly potassium and nitrogen.

Importantly, several of the reported values meet or approach Mexican compost quality standards (NMX-AA-180-SCFI-2018; NADF-020-AMBT-2011), supporting their use for sustainable smallholder systems [37,38]. However, gaps remain in the systematic reporting of some parameters which limit direct comparability and standardization.

4. Discussion

4.1. Global Landscape and Regional Gaps

The global analysis of *Opuntia* spp. research between 2014 and 2024 shows a rapidly expanding and thematically diversified field. Publication output has grown at an annual rate of 16.8%, and collaboration networks reveal active hubs in Africa, the Mediterranean, and Latin America. Yet this expansion is not evenly distributed. Distinct regional agendas have emerged: African groups focus on drought resilience and nutritional use of cladodes; European and Mediterranean teams, particularly in Italy, emphasize high-value bioactive compounds and industrial applications; while Latin American efforts, including Brazil and Mexico, remain strongly agronomic and production-oriented.

The temporal evolution of keywords indicates a clear global shift after 2020: earlier work focused on descriptive aspects—pigments, mucilage, basic agronomy—whereas recent research increasingly emphasizes sustainability, circular economy, and bioactive compounds. This transition reflects a maturing field seeking to move beyond simple characterization toward integrated systems that convert residues into high-value products and sustainable inputs [15–18].

In contrast, Mexican research has remained largely focused on compost, organic amendments, soil improvement, and fertilization. This emphasis reveals a strong orientation toward technical feasibility and localized agronomic performance but shows limited engagement with the nutraceutical, bio industrial, policy, and sustainability-driven perspectives that shape the global conversation. As a result, the relatively weak presence of Mexican studies within clusters linked to industrial applications represents a missed opportunity to move *Opuntia* spp. research beyond farm-level soil enhancement toward higher-value innovation pathways [14].

This gap is particularly striking given Mexico's historical and biological leadership as the center of Opuntia domestication. Although bibliometric results confirm that Mexico leads in the total number of publications, they also reveal a lag in diversification and thematic integration. Collaboration networks further highlight this imbalance: Mexican authors are strongly connected within Latin America and North America but maintain limited links with Mediterranean and African research groups where advances in circular economy practices and value-added products are progressing most dynamically.

4.2. Valorization Processes of Cladodes Residues

The analysis of the five studies revealed three distinct strategies for the valorization of cladodes residues: aerobic composting, anaerobic fermentation, and direct soil application.

The most common strategy was aerobic composting, tested in three studies, though with different system designs and operational scales. Alvarado-Raya et al. [35] and Escamilla-García et al. (2025) [31] applied windrow systems with manual turning but varied in design, duration, and complementary materials. For instance, Alvarado-Raya et al. [35]

combined cattle manure, pruned cladodes, and garden waste, producing compost with a C:N ratio of 25.5 after three months, while Escamilla-García et al. (2025) [31] incorporated regional grasses and extended the process to 120 days, obtaining a mature product and a nutrient-rich leachate proposed as a liquid fertilizer. By contrast, Rodas-Gaitán et al. (2019) [33] implemented static piles with passive aeration and biodynamic preparations, achieving greater mineralization and microbial activity after a one-year process. This methodological variety underscores both the adaptability of Opuntia as a composting substrate and the trade-offs between labor intensity, process duration, and final product quality.

Anaerobic fermentation was tested by Cruz-Méndez et al. (2021) [32], who developed a liquid biofertilizer from cladodes and rabbit manure. Conducted in batch reactors for 50 days, this process yielded improvements in water retention and phosphorus availability when applied to tomatillo crops, offering a complementary pathway for soils with low buffering capacity.

Finally, González-Torres et al. (2024) [34] explored the direct incorporation of raw cladodes into the soil, combined with goat manure. Results showed significant yield increases—up to 3.3 times for carrot and 3.9 kg per tomato plant—demonstrating that even minimally processed residues can improve soil fertility and crop productivity.

Taken together, the three strategies [31–35] delineate a practical menu of options with distinct cost–labor profiles. Aerobic windrows are accessible and adaptable but labor-intensive; passive static piles reduce labor at the expense of longer maturation; and anaerobic fermentation generates complementary liquid inputs suited to alkaline soils. Direct soil incorporation, while the least processed, can deliver yield gains when integrated with manures. Across strategies, three operational levers consistently emerge: (i) targeting balanced C/N ratios through locally available co-substrates; (ii) choosing aeration regimes aligned with labor and time constraints; and (iii) valorizing process by-products (e.g., leachates) to improve overall nutrient circularity [31–35].

4.3. Quality and Properties of the Products

The physicochemical properties reported in the analyzed Mexican studies reveal considerable variation depending on processing method, treatment duration, and the composition of the mixtures. These parameters are critical because they determine both the stability of the product and its suitability for agricultural use. Using Mexican regulatory standards such as NMX-AA-180-SCFI-2018 and NADF-020-AMBT-2011 provides a useful benchmark for assessing the compliance and potential applicability of these bio-inputs in local production systems [37,38]. Interpreting product quality against NMX-AA-180-SCFI-2018 and NADF-020-AMBT-2011 offers not only a compliance lens but also a decision framework for producers—e.g., maturity indicators (C/N) to time field application, pH/EC thresholds to manage salinity risks, and nutrient panels to align product use with specific crop stages [37,38].

In the study by Alvarado-Raya et al. (2016) [35], compost produced through three months of windrow composting with weekly turning reached a C/N ratio of 25.5:1. This value slightly exceeds the maximum threshold of 20:1 established by NADF-020-AMBT-2011, suggesting that additional maturation may be necessary before direct application to crops such as Opuntia [37,38] The pH was slightly alkaline (7.8–9.0), a property that can be beneficial in correcting acidic soils but requires caution when applied to already alkaline soils [37]. The compost was also characterized by its high organic matter content, derived from a mixture of fresh manure, cladodes residues, and grasses, which contributes to soil health by enhancing water retention and gradual nutrient release [37]. Although macronutrients and micronutrients were not directly quantified, the presence of nitrogen

(N), phosphorus (P), potassium (K), calcium (Ca), and magnesium (Mg) from cattle manure is assumed.

Escamilla-García et al. (2025) obtained standardized compost after 120 days, with a C/N ratio of 26.5:1 and 32.9% organic matter [31]. Despite the C/N ratio being marginally above recommended limits, the compost displayed favorable stability and nutrient richness [37,38,41]. The measured pH (8.6) and electrical conductivity (2.89 mS/cm) suggested a low risk of salinization. Importantly, the process also generated a nutrient-rich leachate, with high concentrations of nitrogen, phosphorus, and potassium, which was proposed as a liquid biofertilizer. This dual output (solid compost and liquid leachate) demonstrates a circular approach that enhances the value of cladodes residues for integrated nutrient management [31,37–40].

In contrast, Rodas-Gaitán et al. (2019) reported highly stable compost from static piles with passive aeration over one year [33]. The final product exhibited a C/N ratio between 12:1 and 14:1, indicating advanced mineralization and compost maturity, which is ideal for use in perennial crops like Opuntia [36,37,41]. The pH was alkaline (estimated between 7.8 and 9.0), and the nutrient profile showed increased concentrations of phosphorus (P), calcium (Ca), magnesium (Mg), iron (Fe), zinc (Zn), and manganese (Mn), though nitrogen (N), potassium (K), and copper (Cu) showed slight decreases. This compost is highly stable and suitable for direct application in crops like Opuntia, though further quantitative data on organic matter and electrical conductivity (EC) would be useful for complete regulatory compliance [40,41].

The anaerobic fermentation process studied by Cruz-Méndez et al. (2021) generated a liquid biofertilizer with distinct properties [32]. With a slightly acidic pH (5.9), this product could be useful in correcting alkaline soils or improving phosphorus solubility [36,41]. Its electrical conductivity (3.83 mS/cm) fell within acceptable ranges, and the total organic carbon content was reported at 1.8%. The biofertilizer was particularly rich in potassium (5568 mg/L), alongside nitrogen, phosphorus, and iron, making it a promising input for foliar applications during critical growth stages such as flowering and fruiting [32,38,40,41].

In practical terms, two quality-by-design implications follow from the Mexican cases: (i) when C/N remains >20 at curing, either extend maturation or apply to non-sensitive crops/soils to avoid transient N immobilization; and (ii) leverage liquid fractions (leachates/biofertilizers) in phenological windows where K and readily available N/P are most effective (e.g., flowering, fruit set), particularly in alkaline soils where mildly acidic inputs aid P availability.

Taking together, these findings demonstrate that cladodes residues can be transformed into high-quality composts and biofertilizers that meet or approach regulatory standards [31–35,37,38]. The key properties—C/N ratio, pH, organic matter, and nutrient content—not only validate their technical feasibility but also highlight their potential to reduce dependence on synthetic fertilizers, improve soil health, and increase crop productivity [31–35,37,38]. These results are particularly relevant for smallholder farmers in Mexico, who often face limited access to external inputs. By valorizing locally available residues, these practices foster economic self-sufficiency and align with circular economy principles, turning waste management challenges into opportunities for sustainable agricultural development.

4.4. Why Mexican Research Lags Behind

The gap between Mexico's potential leadership [6] and its current focus on small-scale feasibility studies stems from interconnected structural and institutional barriers.

Funding limitations remain a primary constraint. Most agri-environmental research in Mexico is financed through short-term, project-based schemes. These are adequate for exploratory composting experiments—such as the five Mexican studies analyzed [31–35]—which typically assessed process parameters and immediate soil or yield effects but seldom support multi-year field validation or the development of scalable, standardized protocols. In contrast, countries like Italy and South Africa maintain long-term research programs that integrate agronomy, chemistry, and industrial valorization, allowing the generation of robust and replicable results [1,16].

Weak policy incentives further limit adoption and technological refinement. Although national strategies mention sustainable agriculture and organic production, there is no dedicated policy framework to promote residue-based bio-inputs for Opuntia [34]. Economic incentives, certification schemes, or subsidies that would make compost and biofertilizer production competitive against synthetic fertilizers are lacking. This absence of support discourages farmers and private actors from investing in valorization technologies beyond the experimental stage.

Technology transfer gaps also contribute to the lag. The reviewed Mexican studies yielded technically promising results, yet these outputs largely remain within academic settings [31–35]. Extension services and farmer training networks are underdeveloped, leading to limited dissemination and adoption of protocols that could otherwise be cost-effective and locally adapted.

Finally, fragmented research agendas hinder the creation of synergies. Much of the existing work is driven by isolated initiatives rather than coordinated, multi-institutional programs. In the absence of national or regional strategies that link agronomic research, policy frameworks, and producer networks, innovations remain scattered and difficult to scale.

Addressing these systemic barriers would help align Mexican research with global momentum toward circular economy and sustainable value chains. Coordinated funding schemes, policy-driven incentives for compost and biofertilizer use, and stronger researcher–farmer interactions could accelerate technology adoption, reduce the environmental footprint of *Opuntia* spp. production, and enhance the resilience of smallholder systems.

5. Conclusions

This study offers an integrated perspective on international *Opuntia* spp. research and the emerging practices of residue valorization in Mexico. The findings reveal a rapidly expanding and diversified international research field increasingly focused on sustainability, bioactive compounds, and circular economy strategies. By contrast, Mexican research remains concentrated on small-scale agronomic feasibility studies and shows limited integration into global innovation networks.

Bridging this gap will require long-term funding, stronger policy incentives, and improved technology transfer mechanisms to scale up low-cost residue-based bio-inputs. Expanding research efforts beyond localized feasibility studies toward standardized protocols and diversified applications could position Mexico as a leader in sustainable *Opuntia* spp. value chains.

This work has certain limitations, including the small number of empirical studies available for analysis and the absence of long-term field validations, which limit the generalizability of current findings.

Future research should focus on developing and testing scalable models for the valorization of cladodes residues, as well as stepwise approaches to ensure adoption. Interdisciplinary studies are also needed to address socio-economic barriers, identify enabling incentives, and inform policies that foster circular economy practices and cross-regional knowledge exchange.

Overall, residue valorization emerges as a practical and socially acceptable strategy to reduce waste, improve soil health, and increase the resilience and profitability of smallholder systems. Strengthening the linkages between research, policy, and producer networks is essential to fully realize the sustainability potential of Opuntia cultivation.

Supplementary Materials: The following supporting information can be downloaded at: https://www.mdpi.com/article/10.3390/agronomy15102311/s1, Table S1, dataset of the global research analysis (789 records); and Table S2, dataset of the empirical evidence screening (721 records).

Author Contributions: Conceptualization, A.J.T.-S., Y.D.O.-H., M.A.A.-O. and M.E.T.-C.; Data curation, A.J.T.-S. and M.A.A.-O.; Funding acquisition, M.E.T.-C. and Y.D.O.-H.; Investigation, M.A.A.-O. and A.J.T.-S.; Methodology, M.A.A.-O. and A.J.T.-S.; Project administration, M.E.T.-C. and Y.D.O.-H.; Resources, A.J.T.-S. and M.E.T.-C.; Supervision, Y.D.O.-H., M.A.A.-O., G.L.-E. and M.E.T.-C.; Validation, M.A.A.-O., G.L.-E. and Y.D.O.-H.; Writing—original draft, A.J.T.-S. and M.A.A.-O.; Writing—review & editing, A.J.T.-S., M.A.A.-O., G.L.-E. and Y.D.O.-H. All authors have read and agreed to the published version of the manuscript.

Funding: This research was funded by the Instituto Politécnico Nacional through the project SIP-IPN 20253717.

Data Availability Statement: The data supporting this study are available in the Supplementary Materials (Tables S1 and S2), which include the complete datasets of the global research analysis and the empirical evidence screening. Any additional questions may be addressed to the corresponding authors.

Acknowledgments: The authors would like to thank the Instituto Politécnico Nacional (IPN), Mexico, and Centro Interdisciplinario de Investigación para el Desarrollo Integral Regional, Unidad Oaxaca for their support. Appreciation is also extended to the Secretaría de Ciencias, Humanidades, Tecnología e Innovación (SECIHTI) for granting a scholarship to the first author to pursue postgraduate studies, as well as to the Sistema Nacional de Investigadoras e Investigadores (SNII-SECIHTI). During the preparation of this manuscript, the authors used ChatGPT (OpenAI, GPT-5 version) in a manner comparable to the assistance provided by a language editor, specifically to refine the English translation of the original text, improving grammar, style, and overall language consistency throughout the document. All AI-assisted content was thoroughly reviewed, edited, and validated by the authors to ensure its accuracy, precision, and faithful representation of the original ideas and intended meaning. The authors are fully responsible for the final content.

Conflicts of Interest: The authors declare no conflicts of interest.

References

1. Sáenz, C.; Berger, H.; García, J.; Galletti, L.; de Cortázar, V.; Higuera, I.; Mondragón, C.; Rodriguez-Félix, A.; Sepúlvida, E.; Varnero, M. *Utilización Agroindustrial del Nopal*, 1st ed.; FAO: Rome, Italy, 2006; p. 182.
2. Inglese, P.; Saenz, C.; Mondragon, C.; Nefzaoui, A.; Louhaichi, M. *Crop Ecology, Cultivation and Uses of Cactus Pear*, 1st ed.; Inglese, P., Mondragon, C., Nefzaoui, A., Sáenz, C., Eds.; FAO: Rome, Italy, 2017; p. 244.
3. Tenorio-Escandón, P.; Ramírez-Hernández, A.; Flores, J.; Juan-Vicedo, J.; Martínez-Falcón, A.P. A Systematic Review on Opuntia (Cactaceae; Opuntioideae) Flower-Visiting Insects in the World with Emphasis on Mexico: Implications for Biodiversity Conservation. *Plants* **2022**, *11*, 131. [CrossRef]
4. Dirección General del Servicio de Información Agroalimentaria y Pesquera. ¡Y el Nopal Ingreso en la Dieta Simbólica y Real de los Mexicanos! Available online: https://www.gob.mx/agricultura%7Cdgsiap/articulos/y-el-nopal-ingreso-en-la-dieta-simbolica-y-real-de-los-mexicanos (accessed on 15 May 2025).
5. Secretaría de Agricultura y Desarrollo Rural. Crece en México el Consumo y Producción de Nopal: Agricultura. Available online: http://www.gob.mx/agricultura/prensa/crece-en-mexico-el-consumo-y-produccion-de-nopal-agricultura?idiom=es (accessed on 10 September 2025).
6. Carmona Motolinia, J.R.; Tetreault, D. Pueblos originarios, formas de comunalidad y resistencia en Milpa Alta. *Rev. Mex. Cienc. Políticas Y Soc.* **2020**, *66*, 155–180. [CrossRef]

7. Wacher-Rodarte, M.M. *Nahuas de Milpa Alta. Gobierno de la Ciudad de México*, 1st ed.; Comisión Nacional para el Desarrollo de los Pueblos Indígenas: Mexico City, Mexico, 2006; Volume 1, p. 58.
8. Barboza-Palomino, M. Una guía para desarrollar estudios bibliométricos en psicología. *Rev. Peru. Hist. Psicol.* **2021**, *7*, 37–43.
9. Manterola, C.; Rivadeneira Dueñas, J.; Salgado Castillo, C. Estudios bibliométricos. Una opción para desarrollar investigación en cirugía y disciplinas afines. *Rev. Cir.* **2024**, *76*, 147–156. [CrossRef]
10. Donthu, N.; Kumar, S.; Mukherjee, D.; Pandey, N.; Lim, W.M. How to conduct a bibliometric analysis: An overview and guidelines. *J. Bus. Res.* **2021**, *133*, 285–296. [CrossRef]
11. Aria, M.; Cuccurullo, C. Bibliometrix: An R-tool for comprehensive science mapping analysis. *J. Informetr.* **2017**, *11*, 959–975. [CrossRef]
12. Van Eck, N.J.; Waltman, L. Software survey: VOSviewer, a computer program for bibliometric mapping. *Scientometrics* **2010**, *84*, 523–538. [CrossRef]
13. Page, M.J.; McKenzie, J.E.; Bossuyt, P.M.; Boutron, I.; Hoffmann, T.C.; Mulrow, C.D.; Shamseer, L.; Tetzlaff, J.M.; Akl, E.A.; Brennan, S.E.; et al. The PRISMA 2020 statement: An updated guideline for reporting systematic reviews. *BMJ* **2021**, *372*, n71. [CrossRef]
14. Shackleton, R.T.; Shackleton, C.M.; Kull, C.A. The role of invasive alien species in shaping local livelihoods and human well-being: A review. *J. Environ. Manag.* **2019**, *229*, 145–157. [CrossRef]
15. De Wit, M.; Hugo, A.; Shongwe, N. Quality Assessment of Seed Oil from Selected Cactus Pear Cultivars (*Opuntia ficus-indica* and *Opuntia robusta*). *J. Food Process. Preserv.* **2017**, *41*, e12898. [CrossRef]
16. De Wit, M.; Hugo, A.; Shongwe, N. South African Cactus Pear Seed Oil: A Comprehensive Study on 42 Spineless Burbank *Opuntia ficus-indica* and *Opuntia robusta* Cultivars. *Euro J. Lipid Sci. Technol.* **2018**, *120*, 1700343. [CrossRef]
17. Du Toit, A.; Mpemba, O.; De Wit, M.; Venter, S.L.; Hugo, A. The effect of size, cultivar and season on the edible qualities of nopalitos from South African cactus pear cultivars. *S. Afr. J. Bot.* **2021**, *142*, 459–466. [CrossRef]
18. De Wit, M.; Du Toit, A.; Fouché, H.J.; Hugo, A.; Venter, S.L. Screening of cladodes from 42 South African spineless cactus pear cultivars for morphology, mucilage yield and mucilage viscosity. *Acta Hortic.* **2019**, *1247*, 47–56. [CrossRef]
19. Donato, P.E.R.; Donato, S.L.R.; Silva, J.A.; Pires, A.J.V.; Silva Junior, A.A.E. Extraction/exportation of macronutrients by cladodes of 'Gigante' cactus pear under different spacings and organic fertilization. *Rev. Bras. Eng. Agríc. Ambient.* **2017**, *21*, 238–243. [CrossRef]
20. Fonseca, V.A.; Donato, S.L.R.; Santos, M.R.D.; Silva, J.A.D.; Oliveira, C.M.; Batista, R.D.S. Morphometry and yield of 'Gigante' forage cactus pear under irrigation and different planting densities. *Rev. Caatinga* **2023**, *36*, 690–701. [CrossRef]
21. Fouché, H.J.; Coetzer, G.M.; Smith, M.F. Genotype × environmental interactions of cactus pear (*Opuntia ficus-indica*) in semi-arid regions of South Africa: Cladode production. *Acta Hortic.* **2019**, *1247*, 81–90. [CrossRef]
22. Choudhary, M.; Kumar, R.; Neogi, S. Activated biochar derived from *Opuntia ficus-indica* for the efficient adsorption of malachite green dye, Cu^{+2} and Ni^{+2} from water. *J. Hazard. Mater.* **2020**, *392*, 122441. [CrossRef]
23. Ribeiro, R.; Barreto, S.; Ostrosky, E.; Rocha-Filho, P.; Veríssimo, L.; Ferrari, M. Production and Characterization of Cosmetic Nanoemulsions Containing *Opuntia ficus-indica* (L.) Mill Extract as Moisturizing Agent. *Molecules* **2015**, *20*, 2492–2509. [CrossRef]
24. Saleem, M.; Bachmann, R.T. A contemporary review on plant-based coagulants for applications in water treatment. *J. Ind. Eng. Chem.* **2019**, *72*, 281–297. [CrossRef]
25. Elazzouzi, M.; Haboubi, K.; Elyoubi, M.S. Electrocoagulation flocculation as a low-cost process for pollutants removal from urban wastewater. *Chem. Eng. Res. Des.* **2017**, *117*, 614–626. [CrossRef]
26. Treviño-Garza, M.Z.; García, S.; Heredia, N.; Alanís-Guzmán, M.G.; Arévalo-Niño, K. Layer-by-layer edible coatings based on mucilages, pullulan and chitosan and its effect on quality and preservation of fresh-cut pineapple (*Ananas comosus*). *Postharvest Biol. Technol.* **2017**, *128*, 63–75. [CrossRef]
27. Yang, L.; Lu, M.; Carl, S.; Mayer, J.A.; Cushman, J.C.; Tian, E.; Lin, H. Biomass characterization of Agave and Opuntia as potential biofuel feedstocks. *Biomass Bioenergy* **2015**, *76*, 43–53. [CrossRef]
28. Panda, S.K.; Behera, S.K.; Witness Qaku, X.; Sekar, S.; Ndinteh, D.T.; Nanjundaswamy, H.M.; Ray, R.C.; Kayitesi, E. Quality enhancement of prickly pears (*Opuntia* sp.) juice through probiotic fermentation using Lactobacillus fermentum—ATCC 9338. *LWT* **2017**, *75*, 453–459. [CrossRef]
29. Radhakrishnan, K.; Panneerselvam, P. Green synthesis of surface-passivated carbon dots from the prickly pear cactus as a fluorescent probe for the dual detection of arsenic(iii) and hypochlorite ions from drinking water. *RSC Adv.* **2018**, *8*, 30455–30467. [CrossRef]
30. Ramaswamy, R.; Gurupranes, S.V.; Kaliappan, S.; Natrayan, L.; Patil, P.P. Characterization of prickly pear short fiber and red onion peel biocarbon nanosheets toughened epoxy composites. *Polym. Compos.* **2022**, *43*, 4899–4908. [CrossRef]
31. Escamilla-García, P.E.; Alvarado-Raya, H.E.; Caire Alfaro, C.; Dorantes Hernández, P.I.; Ángeles Tovar, L.C. Standardization Process for the Production of Nopal-Based (*Opuntia ficus-indica* (L.) Mill.) Compost: A Technical and Economic Analysis. *Compost. Sci. Util.* **2025**, 1–16. [CrossRef]

32. Cruz-Méndez, A.S.; Ortega-Ramírez, E.; Lucho-Constantino, C.A.; Arce-Cervantes, O.; Vázquez-Rodríguez, G.A.; Coronel-Olivares, C.; Beltrán-Hernández, R.I. Bamboo Biochar and a Nopal-Based Biofertilizer as Improvers of Alkaline Soils with Low Buffer Capacity. *Appl. Sci.* **2021**, *11*, 6502. [CrossRef]
33. Rodas-Gaitán, H.A.; Palma-García, J.M.; Olivares-Sáenz, E.; Gutiérrez-Castorena, E.V.; Vázquez-Alvarado, R. Biodynamic preparations on static pile composting from prickly pear cactus and moringa crop wastes. *Open Agric.* **2019**, *4*, 247–257. [CrossRef]
34. González-Torres, G.Y.; Bernardino-Nicanor, A.; Fernández-Avalos, S.; Acosta-García, G.; Juárez-Goiz, J.M.S.; González-Cruz, L. Effects of Nopal and Goat Manure on Soil Fertility and the Growth, Yield and Physical Characteristics of Tomato and Carrot Plants. *Agronomy* **2024**, *14*, 1221. [CrossRef]
35. Alvarado Raya, H.E.; Salinas Callejas, E.; Ortiz Huerta, G. Peso fresco y calidad de nopalito (*Opuntia ficus-indica* L.) fertilizado con composta de estiércol de vaca: Fresh weight and quality of nopalito (*Opuntia ficus-indica* L.) fertilized with cow manure compost. *TECNOCIENCIA Chih.* **2016**, *10*, 13–22. [CrossRef]
36. Abiven, S.; Menasseri, S.; Chenu, C. The effects of organic inputs over time on soil aggregate stability—A literature analysis. *Soil Biol. Biochem.* **2009**, *41*, 1–12. [CrossRef]
37. *NMX-AA-180-SCFI-2018*; Norma Mexicana NMX-AA-180-SCFI-2018, Que Establece los Métodos y Procedimientos Para el Tratamiento Aerobio de la Fracción Orgánica de los Residuos Sólidos Urbanos y de Manejo Especial, Así Como la Información Comercial y de Sus Parámetros de Calidad de los Productos Finales. Secretaría de Economía: Mexico City, Mexico, 2018; p. 52.
38. *NADF-020-AMBT-2011*; Norma Ambiental Para el Distrito Federal NADF-024-AMBT-2013, Que Establece los Criterios y Especificaciones Técnicas Bajo los Cuales se Deberá Realizar la Separación, Clasificación, Recolección Selectiva y Almacenamiento de los Residuos del Distrito Federal. Secretaría del Medio Ambiente: Mexico City, Mexico, 2012; p. 15.
39. Andreu, L. Valorization of Prickly Pear (*Opuntia ficusindica* (L.) Mill): Study of Its Phytochemical, Nutraceutical, and Functional Properties. Ph.D. Thesis, Universidad Miguel Hernández de Elche, Alicante, Spain, 2022.
40. Gajender, R.; Yadav, J.; Dagar, J.; Lal, K.; Gurbachan, S. Growth and fruit characteristics of edible cactus (*Opuntia ficus-indica*) under salt stress environment. *J. Soil Salin. Water Qual.* **2013**, *2013*, 135–142.
41. Bernal, M.P.; Alburquerque, J.A.; Moral, R. Composting of animal manures and chemical criteria for compost maturity assessment. A review. *Bioresour. Technol.* **2009**, *100*, 5444–5453. [CrossRef] [PubMed]

agronomy

MDPI

Article

Self-Organizing Map-Based Assessment of Compost Maturity and Cu/Zn Passivation in Biochar-Amended Pig Manure

Hongqiong Zhang [1], Xinlong Yu [1], Lina Luo [1], Yong Sun [1], Ling Zhou [2,3,*] and Haimei Fu [4]

[1] Key Laboratory of Swine Facilities Engineering, Ministry of Agriculture and Rural Affairs, College of Engineering, Northeast Agricultural University, Harbin 150030, China; zhhqiong@neau.edu.cn (H.Z.); 15552176185@163.com (X.Y.); luolina@neau.edu.cn (L.L.); sunyong@neau.edu.cn (Y.S.)

[2] Modern Agricultural Engineering Key Laboratory at Universities of Education Department of Xinjiang Uygur Autonomous Region, Alar 843300, China

[3] School of Mechanical Electrification Engineering, Tarim University, Alaer 843300, China

[4] Institute of Analysis and Testing, Beijing Academy of Science and Technology (Beijing Center for Physical and Chemical Analysis), Beijing 100089, China; dena1988@sina.com

* Correspondence: zhoul-007@163.com

Abstract: The accumulation of copper (Cu) and zinc (Zn) from piglet feed, coupled with inadequate compost maturation, hinders the safe land application of pig manure (PM). This study employed self-organizing maps (SOMs) integrated with three-dimensional excitation–emission matrix fluorescence spectroscopy (3D-EEM) and parallel factor analysis (PARAFAC) to evaluate PM compost maturity and Cu/Zn passivation under different biochar (BC) dosages (0%, 8%, 10%, and 12%). The results revealed that SOM clustering effectively distinguished composting phases and organic matter transformation trends, while network analysis identified key microbial modules (M5, M6) linked to Cu/Zn passivation. Moreover, 12% BC accelerated compost maturation, maximizing humic content (C1: anthropogenic; C4: terrestrial) by increasing *Luteimonas* abundance (241.98%) and reducing *Terrisporobacter* (92%). It also achieved the highest Cu (36.36%) and Zn (32.34%) passivation. Although 10% BC promoted C4 synthesis but inhibited C1 formation, it ultimately reached a similar maturity level to 12% BC. Additionally, 10% BC demonstrated comparable Cu (34.85%) and Zn (27.89%) passivation, making it a more cost-effective alternative. These findings highlight SOM as a robust tool for compost evaluation, optimizing BC application and improving composting efficiency.

Keywords: compost maturity assessment; biochar; humification; Cu/Zn passivation; self-organizing map

1. Introduction

Large-scale farms generate about 4.4 billion tons of livestock waste annually, with pig manure (PM) accounting for 45.9% [1]. Rich in organic matter, PM can cause environmental pollution and resource waste if not properly managed. Composting is an eco-friendly approach that transforms PM into high-quality organic fertilizer, improving soil fertility and plant growth [2]. However, PM contains high levels of heavy metals (HMs), especially copper (Cu) and zinc (Zn), mainly from pig feed. Pigs absorb less than 5% of these metals, leaving the rest to accumulate in PM [1]. The excessive HMs in PM-based fertilizers contribute to Cu and Zn accumulation in paddy soils, increasing their bioavailability [3]. Reportedly, in severely polluted farmlands, zinc (349–1100 mg/kg) and copper (276–703 mg/kg) concentrations far exceed the recommended limits of 200 mg/kg (Zn) and 150 mg/kg (Cu) [4]. Thus, reducing the bioavailability of HMs is essential for environmental sustainability and agricultural safety. HMs in PM exist mainly in exchangeable (Exc), reducible (Red), oxidizable (Oxi), and residual (Res) forms [5]. Among these, the Exc and Red forms are biologically available and influence HM toxicity and mobility, while the Oxi and Res forms are more stable and less bioavailable [6]. Therefore, limiting HM bioavailability is crucial for enhancing compost quality.

To reduce HM bioavailability, various amendments have been studied. Limestone, zeolite, and clay minerals stabilize metals by modifying pH and enabling ion exchange [7]. Organic amendments, such as humic substances and microbial inoculants, aid metal complexation but often show inconsistent efficiency [7]. Among these, biochar (BC) is particularly effective due to its high surface area and functional groups, passivating Cu and Zn through adsorption, precipitation, and redox reactions [8]. Additionally, BC enhances compost structure, stimulates microbial activity, and accelerates organic matter transformation, improving HM stabilization [9]. BC application increased *Firmicutes* (52.75%), *Bacteroidetes* (28.41%), and *Proteobacteria* (4.94%), further promoting metal immobilization [10]. The effectiveness of BC in HM passivation depends on its dosage [7]. Low doses (2%) promote microbial growth under metal stress, while 6% BC alters Cu/Zn forms, improving stability [11]. A 10% BC amendment achieved the best results, reducing Cu bioavailability by 46.95% and Zn by 56.27% [12]. Higher levels (15%) further enhanced passivation by boosting humic acid formation and optimizing microbial communities [1]. However, 20% BC lowered compost temperature and substrate alkalinity, hindering passivation [1]. Excessive BC also increases costs, limiting its large-scale use. To optimize BC application, this study examines the effects of 0%, 8%, 10%, and 12% BC on Cu/Zn passivation, humic substance formation, compost maturity, and microbial dynamics. The goal is to determine the most effective BC dosage for maximizing metal stabilization, compost efficiency, and cost-effectiveness.

Various methods are used to assess the composition and dynamic changes in organic matter during composting, including ultraviolet–visible (UV-Vis) spectroscopy, chemometrics, and Fourier transform infrared spectroscopy (FTIR) [13,14]. Compared to these traditional techniques, three-dimensional excitation–emission matrix fluorescence spectroscopy (3D-EEM) combined with parallel factor analysis (PARAFAC) offers real-time tracking of organic matter variations, enabling qualitative and quantitative analysis of humic substances and proteins [15]. However, composting involves complex biochemical processes, making traditional methods limited in evaluation. Self-organizing map (SOM), an unsupervised machine learning technique, processes large datasets and reveals hidden patterns in hydrochemistry, environmental risk assessment, and ecosystem analysis [16]. It integrates spectral, microbial, and chemical data to analyze organic matter transformation and microbial metabolism [17,18]. Traditional fluorescence spectroscopy identifies component composition and distribution but fails to capture sample similarities and underlying patterns. In contrast, SOM combined with 3D-EEM maps fluorescence spectral data topologically, intuitively visualizing sample similarities and differences. However, its application in assessing compost maturity and HM passivation remains underexplored, offering potential for compost evaluation improvement. This study hypothesizes that within the 8–12% range, increasing biochar enhances compost maturity and HM passivation, but exceeding the optimal dosage may cause negative effects. SOM combined with 3D-EEM effectively identifies organic matter transformation and heavy metal passivation during composting, optimizing biochar application strategies. Therefore, this study aims to (1) determine the optimal BC dosage for PM composting and assess its effects on humic substance formation and Cu/Zn passivation using SOM; and (2) identify key factors influencing composting dynamics and HM stabilization to enhance compost quality assessment and optimize BC application. By integrating SOM, this study provides a novel method for assessing compost quality and optimizing BC application for HM stabilization, offering a potential approach for waste treatment evaluation.

2. Materials and Methods

2.1. Acquisition of Compost Raw Substrate

This study was conducted from October to December 2024. The raw substrate for composting consisted of PM and corn stover (CS), both sourced from a farm in Harbin, Heilongjiang Province, China. Before further processing, PM and CS were air-dried at 25 °C in a well-ventilated area for approximately 48 h. PM was ground to particles smaller than 0.5 cm, while CS was cut into pieces approximately 2.0 cm long. BC was obtained from Lize Environmental Protection Technology, CO., Ltd., Zhengzhou, Henan Province, China, and produced by pyrolyzing CS at 500 °C for 3 h. The resulting BC had a particle size of less than 0.5 cm. Before composting, BC was thoroughly mixed with PM at the designated ratios to ensure uniform distribution. Detailed characteristics of PM, CS, and BC are provided (see Supplementary Materials).

2.2. Experimental Design and Sampling Procedure

Four composting treatments were established, incorporating different BC percentages based on the dry weight of PM: the treatments were designated as CK (0% BC, control group), F1 (8% BC), F2 (10% BC), and F3 (12% BC). The carbon-to-nitrogen ratio (C/N) was adjusted to 25 using CS, and the moisture content was maintained at 65% to support microbial activity. Composting was conducted in 100 L polyethylene bins, with each bin containing 15–20 kg of composting material [1].

To ensure proper oxygenation, compost was manually turned during each sampling event. Samples were collected at nine time points (1, 8, 15, 22, 29, 35, 42, 49, and 56 days) from three depths (upper, middle, and lower layers). Each sample was divided into three portions: the first portion underwent physicochemical analysis, the second was dried and ground for Cu/Zn assays, and the third was stored at −20 °C for microbial analysis.

2.3. Analysis of Composting Properties and Heavy Metals

The compost temperature was monitored twice daily using a three-point measurement method. Fresh samples were diluted in a 1:10 (w/v) ratio with deionized water for the determination of pH, electrical conductivity (EC), and germination index (GI). pH was measured using a pH meter (INESA, Shanghai, China), EC was determined with a conductivity meter (Mettler Toledo, Greifensee, Switzerland), and the GI was measured and computed using the method described by Zhang et al. [19]:

$$GI = \frac{\text{Seed germination rate in treatment} \times \text{Root length in treatment}}{\text{Seed germination rate in control} \times \text{Root length in control}} \quad (1)$$

The dried samples were treated with nitric acid and H_2O_2 to analyze the total Cu and Zn content [1]. The chemical states of Cu and Zn were assessed using the modified four-step sequential extraction method [9]. The passivation rate of Cu and Zn was calculated using the equation:

$$\text{Passivation Rate} = \frac{\text{Distribution rate of Exc before composting} - \text{Distribution rate of Exc after composting}}{\text{Distribution rate of Exc before composting}} \quad (2)$$

2.4. Fluorescence Spectrometry and EEM-PARAFAC Analysis

Three-dimensional excitation–emission matrix fluorescence spectroscopy was employed to identify and analyze fluorescent substances in the compost samples. Samples were mixed with deionized water in a 1:20 (*w*/*v*) ratio and allowed to equilibrate at room temperature. Subsequently, the mixture underwent filtration through a 0.45 μm membrane. To reduce potential spectral interference, the pH of the filtrate was adjusted to 3.0, and the filtrate was diluted to an optical density of 0.05 cm^{-1} using ultrapure water (18.2 MΩ·cm resistivity). The diluted extract was scanned using a fluorescence spectrophotometer (F-7000, Hitachi, Tokyo, Japan) following optimized scanning parameters [20]. In detail, the excitation (Ex) wavelength spanned from 200 nm to 500 nm in 5 nm increments, and the emission (Em) wavelength covered the range from 250 nm to 600 nm in 2 nm increments. The DOMFluor Toolbox in MATLAB R2021b was employed for the excitation–emission matrix combined with parallel factor (EEM-PARAFAC) analysis of the dissolved organic matter (DOM). The Tucker convergence coefficient value surpassed 95% for the four components [21].

2.5. Integration of Self-Organizing Maps with PARAFAC

SOM is an unsupervised artificial neural network technique used for pattern recognition and dimensionality reduction [22]. During the training phase, the map vector is iteratively adjusted to approximate the sample vector, projecting high-dimensional data onto a lower-dimensional grid while preserving its inherent structure [21]. After training, SOM identifies optimal adjacent map vectors. The weights of these map vectors and neighboring neurons are fine-tuned to better align with sample characteristics. In this study, fluorescence data were processed using the SOM toolbox (version 2.0) in MATLAB R2021b. Before analysis, the 3D-EEM data were converted into a 2D-EEM format [20].

In addition, a hybrid PARAFAC-SOM method was employed for data analysis. This method integrates the strengths of both PARAFAC and SOM to enable more nuanced data classification and visualization [23]. In this work, the 3D-EEM of 41 samples was analyzed with DOMFluor, identifying four distinct fluorescent components (C1-C4). The percentage of fluorescence (Fmax%) was then used to train the SOM (referred to as Fmax%-SOM), allowing for the monitoring of mass changes in Cu and Zn [23]. The quality of the resulting map was evaluated based on the average quantization error (Fmax%-SOM: 0.691) and the final topographic error (Fmax%-SOM: 0.029).

2.6. Microbial Analysis

Genomic DNA was extracted from compost samples (filtered through a 0.22 μm filter) using a FastDNA Spin Kit for Soil (MP Biomedicals, Santa Ana, CA, USA), and 16S rRNA gene sequencing was performed to analyze microbial community composition [1]. Polymerase chain reaction (PCR) amplification targeted the V3–V4 hypervariable region of the 16S rRNA gene using primers 338F (5′-ACTCCTACGGGAGGCAGCAG-3′) and 806R (5′-GGACTACHVGGGTWTCTAAT-3′) [20]. Sequencing was conducted on an Illumina NovaSeq 6000 platform (Illumina, San Diego, CA, USA), and microbial community diversity was analyzed using QIIME2 (v2021.4) with the SILVA 138 database [24].

2.7. Data Analysis and Software Tools

All data were processed using Microsoft Excel 2020, and SPSS 23.0 was used for statistical analysis. Each experiment was conducted in biological triplicate, and during the analysis of each parameter, samples underwent three technical replicates to ensure measurement accuracy and reproducibility. Graphs were generated using ORIGIN 2021b. Microbial correlation analysis was conducted using the WEKEMO Bioincloud Platform (https://www.bioincloud.tech) and Circos software (http://circos.ca/) to assess relationships between microbial taxa and environmental variables (pH, temperature, EC, Cu, Zn, and UV_{254} absorbance). A community network was constructed to visualize microbial interactions.

3. Results and Discussion

3.1. Physicochemical Properties

3.1.1. Temperature Dynamics

Temperature is a crucial indicator of microbial activity and organic matter decomposition during composting [13]. Figure 1a presents the temperature evolution across four treatments, including the heating, thermophilic, cooling, and maturity phases. All treatments exhibited a typical temperature progression: a rapid increase in the initial stage, followed by a thermophilic plateau, and a gradual decline toward stabilization. The initial temperature rise was attributed to the breakdown of low-molecular-weight compounds such as lipids, specific sugars, and proteins [25]. After day 47, all treatments reached ambient temperatures, signaling the depletion of easily degradable organic matter. The CK group peaked at 54.37 °C on day 17, maintaining a thermophilic phase (≥50 °C) for 7 days [6]. In contrast, F1 and F2 maintained this phase for 6 and 10 days, respectively, peaking at 53.90 °C and 61.7 °C. F3 reached a significantly higher peak of 65.37 °C on day 9, with an extended thermophilic phase of 23 days, which was notably longer than CK. This suggests that BC facilitated higher and prolonged thermophilic conditions, promoting enhanced organic matter degradation [9,26].

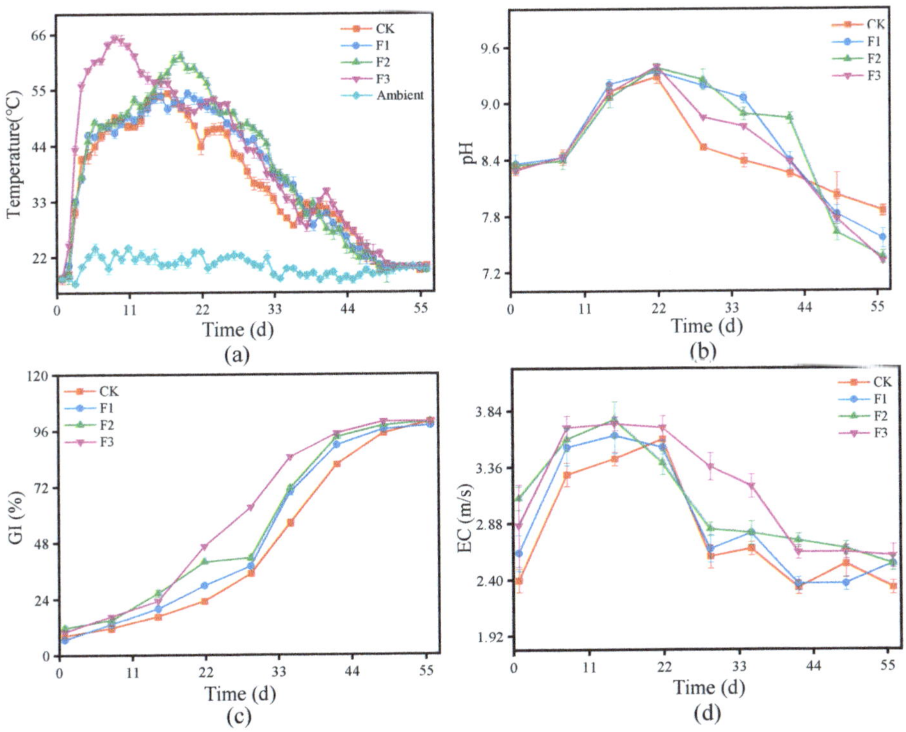

Figure 1. Physicochemical properties of all treatments: (**a**) temperature variation; (**b**) pH fluctuation; (**c**) seed germination index change; (**d**) electrical conductivity change. CK is the control without BC; F1, F2, and F3 contain 8%, 10%, and 12% BC, respectively.

3.1.2. Changes in pH

pH fluctuations followed a typical trend, initially increasing and then stabilizing, as shown in Figure 1b. This trend is attributed to the rapid degradation of organic acids and mineralization of organic nitrogen, followed by ammonia volatilization and nitrification [27,28]. Throughout composting, pH remained between 7.31 and 9.38, supporting microbial growth and activity [29]. pH peaked on day 22 at 9.28 (CK), 9.34 (F1), 9.38 (F2), and 9.40 (F3), then declined to 7.85, 7.56, 7.36, and 7.32, respectively, at the final stage. These fluctuations are linked to alkaline bacterial proliferation [19]. Specifically, the pH values recorded were as follows: 9.28 for CK, 9.34 for F1, 9.38 for F2, and 9.40 for F3. By the end of the composting process, the pH values had shifted to 7.85 for CK, 7.56 for F1, 7.36 for F2, and 7.32 for F3. These fluctuations are attributed to alkaline bacterial growth in compost [24]. Notably, all values met regulatory limits ensuring compliance with agricultural safety standards [28].

3.1.3. Germination Index Variations

GI assesses compost toxicity and quality, with values above 80% indicating plant-safe compost [30,31]. As shown in Figure 1c, GI gradually increased in all treatments, particularly during the thermophilic phase [26]. Initially, GI ranged from 6.77% to 11.67%, indicating high phytotoxicity. By the end of composting, all treatments exceeded 80%, meeting plant safety criteria [28]. Notably, F3, with the highest BC addition (12%), reached 80% GI seven days earlier than other treatments, suggesting that increased BC accelerated compost maturation. However, F2 (10% BC) achieved a similar GI by day 55.

3.1.4. Electrical Conductivity Variations

EC indicates the soluble salt and ion levels in compost [32]. Figure 1d illustrates that all treatments experienced an initial increase in EC, most notably during the thermophilic phase of composting, followed by a progressive decline. Specifically, a discernable decreasing trend in EC began on day 22 for most treatments, whereas in F1, it became evident as early as day 15. Upon completion of the composting process, the EC values stabilized at 2.33 for CK, 2.53 for F1, 2.53 for F2, and 2.60 for F3.

3.2. Spectral Analysis of Dissolved Organic Matter

3.2.1. Identification of Fluorescent Components

The 3D-EEMs of 41 samples were analyzed using a PARAFAC model, identifying four primary fluorescent components, as reported in previous studies. The EEM characteristics of these components are presented, detailing fluorescence intensities across different wavelengths (see Supplementary Materials). These components comprise two humic substances (C1: anthropogenic humic, Ex/Em 330/422 nm; C4: terrestrial humic, Ex/Em 370/486 nm) and two protein substances (C2: tryptophan, Ex/Em 290/366 nm; C3: tyrosine, Ex/Em 280/310 nm), similar to those found in coastal waters and lakes [33–35].

3.2.2. Analysis of Fluorescent Components

The PARAFAC-SOM Fmax model provides an in-depth characterization of organic matter evolution (Figure 2). The U-matrix (Figure 2a) visualizes fluorescence variations, with similar samples clustering in the lower region. The BMU distribution (Figure 2b) validates the neural network's classification efficiency, highlighting distinct fluorescence clusters. CK-1, F1-1, F2-1, and F3-1 (where "−1" denotes day 1, similarly for subsequent labels) resided in the lower left corner, indicating similar initial conditions across all treatments. As composting progressed, maturity levels diverged, with CK-49, F1-49, and F2-42 forming one cluster and F1-42, F2-35, and F3-29 another. Compared to CK, 12% BC accelerated maturation the most, followed by 10% and 8% BC. The multiple-hit histogram (Figure 2c) confirms fluorescence clustering among the treatments. The component planes (Figure 2d–g) illustrate the temporal changes in fluorescence intensity. C1 and C4 increased, while C2 and C3 degraded, signifying protein decomposition and humification. Studies have shown that amino acids are key precursors in humus formation [36]. During protein hydrolysis, released amino acids undergo further transformation and participate in humification processes, thereby enhancing humus accumulation and improving compost maturity. This mechanism is consistent with the findings of this study. Samples F3-56, F3-49, and F1-56 clustered in the upper right, exhibiting the highest C1 content. The similarity between F1-56 and F3-56 suggests that 8% and 12% BC had comparable effects on C1 formation, though F3 reached maturation by day 49, while F1 required more time.

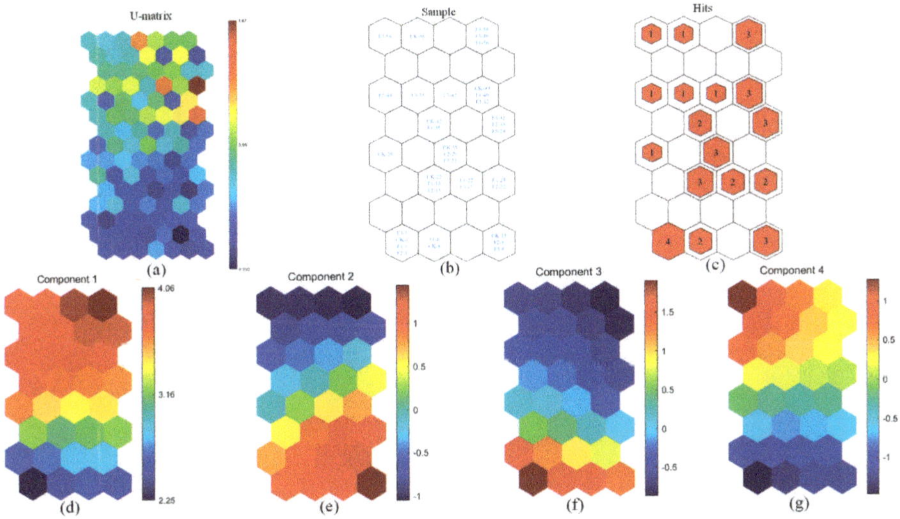

Figure 2. PARAFAC-SOM analysis of four organic components: (**a**) U-matrix; (**b**) sample distribution; (**c**) multiple-hit histogram; (**d**–**g**) fluorescence intensity maps of Components 1–4. The U-matrix visualizes sample similarity, where identical neurons indicate similar fluorescence properties. Color shifts from blue to red denote increasing fluorescence signal variation. In intensity maps, redder hues indicate higher fluorescence levels. CK-1 represents day 1 of the control group, with other labels following the same format.

At the end of composting, C1 production followed the order F3-56 ≈ F1-56 > CK-56 > F2-56 (Figure 2d), while F2-56 had the highest C4 content (Figure 2g), suggesting 10% BC favored C4 accumulation. This indicates that an increase in biochar concentration is more conducive to humic acid accumulation, which is consistent with the findings of other studies [37]. C2 and C3 showed a stepwise decline (Figure 2e,f), peaking in CK-15, F2-8, and F3-8 before progressively degrading. CK-56 retained the highest residual protein content, whereas F2-56, F3-49, F3-56, and F1-56 had the lowest, highlighting BC's role in enhancing protein degradation, with 10% and 12% BC exhibiting the strongest effects. In contrast, despite lower BC input, F2 demonstrated 0 formation efficiency comparable to F3, providing a cost-effective alternative.

3.3. Biochar-Induced Cu and Zn Passivation

3.3.1. Passivation of Cu

After the composting process, the Cu content in the four experimental groups showed distinct increases: CK increased by 12.00% to reach 108.54 mg/kg, F1 by 10.84% to 117.44 mg/kg, F2 by 13.66% to 114.18 mg/kg, and F3 by 9.63% to 121.45 mg/kg (see Supplementary Materials). The post-composting concentrations of Cu, in descending order, were F3, F1, F2, and CK. The variations in Cu levels result from the differing BC dosages, which subsequently affected the duration of the thermophilic phase during composting [6]. Figure 3 illustrates the distribution of Cu species, with initial bioavailable Cu, comprising

Exc-Cu and Red-Cu, ranging from 61.24% to 62.82%, indicating high Cu toxicity in raw materials [1].

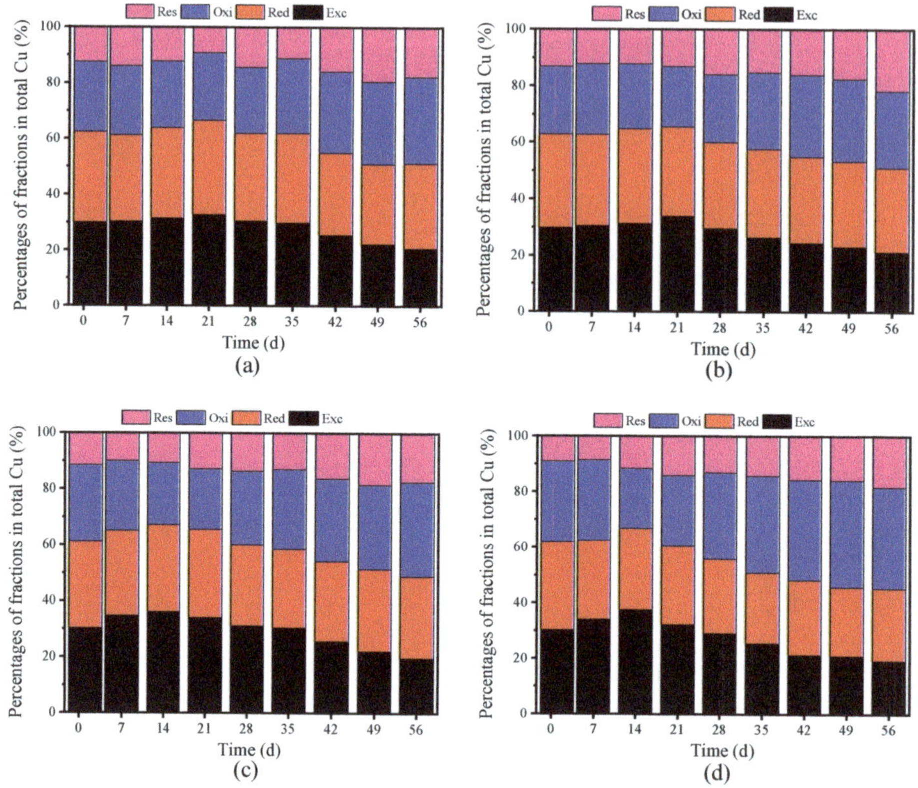

Figure 3. Cu speciation under different treatments: (**a**) CK (control without BC); (**b**) F1 (8% BC); (**c**) F2 (10% BC); (**d**) F3 (12% BC).

The compost samples were divided into seven regions using SOM combined with 3D-EEM; the U-matrix (Figure 4a) visualizes fluorescence variations. Heating-phase samples were located in the map's upper section, while maturity-phase samples mainly appeared in the lower right corner (Figure 4b). A transition zone between these phases included samples from the thermophilic and cooling stages. Figure 4 shows a consistent downward trend in fluorescence intensity from the upper to the lower areas of the grid. This pattern suggests that the relative abundance of bioavailable Cu forms decreased as composting progressed, reducing Cu migration. This shift correlates with rising pH, accelerating organic matter degradation and increasing humic substance production [38]. As shown in Figure 4c, among the evaluated samples, F2-7, F2-14, F2-21, and F3-14 exhibited the highest distribution ratios for Exc-Cu. In contrast, samples CK-56, F1-49, F1-56, F2-49, F2-56, F3-42, and F3-56 were characterized by the lowest Exc-Cu distribution ratios. This trend suggests that the concentration of Exc-Cu peaks during the heating phase undergoes a rapid decline in the thermophilic phase, and stabilizes in the maturity phase. In terms of spatial distribution, a lateral decrease in fluorescence intensity was observed from left to right on the map. Moreover, neurons located in the upper left and upper right corners show a direct correlation between the quantity of added BC and peak Exc-Cu values; greater BC incorporation results in elevated peak values. In Figure 4d, in the vertical dimension, the fluorescence intensity corresponding to Red-Cu decreased, indicating a declining distribution ratio, as composting progressed. In addition, the lowest-position neuron showed increased fluorescence, indicating a slight rise in the Red-Cu ratio in F3 during maturation. As shown in Figure 4e,f, Oxi-Cu and Res-Cu fluorescence intensities exhibited a stepwise increase from the top to the bottom of the grid, highlighting effective Cu passivation. Within the sample set, CK-7, CK-14, F1-7, F1-14, and F1-21 exhibited the lowest concentrations of Oxi-Cu, while F3-49, F2-56, F3-42, and F3-56 presented the highest concentrations, indicating a progressive increase in Oxi-Cu content during the composting.

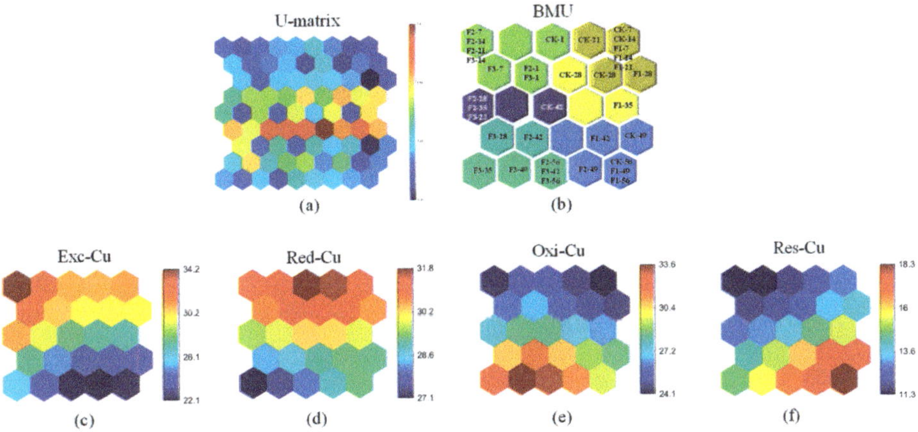

Figure 4. Fmax%-SOM analysis of Cu: (**a**) U-matrix; (**b**) best matching unit (BMU); (**c**) Exc-Cu; (**d**) Red-Cu; (**e**) Oxi-Cu; (**f**) Res-Cu fluorescence intensity mapping. The U-matrix represents sample similarity, where samples in the same neuron share fluorescence characteristics. Color transitions from blue to red indicate increasing fluorescence signal differences, with redder hues signifying lower similarity. In fluorescence maps, blue-to-red shifts denote rising fluorescence intensity, reflecting higher Cu concentrations. CK-1 represents the day 1 sample from the control without BC; all other labels follow this format.

As shown in Figure 3, the initial distribution ratios of Oxi-Cu were 25.04% for CK, 24.04% for F1, 27.24% for F2, and 29.01% for F3. Upon completion of composting, the initial increased by 24.12%, 25.72%, 23.79%, and 25.72%, respectively, revealing a significant enhancement in Oxi-Cu distribution, most notably in the F1 and F3 treatments, which exhibited similar effects. By the end of composting, the distribution ratios for Res-Cu were 17.77% for CK, 21.76% for F1, 17.43% for F2, and 18.31% for F3. These ratios represented increases of 42.73%, 65.60%, 51.30%, and 100.77%, respectively, compared to the initial levels. The inclusion of 12% BC appeared to significantly enhance the conversion of Cu into its Res-Cu form. After composting, the Cu passivation rates were 30.83% for CK, 29.11% for F1, 34.85% for F2, and 36.36% for F3. In a previous study involving a 10% BC and stover mixture added to PM for composting, the Cu passivation rate was reported to be 21.52% [38]. In comparison, the utilization of 12% BC in the current experiment notably improved Cu passivation.

3.3.2. Passivation of Zn

In contrast to Cu, Zn demonstrated enhanced characteristics. After composting, the total Zn concentrations for the four treatment groups were as follows: 719.38 mg/kg (CK), 795.79 mg/kg (F1), 818.89 mg/kg (F2), and 714.56 mg/kg (F3). These values represent respective increases of 7.17%, 8.98%, 2.96%, and 2.88% in comparison to the initial day of composting (see Supplementary Materials). These findings suggest a direct correlation between pH and total Zn content; specifically, as the pH levels rise, the Zn concentration also increases. Variations in composting conditions across different treatment groups lead to disparate rates of organic matter degradation, which in turn results in differing levels of Zn accumulation [39].

Figure 5 illustrates the distribution of Zn species. During the initial phase, bioavailable Zn rates were 65.22% for CK, 61.68% for F1, 58.67% for F2, and 59.51% for F3, indicating high initial Zn toxicity [1]. As composting progressed, bioavailable Zn levels declined, with Exc-Zn and Red-Zn primarily converting into Oxi-Zn and Res-Zn (Figure 6). However, bioavailable Zn rates were still high, meaning Zn was less easily stabilized compared to Cu (Figure 6c,d). Except for Exc-Zn remained relatively high by the end, suggesting that Zn was less readily stabilized than Cu. Notably, while Exc-Zn in F3 continued to decrease, the other treatments exhibited an initial increase followed by a decline (Figure 5). Oxi-Zn content increased by 6.91% to 18.70%, with minimal change observed in CK, indicating that BC facilitated the transformation of bioavailable Zn into Oxi-Zn, particularly in F1 and F3. A greater proportion of bioavailable Zn was converted into Res-Zn rather than Oxi-Zn. By the end of composting, Res-Zn levels increased across all treatments, ranging from an initial 9.74–14.28% to 17.67–20.64%. The relative growth rates were 81.42% in CK, 44.54% in F1, 48.01% in F2, and 76.49% in F3. The overall Zn stabilization rates reached 22.56% for CK, 22.35% for F1, 27.89% for F2, and 32.34% for F3, with F3 demonstrating the highest efficiency in Zn stabilization. This suggests that the higher BC content in F3 enhanced microbial activity, promoting the secretion of bioactive compounds that contributed to Cu/Zn passivation [38].

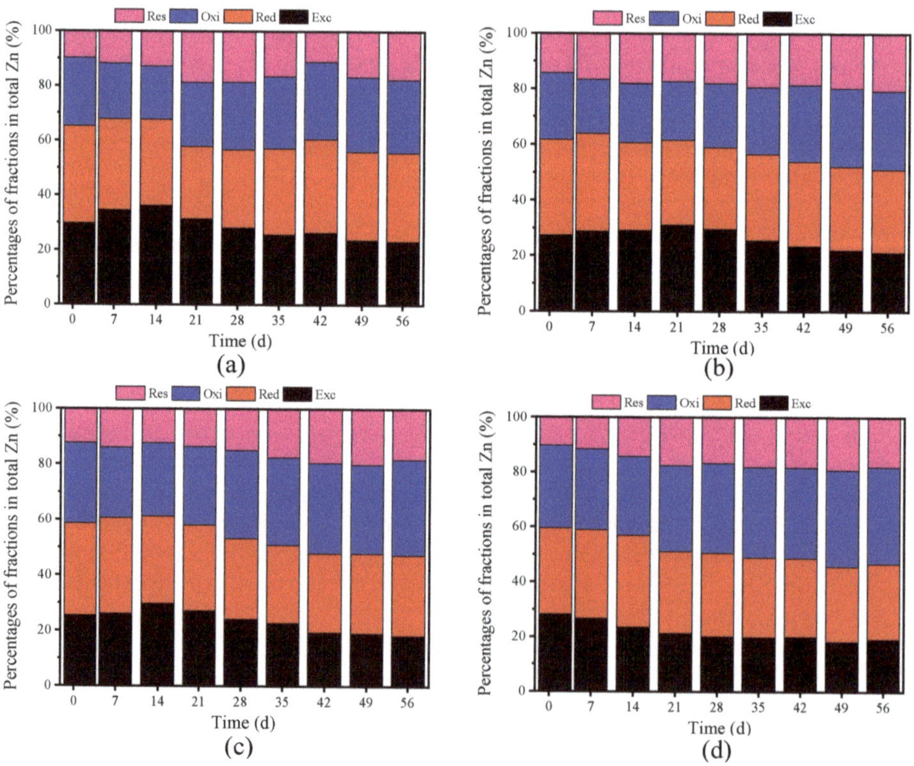

Figure 5. Zn speciation under different treatments: (**a**) CK (control without BC); (**b**) F1 (8% BC); (**c**) F2 (10% BC); and (**d**) F3 (12% BC).

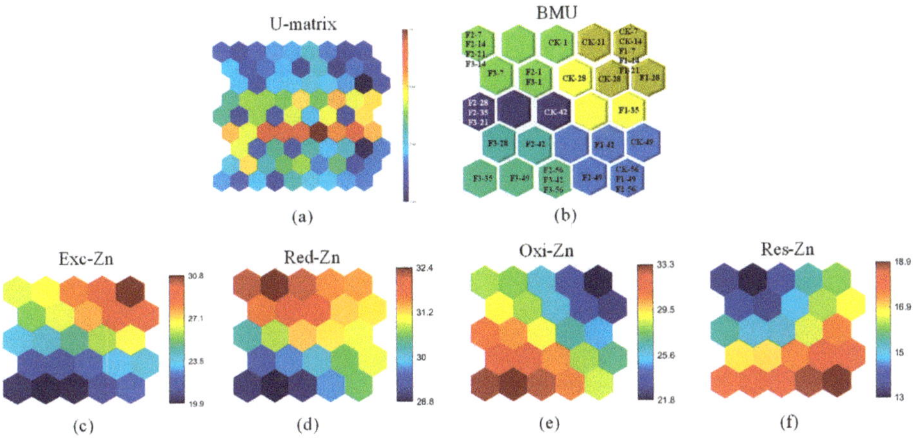

Figure 6. Fmax%-SOM analysis of Zn: (**a**) U-matrix; (**b**) best matching unit (BMU); and fluorescence intensity maps for (**c**) Exc-Zn; (**d**) Red-Zn; (**e**) Oxi-Zn; and (**f**) Res-Zn. The U-matrix illustrates sample similarity, where samples in the same neuron exhibit comparable fluorescence characteristics. Color shifts from blue to red indicate increasing fluorescence differences, with redder hues denoting lower similarity. In intensity maps, a blue-to-red transition signifies enhanced fluorescence, reflecting higher Zn content. CK-1 represents the day 1 sample from the control without BC, with all other labels following this format.

3.4. Microbial Composition Analysis

High-throughput 16S rRNA gene sequencing targeting the V3-V4 regions was employed to assess the impact of BC addition on PM composting. Samples for bacterial genomic analysis were collected from the thermophilic and thermotolerant phases. A total of 2030 unique operational taxonomic units (OTUs) were identified, with a sequence similarity threshold of 97%. At the phylum level (Figure 7a), nine dominant bacterial phyla were identified, primarily *Bacteroidetes* (2.49–62.66%), *Firmicutes* (9.51–85.81%), *Proteobacteria* (1.36–34.71%), and *Actinobacteria* (3.99–17.15%). These findings align well with previous research, highlighting the crucial role these bacterial phyla play in compost degradation [40]. Moreover, the results indicate that BC addition did not significantly alter the distribution of these dominant microbial groups. At the start of composting, *Firmicutes* were the most abundant phylum, with relative abundances of 44.27% (CK-0), 45.43% (F1-0), 60.06% (F2-0), and 85.81% (F3-0). Significantly, *Firmicutes* levels increased with rising BC content. During the thermophilic phase, *Bacteroidetes* emerged as the dominant phylum in CK-29 (42.23%), F1-29 (45.78%), and F2-29 (51.27%), whereas *Firmicutes* remained dominant in F3-29 (40.54%). In the maturity phase, *Proteobacteria* became predominant in F2-25 (34.32%), while *Bacteroidetes* remained dominant in CK-25 (46.63%), F1-25 (62.66%), and F3-25 (39.45%).

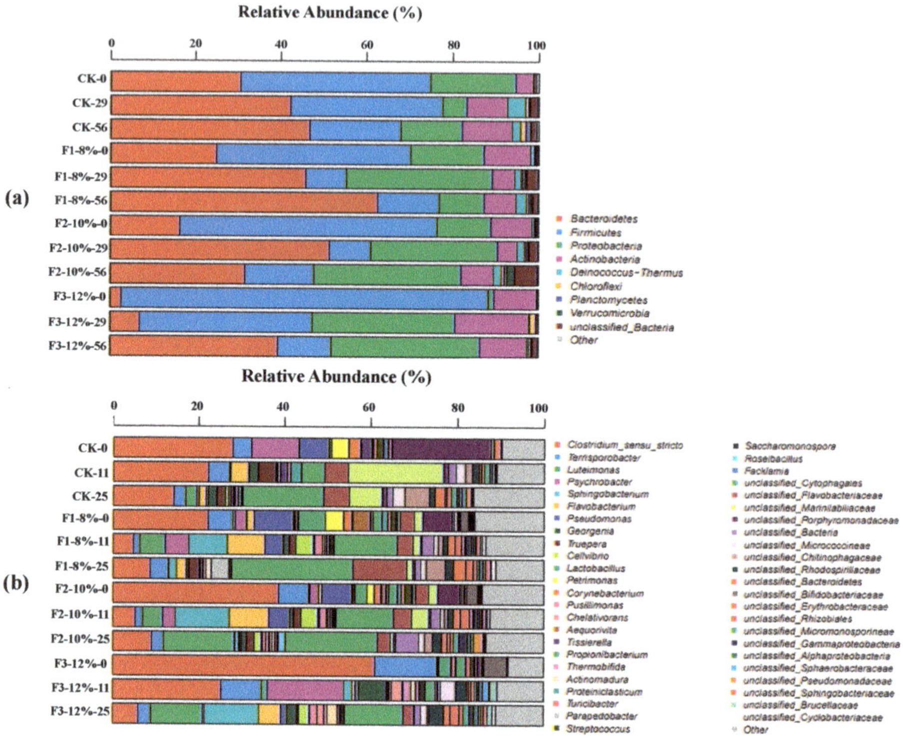

Figure 7. Relative abundance of microbial communities: (**a**) phylum level; (**b**) genus level. CK-0 represents the day 0 sample from the control without BC, with all other labels following this format.

Some studies have found that *Firmicutes* are not heat-tolerant, but they play a crucial role in humic substance formation by utilizing cellulose and hydrolyzed sugars as substrates [41]. By the end of composting, Firmicutes abundance declined by 51.72% (CK), 68.63% (F1), 73.16% (F2), and 85.69% (F3), which is consistent with the findings of other studies [42]. These findings suggest that 12% BC enhances composting temperature, accelerates organic matter degradation, and improves compost quality. Compared to CK, BC-treated groups exhibited a marked increase in *Proteobacteria* abundance. Both *Actinobacteria* and *Bacteroidetes* are thermosensitive. In the maturity phase, *Actinobacteria* abundance in CK and F3 exceeds initial levels, whereas F1 and F2 had lower final abundances. *Bacteroides,* a key decomposer of organic macromolecules, is widely recognized as an indicator of compost maturity [43]. Throughout the composting process, its abundance gradually increased, eventually becoming the dominant bacterial phylum across all treatments. At the genus level (Figure 7b), microbial succession exhibited distinct patterns across different composting phases. In the initial stage, *Clostridium_sensu_stricto* was the predominant genus, accounting for 27.83% (CK-0), 22.26% (F1-0), 38.75% (F2-0), and 61.09% (F3-0). However, with increasing temperatures, the abundance of *Clostridium_sensu_stricto* significantly declined, reaching 5.81–14.22% in the thermophilic phase. This indicates that pathogenic bacteria can be effectively eliminated at high temperatures [44]. *Terrisporobacter*, a known pathogenic genus found in livestock manure, demonstrated low heat resistance, with its abundance declining as composting temperature increased, indicating that composting effectively mitigates PM-associated pathogens [44]. *Luteimonas,* a genus linked to plant growth promotion and disease resistance [45], exhibited increased abundance during the thermophilic phase, with relative rises of 28.15% (CK), 58.76% (F2), and 241.98% (F3), whereas F1 showed a slight reduction (0.66%). These findings suggest that incorporating 12% BC enhances compost quality and plant-beneficial microbial activity. Additionally, *Unclassified_Cytophagales* showed a positive correlation with NO_3-N and total nitrogen concentrations (n = 0.30, $p < 0.01$), indicating its involvement in nitrogen fixation [46,47]. It was most abundant during the thermophilic phase, accounting for 18.74% (CK-25), 28.16% (F1-25), 21.33% (F2-25), and 13.71% (F3-25), highlighting the significant role of 8% BC in nitrogen transformation.

3.5. Bacterial Community Network Analysis

Principal components analysis (PCA) was employed for dimensionality reduction, classifying composting microorganisms into six primary modules (M1, M2, M3, M4, M5, M6). Network analysis (Figure 8) illustrates correlations between microbial modules and compost physicochemical properties, HMs, UV_{254}, UV_{280}, and other composting factors. UV_{254} symbolizes humic substance content and UV_{280} indicates protein content in the compost matrix [43]. The percentage of correlated genera is provided (see Supplementary Materials). In M1, M4, M5, and M6, 13.29–29.37% of genera exhibit a positive correlation with pH, whereas M2 and M3 display a negative correlation. A similar trend has been reported in previous studies, where microbial community shifts were associated with pH changes due to microbial metabolic activity and organic matter degradation [1]. M6 is positively correlated with EC, while M5 shows limited correlation. M1-M4 exhibit negative correlations with EC, with M2 demonstrating the strongest correlation (40% of correlated genera). However, the correlation between pH and EC is not statistically significant, suggesting that compost microorganisms exert limited influence on these factors and are inadequate for determining the compost maturation period (Figure 8). M6 demonstrates a strong negative correlation with GI (62.24% of correlated genera), whereas the remaining modules exhibit positive correlations, especially M3 (77.62%), followed by M2 (70.63%) and M5 (58.04%) (see Supplementary Materials). This underscores the substantial role of microbial communities in influencing GI. M5 exhibits a strong positive correlation with temperature, while M1 and M3 also show significant positive correlations, suggesting that these modules contribute to microbial thermophilic activity and enzymatic degradation of organic matter [14]. Conversely, M3 is negatively correlated with moisture (76.84% of correlated genera). Moisture is also negatively correlated with M2 but positively correlated with M6, supporting findings that microbial community composition is highly sensitive to moisture levels, influencing metabolic pathways and enzymatic reactions.

Bioavailable Zn exhibits a notable positive correlation with M2 (58.74%), and bioavailable Cu demonstrates a significant positive correlation with M3 (60.84%). This suggests that microorganisms in M2 and M3 do not promote of Cu and Zn passivation. Conversely, M4, M5, and M6 exhibit negative correlations with Cu and Zn, indicating a potential role in mitigating HM toxicity [1]. Among environmental factors, UV_{254} is highlighted in orange with the largest area, emphasizing its dominant role in shaping microbial communities. UV_{280} displays a significant correlation with M3, while M2 shows the strongest response to UV_{254} and UV_{280} (80.42% of genera linked), indicating its profound impact on nutrient transformation and microbial diversity in compost. These results are consistent with previous reports, which suggest that microbial communities drive the decomposition of proteinaceous materials and humification, directly influencing compost quality [48].

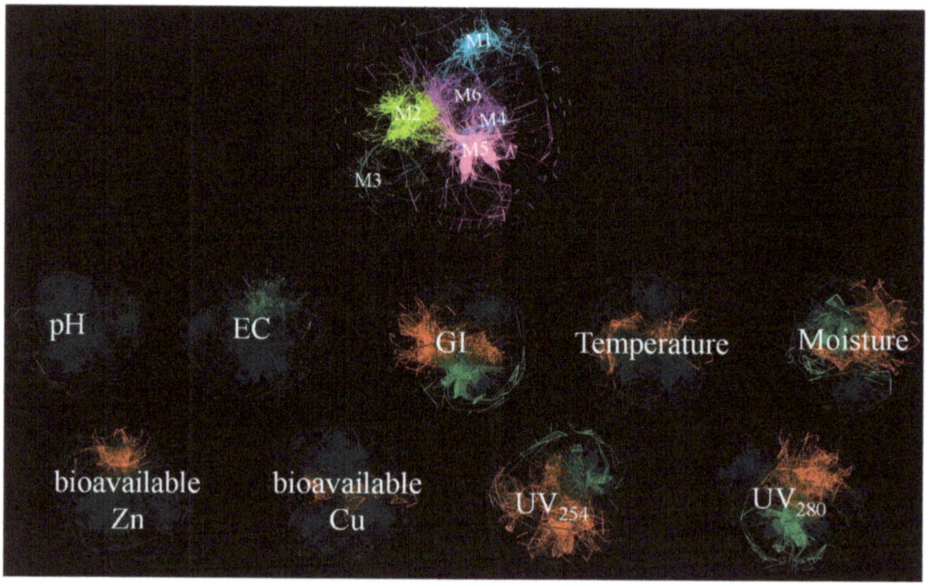

Figure 8. Network analysis of microbial modules and composting parameters. Modules are color-coded: cyan (M1), yellow (M2), green (M3), blue (M4), pink (M5), and purple (M6). Orange indicates a positive correlation, green a negative correlation, and gray indicates the absence of correlation.

4. Conclusions

This study demonstrated the effectiveness of SOM in evaluating PM compost maturation and Cu/Zn passivation. By integrating SOM with 3D-EEM and PARAFAC, distinct fluorescence characteristics and HM bioavailability patterns were identified. SOM clustering effectively differentiated composting phases, revealing organic matter transformation trends and Cu/Zn stabilization pathways. The results showed that 12% BC accelerated humification, achieving maturity in 49 days, while 10% BC reached similar maturity at 55 days, offering a cost-effective alternative. For HM passivation, 12% BC achieved the highest Cu (36.36%) and Zn (32.34%) passivation, while 10% BC showed comparable performance (34.85% Cu, 27.89% Zn), balancing cost and efficiency. Therefore, in large-scale composting projects, using 10% BC instead of 12% BC is a feasible strategy that balances cost control and economic benefits. Network analysis confirmed strong correlations between humic substances and microbial modules (M5, M6), emphasizing contributions to Cu/Zn passivation. The integration of SOM with fluorescence analysis enabled precise compost quality tracking, offering a data-driven approach for real-time assessment. In future studies, soil incubation experiments will be conducted to investigate the stability and potential mobility of Cu/Zn heavy metals in soil. Additionally, SOM models will be further refined to expand their applications in compost evaluation and organic waste treatment.

Supplementary Materials: The following supporting information can be downloaded at https://www.mdpi.com/article/10.3390/agronomy15040778/s1, Figure S1: Fluorescence intensity of four organic components at different wavelengths. (a) Anthropogenic humic substance; (b) Tryptophan-like substance; (c) Tyrosine-like substance; (d) Terrestrially derived humic substance; Figure S2: EEM-PARAFAC identification of four fluorescence components: (a) Anthropogenic humic substance; (b) Tryptophan-like substance; (c) Tyrosine-like substance; (d) Terrestrially derived humic substance; Figure S3: Cu and Zn contents under different treatments: CK (without BC), F1 (8% BC), F2 (10% BC), and F3 (12% BC); Table S1: Basic physical and chemical properties of compost raw materials; Table S2: Percentage of similarity between microbial genus and composting parameters.

Author Contributions: Conceptualization, H.Z., X.Y., and L.Z.; methodology, L.L.; software, Y.S.; validation, X.Y. and H.Z.; formal analysis, H.F. and Y.S.; investigation, H.F. and L.L.; resources, L.Z. and H.Z.; data curation, H.F. and X.Y.; writing—original draft preparation, H.Z. and X.Y.; writing—review and editing, L.Z., L.L., and Y.S.; visualization, X.Y.; supervision, L.Z. and H.Z.; project administration, H.F.; funding acquisition, H.Z. and L.Z. All authors have read and agreed to the published version of the manuscript.

Funding: This research was supported by the National Key Research and Development Program of China (2023YFD1701601), the Key Research and Development Program of Heilongjiang Province (2022ZXJ05C02), and the Science and Technology Project of the First Division of Alar (YS2022KB226).

Data Availability Statement: Data are contained within this article.

Conflicts of Interest: The authors declare no conflicts of interest.

References

1. Zhou, L.; Xue, J.; Xu, Y.; Tian, W.X.; Huang, G.W.; Liu, L.Q.; Zhang, Y.C. Effect of biochar addition on copper and zinc passivation pathways mediated by humification and microbial community evolution during pig manure composting. *Bioresour. Technol.* **2023**, *370*, 128575. [CrossRef]
2. Cheng, L.; Wang, L.X.; Wang, X.H.; Ou, Y.; Liu, H.P.; Hou, X.; Yan, L.M.; Li, X.Y. The various effect of cow manure compost on the degradation of imazethapyr in different soil types. *Chemosphere* **2023**, *337*, 139325. [CrossRef]
3. Yang, M.; He, L.L.; Rao, Z.X.; Huang, D.Y.; Tian, Y.B.; Zhu, H.H.; Xu, C.; Zhang, Q. Effects of long-term fertilization on accumulation and availability of Cd, Zn and Cu in paddy soil. *Res. Agric. Mod.* **2021**, *42*, 302–310. [CrossRef]
4. Li, X.X.; Wei, Y.F.; Wang, L.J.; Jin, S.J.; Wang, P.; Chang, J.; Yin, Q.Q.; Liu, C.Q.; Li, M.L.; Liu, Y.Y.; et al. Effects of multi-component passivator on heavy metal passivation, compost quality andplant growth. *Resour. Environ. Sustain.* **2024**, *17*, 100166. [CrossRef]
5. Zheng, X.R.; Wu, K.H.; Sun, P.J.; Zhouyang, S.Y.; Wang, Y.P.; Wang, H.T.; Zheng, Y.M.; Li, Q.B. Effects of substrate types on the transformation of heavy metal speciation and bioavailability in an anaerobic digestion system. *J. Environ. Sci.* **2021**, *101*, 361–372. [CrossRef]
6. Shen, Y.J.; Zhao, L.X.; Meng, H.B.; Hou, Y.Q.; Zhou, H.B.; Wang, F.; Cheng, H.S.; Liu, H.B. Effect of aeration rate, moisture content and composting period on availability of copper and lead during pig manure composting. *Waste Manag. Res.* **2016**, *34*, 578–583. [CrossRef] [PubMed]
7. Chen, K.L.; Li, J.L.; Lin, L.; Qin, W.K.; Gao, Y.C.; Hu, E.D.; Jiang, J.G. Occurrence, fate and control strategies of heavy metals and antibiotics in livestock manure compost land application: A review. *Sci. Total Environ.* **2024**, *957*, 177381. [CrossRef] [PubMed]
8. Song, H.R.; Chen, S.F.; Si, G.L.; Bhatt, K.; Chen, S.H.; Chen, W.J. Removal of environmental pollutants using biochar: Current status and emerging opportunities. *Environ. Geochem. Health* **2024**, *46*, 384. [CrossRef]
9. Li, Y.; Awasthi, M.K.; Sindhu, R.; Binod, P.; Zhang, Z.Q.; Taherzadeh, M.J. Biochar preparation and evaluation of its effect in composting mechanism: A review. *Bioresour. Technol.* **2023**, *384*, 129329. [CrossRef]
10. Duan, Y.M.; Yang, J.F.; Guo, Y.R.; Wu, X.P.; Tian, Y.L.; Li, H.K.; Awasthi, M.K. Pollution control in biochar-driven clean composting: Emphasize on heavy metal passivation and gaseous emissions mitigation. *J. Hazard. Mater.* **2021**, *420*, 126635. [CrossRef]
11. Zhou, Y.W.; Awasthi, S.K.; Liu, T.; Verma, S.; Zhang, Z.Q.; Pandey, A.; Varjani, S.; Li, R.H.; Taherzadeh, M.J.; Awasthi, M.K. Patterns of heavy metal resistant bacterial community succession influenced by biochar amendment during poultry manure composting. *J. Hazard. Mater.* **2021**, *420*, 126562. [CrossRef] [PubMed]
12. Cui, H.; Ou, Y.; Wang, L.X.; Yan, B.X.; Li, Y.X.; Ding, D.W. The passivation effect of heavy metals during biochar-amended composting: Emphasize on bacterial communities. *Waste Manag.* **2020**, *118*, 360–368. [CrossRef]
13. Kong, Y.L.; Zhang, J.; Zhang, X.S.; Gao, X.; Yin, J.; Wang, G.Y.; Li, J.M.; Li, G.X.; Cui, Z.L.; Yuan, J. Applicability and limitation of compost maturity evaluation indicators: A review. *Chem. Eng. J.* **2024**, *489*, 151386. [CrossRef]
14. Zhu, Y.P.; Cao, Y.X.; Fu, B.B.; Wang, C.J.; Shu, S.H.; Zhu, P.J.; Wang, D.F.; Xu, H.; Zhong, N.Q.; Cai, D.Q. Waste milk humification product can be used as a slow release nano-fertilizer. *Nat. Commun.* **2024**, *15*, 1–13. [CrossRef] [PubMed]
15. Rajapaksha, A.U.; Ok, Y.S.; El-Naggar, A.; Kim, H.; Song, F.H.; Kang, S.; Tsang, Y.F. Dissolved organic matter characterization of biochars produced from different feedstock materials. *J. Environ. Manag.* **2019**, *233*, 393–399. [CrossRef]
16. Zhang, H.Y.; Ni, J.Z.; Wei, R.; Chen, W.F. Water-soluble organic carbon (WSOC) from vegetation fire and its differences from WSOC in natural media: Spectral comparison and self-organizing maps (SOM) classification. *Sci. Total Environ.* **2023**, *895*, 165180. [CrossRef]

17. Yan, C.X.; Wang, W.Y.; Nie, M.H.; Ding, M.J.; Wang, P.; Zhang, H.; Huang, G.X. Characterization of copper binding to biochar-derived dissolved organic matter: Effects of pyrolysis temperature and natural wetland plants. *J. Hazard. Mater.* **2023**, *442*, 130076. [CrossRef] [PubMed]
18. Wang, R.; Shi, T.; Zhang, X.M.; Wei, J.S.; Lu, J.; Zhu, J.X.; Wu, Z.H.; Liu, Q.; Liu, M. Implementing in-situ self-organizing maps with memristor crossbar arrays for data mining and optimization. *Nat. Commun.* **2022**, *13*, 1–10. [CrossRef]
19. Zhang, H.Y.; Li, G.X.; Gu, J.; Wang, G.Q.; Li, Y.Y.; Zhang, D.F. Influence of aeration on volatile sulfur compounds (VSCs) and NH3 emissions during aerobic composting of kitchen waste. *Waste Manag.* **2016**, *58*, 369–375. [CrossRef]
20. Bieroza, M.; Baker, A.; Bridgeman, J. Exploratory analysis of excitation–emission matrix fluorescence spectra with self-organizing maps—A tutorial. *Educ. Chem. Eng.* **2012**, *7*, 22–31. [CrossRef]
21. Ke, W.S.; Liu, Z.; Zhu, F.; Xie, Y.; Hartley, W.; Li, X.; Wu, H.; Xue, S.G. Remediation potential of magnetic biochar in lead smelting sites: Insight from the complexation of dissolved organic matter with potentially toxic elements. *J. Environ. Manag.* **2023**, *344*, 118556. [CrossRef]
22. Yu, S.H.; Zhang, H.Y.; Ni, J.Z.; Xiang, Y.; Wei, R.; Qian, W.; Chen, W.F. Spectral characteristics coupled with self-organizing maps analysis on different molecular size-fractionated water-soluble organic carbon from biochar. *Sci. Total Environ.* **2023**, *857*, 159424. [CrossRef] [PubMed]
23. Fleming, M.; Tai, Y.P.; Zhuang, P.; McBride, M.B. Extractability and bioavailability of Pb and As in historically contaminated orchard soil: Effects of compost amendments. *Environ. Pollut.* **2013**, *177*, 90–97. [CrossRef] [PubMed]
24. Xu, Y.; Bi, Z.T.; Zhang, Y.C.; Wu, H.; Zhou, L.; Zhang, H.Q. Impact of wine grape pomace on humification performance and microbial dynamics during pig manure composting. *Bioresour. Technol.* **2022**, *358*, 127380. [CrossRef]
25. Liu, H.X.; Yang, Y.J.; Yang, Y.Z.; Zhong, X.B.; Lv, J.L. Dynamics of fungal and bacterial communities in different types of soil ageing with different dosages of cadmium. *Ecotox. Environ. Saf.* **2022**, *242*, 113860. [CrossRef]
26. Liu, Y.; Ma, R.N.; Li, D.Y.; Qi, C.R.; Han, L.N.; Chen, M.; Fu, F.; Yuan, J.; Li, G.X. Effects of calcium magnesium phosphate fertilizer, biochar and spent mushroom substrate on compost maturity and gaseous emissions during pig manure composting. *J. Environ. Manag.* **2020**, *267*, 110649. [CrossRef]
27. Wang, G.Y.; Kong, Y.L.; Yang, Y.; Ma, R.N.; Shen, Y.J.; Li, G.X.; Yuan, J. Superphosphate, biochar, and a microbial inoculum regulate phytotoxicity and humification during chicken manure composting. *Sci. Total Environ.* **2022**, *824*, 153958. [CrossRef]
28. Chen, Y.X.; Huang, X.D.; Han, Z.Y.; Huang, X.; Hu, B.; Shi, D.Z.; Wu, W.X. Effects of bamboo charcoal and bamboo vinegar on nitrogen conservation and heavy metals immobility during pig manure composting. *Chemosphere* **2010**, *78*, 1177–1181. [CrossRef]
29. Li, Y.; Luo, W.H.; Li, G.X.; Wang, K.; Gong, X.Y. Performance of phosphogypsum and calcium magnesium phosphate fertilizer for nitrogen conservation in pig manure composting. *Bioresour. Technol.* **2018**, *250*, 53–59. [CrossRef]
30. Bernal, M.P.; Alburquerque, J.A.; Moral, R. Composting of animal manures and chemical criteria for compost maturity assessment. A review. *Bioresour. Technol.* **2009**, *100*, 5444–5453. [CrossRef]
31. Kong, Y.L.; Wang, G.Y.; Chen, W.J.; Yang, Y.; Ma, R.N.; Li, D.Y.; Shen, Y.J.; Li, G.X.; Yuan, J. Phytotoxicity of farm livestock manures in facultative heap composting using the seed germination index as indicator. *Ecotox. Environ. Saf.* **2022**, *247*, 114251. [CrossRef]
32. Ji, Z.Y.; Zhang, L.Y.; Liu, Y.W.; Li, X.Q.; Li, Z.J. Evaluation of composting parameters, technologies and maturity indexes for aerobic manure composting: A meta-analysis. *Sci. Total Environ.* **2023**, *886*, 163929. [CrossRef]
33. Ravindran, B.; Awasthi, M.K.; Karmegam, N.; Chang, S.W.; Chaudhary, D.K.; Selvam, A.; Nguyen, D.D.; Milon, A.R.; Munuswamy-Ramanujam, G. Co-composting of food waste and swine manure augmenting biochar and salts: Nutrient dynamics, gaseous emissions and microbial activity. *Bioresour. Technol.* **2022**, *344*, 126300. [CrossRef]
34. Liu, C.; Du, Y.H.; Yin, H.B.; Fan, C.X.; Chen, K.N.; Zhong, J.C.; Gu, X.Z. Exchanges of nitrogen and phosphorus across the sediment-water interface influenced by the external suspended particulate matter and the residual matter after dredging. *Environ. Pollut.* **2019**, *246*, 207–216. [CrossRef]
35. Lu, K.T.; Gao, H.J.; Yu, H.B.; Liu, D.P.; Zhu, N.M.; Wan, K.L. Insight into variations of DOM fractions in different latitudinal rural black-odor waterbodies of eastern China using fluorescence spectroscopy coupled with structure equation model. *Sci. Total Environ.* **2022**, *816*, 151531. [CrossRef]
36. Li, J.Y.; Wu, S.W.; Zheng, J.X.; Sun, X.C.; Hu, C.X. Combining citrus waste-derived function microbes with biochar promotes humus formation by enhancing lignocellulose degradation in citrus waste compost. *Chemosphere* **2024**, *368*, 143754. [CrossRef] [PubMed]
37. Sun, X.W.; Awasthi, M.K.; Syed, A.; Bahkali, A.H. Effect of cyanobacteria biochar addition on humification, fungal dynamics and its mechanism of action in pig manure composting. *J. Environ. Chem. Eng.* **2024**, *12*, 113755. [CrossRef]
38. Catalá, T.S.; Reche, I.; Fuentes, L.A.; Romera-Castillo, C.; Nieto-Cid, M.; Ortega-Retuerta, E.; Calvo, E.; Álvarez, M.; Marrasé, C.; Stedmon, C.A.; et al. Turnover time of fluorescent dissolved organic matter in the dark global ocean. *Nat. Commun.* **2015**, *6*, 5986. [CrossRef]

39. Kong, Y.L.; Ma, R.N.; Li, G.X.; Wang, G.Y.; Liu, Y.; Yuan, J. Impact of biochar, calcium magnesium phosphate fertilizer and spent mushroom substrate on humification and heavy metal passivation during composting. *Sci. Total Environ.* **2022**, *824*, 153755. [CrossRef]
40. Chen, L.J.; Lin, Y.; Liu, C.W.; Zhang, H.; Lin, C.Q. Improving the Utilization of Flammulina velutipes Waste during Biochar-Amended Composting: Emphasis on Bacterial Communities. *Agronomy* **2024**, *14*, 1046. [CrossRef]
41. Jiang, K.H.; Jiang, D.M.; Li, S.; Guo, Z.Z.; Zhao, L.B.; Wang, J.; Hao, X.X.; Bai, L.; Qiu, S.X.; Kang, B. Impacts of mixed ferrous sulfate-biochar additives on humification and bacterial community during electric field-assisted aerobic composting. *Bioresour. Technol.* **2024**, *404*, 130901. [CrossRef] [PubMed]
42. Zhang, Y.C.; Liu, L.Q.; Huang, G.W.; Yang, C.H.; Tian, W.X.; Ge, Z.Y.; Zhang, B.H.; Wang, S.F.; Zhang, H.Q. Enhancing humification and microbial interactions during co-composting of pig manure and wine grape pomace: The role of biochar and Fe_2O_3. *Bioresour. Technol.* **2024**, *393*, 130120. [CrossRef] [PubMed]
43. Zhang, Y.C.; Xu, Y.; Yu, X.L.; Li, J.K.; Chen, G.; Wang, S.J.; Xu, Y.P.; Xu, R.; Zhang, B.H.; Zhang, H.Q. Microbial metabolism and humic acid formation in response to enhanced copper and zinc passivation during composting of wine grape pomace and pig manure. *Bioresour. Technol.* **2023**, *384*, 129226. [CrossRef]
44. Zhang, Z.C.; Zhao, Y.; Wang, R.X.; Lu, Q.; Wu, J.Q.; Zhang, D.Y.; Nie, Z.F.; Wei, Z.M. Effect of the addition of exogenous precursors on humic substance formation during composting. *Waste Manag.* **2018**, *79*, 462–471. [CrossRef]
45. Wei, Y.H.; Liang, Z.W.; Zhang, Y. Evolution of physicochemical properties and bacterial community in aerobic composting of swine manure based on a patent compost tray. *Bioresour. Technol.* **2022**, *343*, 126136. [CrossRef]
46. Ulrich, K.; Becker, R.; Behrendt, U.; Kube, M.; Schneck, V.; Ulrich, A. Physiological and genomic characterisation of Luteimonas fraxinea sp. nov., a bacterial species associated with trees tolerant to ash dieback. *Syst. Appl. Microbiol.* **2022**, *45*, 126333. [CrossRef] [PubMed]
47. Hur, J.; Lee, B.M. Characterization of binding site heterogeneity for copper within dissolved organic matter fractions using two-dimensional correlation fluorescence spectroscopy. *Chemosphere* **2011**, *83*, 1603–1611. [CrossRef]
48. Lu, M.L.; Lin, B.F.; Zhang, Y.; Hao, Y.H.; Li, K.; Huang, Z.; Li, J.B. Insight into the molecular transformation pathways of humic acid in the co-composting of bagasse and cow manure after adding compound microorganisms. *Process Biochem.* **2024**, *143*, 23–33. [CrossRef]

agronomy

MDPI

Review

State of the Art of Biomethane Production in the Mediterranean Region

Antonio Comparetti [1], Salvatore Ciulla [2], Carlo Greco [1,*], Francesco Santoro [3] and Santo Orlando [1]

[1] Department of Agricultural, Food and Forest Sciences, University of Palermo, Building 4, 90128 Palermo, Italy; antonio.comparetti@unipa.it (A.C.); santo.orlando@unipa.it (S.O.)
[2] Association of Producers SiciliaBio, Via Vittorio Emanuele, 100, 92026 Favara, Italy; agrisinergie@gmail.com
[3] Department of Soil, Plant and Food Sciences (Di.S.S.P.A.), University of Bari Aldo Moro, Piazza Umberto I, 70121 Bari, Italy; francesco.santoro@uniba.it
* Correspondence: carlo.greco@unipa.it

Abstract: The Mediterranean region is increasingly confronted with intersecting environmental, agricultural, and socio-economic challenges, including biowaste accumulation, soil degradation, and high dependency on imported fossil fuels. Biomethane, a renewable substitute for natural gas, offers a strategic solution that aligns with the region's need for sustainable energy transition and circular resource management. This review examines the current state of biomethane production in the Mediterranean area, with a focus on anaerobic digestion (AD) technologies, feedstock availability, policy drivers, and integration into the circular bioeconomy (CBE) framework. Emphasis is placed on the valorisation of regionally abundant feedstocks such as olive pomace, citrus peel, grape marc, cactus pear (*Opuntia ficus-indica*) residues, livestock manure, and the Organic Fraction of Municipal Solid Waste (OFMSW). The multifunctionality of AD—producing renewable energy and nutrient-rich digestate—is highlighted for its dual role in reducing greenhouse gas (GHG) emissions and restoring soil health, especially in areas threatened by desertification such as Sicily (Italy), Spain, Malta, and Greece. The review also explores emerging innovations in biogas upgrading, nutrient recovery, and digital monitoring, along with the role of Renewable Energy Directive III (RED III) and national biomethane strategies in scaling up deployment. Case studies and decentralised implementation models underscore the socio-technical feasibility of biomethane systems across rural and insular territories. Despite significant potential, barriers such as feedstock variability, infrastructural gaps, and policy fragmentation remain. The paper concludes with a roadmap for research and policy to advance biomethane as a pillar of Mediterranean climate resilience, energy autonomy and sustainable agriculture within a circular bioeconomy paradigm.

Keywords: anaerobic digestion; Mediterranean agriculture; circular bioeconomy; agricultural residues; OFMSW; renewable energy sources; sustainability

1. Introduction

The Mediterranean basin is widely recognised as one of the most climate-vulnerable regions in the world, facing mounting challenges from prolonged drought periods, rising temperatures, increasing water scarcity, and a persistent dependency on fossil fuel imports [1,2]. These environmental pressures are further compounded by demographic factors such as high population density and the generation of substantial quantities of organic waste from agricultural, livestock, agro-industrial, and urban sectors [3]. In this context, organic waste streams represent an underutilised yet strategic resource for renewable energy production within the framework of the circular bioeconomy (CBE), supporting both environmental sustainability, and economic resilience.

Among the available renewable energy technologies, anaerobic digestion (AD) has emerged as a mature, flexible, and scalable solution for the treatment of organic substrates. Through this biological process, biodegradable waste is transformed into biogas and digestate. Biogas, in turn, can be valorised through Combined Heat and Power (CHP) systems or upgraded to biomethane, a renewable and grid-compatible alternative to natural gas, which can also be used as a sustainable transport fuel [4–8]. The deployment of AD technologies contributes to multiple sustainability objectives, including the reduction in greenhouse gas (GHG) emissions, the valorisation of local waste streams, and the enhancement of energy security in a region highly dependent on energy imports [9,10].

Foundational research in this field includes the study of Comparetti et al. (2012) [11], who evaluated the energy recovery potential from livestock and agro-industrial residues, establishing a baseline for the development of AD-based energy strategies in Mediterranean rural areas. Attard et al. (2017) [7] further demonstrated the viability of AD systems in Mediterranean island contexts, where such technologies align with local renewable energy and waste management objectives. In more recent years, Greco et al. (2022) [12] highlighted the socio-economic and environmental co-benefits associated with small-scale biowaste AD plants, particularly in peri-urban and urban-rural interface zones.

Complementary valorisation pathways have also been investigated. Greco et al. (2019a) [13] examined manure pyrolysis as an alternative route for waste conversion, while Comparetti et al. (2017) [14] and Greco et al. (2019b) [15] explored nutrient recycling from cactus pear residues as a circular approach to agri-food by-product management. Additional studies by Campiotti et al. (2019) [16] have emphasised the role of renewable energy, including biomethane, in reducing the environmental footprint of Mediterranean greenhouse production systems.

In this integrated context, biomethane production is not only a pathway to clean energy but also an enabler of sustainable rural development, especially when implemented in multifunctional farming systems that integrate energy, nutrient cycling, and waste management [8,17]. Furthermore, Greco et al. (2020, 2021) [18,19] demonstrated that compost and vermicompost derived from digestate can successfully replace peat in growing media, thus contributing to closing the organic matter loop and enhancing soil fertility.

Taken together, these studies underscore the role of biomethane production via AD as a strategic driver for sustainable energy transitions, resource circularity, and climate resilience in Mediterranean agri-food systems.

Building on this foundation, the present review analyses the current state of biomethane production across the Mediterranean region. Particular attention is given to the development and deployment of AD technologies, the availability and regional distribution of feedstocks, and the evolving policy and regulatory frameworks supporting their adoption within the circular bioeconomy paradigm. Moreover, the review highlights emerging innovations in biogas upgrading technologies, nutrient recovery, and digital process monitoring, as well as the implications of the Renewable Energy Directive III (RED III) and national biomethane roadmaps for future scaling up.

Table 1 summarises representative examples of AD plants operating in Mediterranean countries, detailing the primary feedstock sources, plant capacities, and biogas valorisation pathways. These examples illustrate how region-specific organic residues are integrated into circular bioeconomy systems, supporting sustainable energy generation and promoting agricultural and rural development.

Table 1. Overview of selected anaerobic digestion (AD) plants in the Mediterranean region, highlighting dominant feedstocks and pathways of biogas valorisation. The listed examples illustrate region-specific integration of agricultural residues into bioenergy systems and the alignment with local circular bioeconomy strategies.

Country	Region	Plant Name/Project	Feedstocks Used	Biogas Valorisation	Notes
Italy	Emilia-Romagna, Sicily	Azienda Agricola Speranza, ILSA Biometano	Livestock manure, olive pomace, grape marc	Upgrading to biomethane gas grid injection	Feed-in tariff incentives under the Biometano Decree
Spain	Catalonia, Murcia	Ecobiogas Murcia, AEMA Biogás	Citrus peel, OFMSW, agro-industrial waste	CHP and partial upgrading	Industrial parks use heat and electricity locally
France	Occitanie, Provence	Agribiométhane, Méthabaz	Livestock manure, pruning biomass, winery waste	Grid injection (GRDF), vehicle fuel	Strong national support through long-term purchase contracts
Greece	Thessaly, Crete	Bioenergy Thessaly pilot, Heraklion Coop	Dairy manure, olive mill waste, fruit waste	Electricity production (CHP)	EU-funded pilot under Horizon 2020 and innovation grants
Tunisia	Cap Bon, Kairouan	Biogas CapBon, PommeVerte Pilot	Tomato waste, cattle slurry, fruit residues	CHP (local use in farms)	GIZ and EIB pilot; digestate used as organic fertiliser in orchards

2. Biomethane Production Technologies

Biomethane production in the Mediterranean region is primarily based on anaerobic digestion (AD), a well-established and versatile biotechnology that enables the biological conversion of organic matter into biogas and digestate under anaerobic conditions (Figure 1). This process is adaptable to various scales and feedstocks, making it particularly suited to the diversified agricultural, livestock, and agro-industrial contexts of Mediterranean countries.

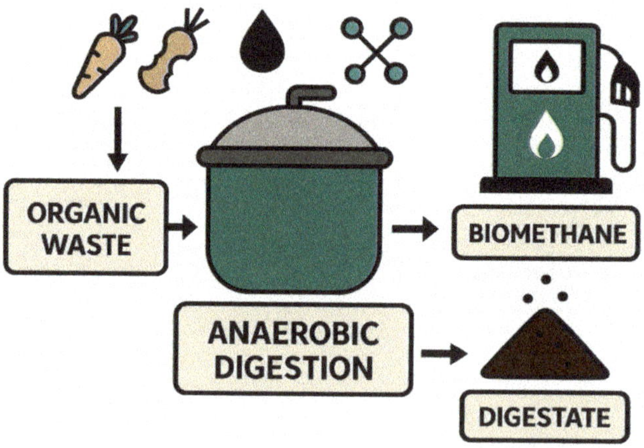

Figure 1. Schematic representation of the anaerobic digestion process aimed at converting organic waste into biomethane and digestate.

The efficiency and stability of the AD process are influenced by a complex interplay of physicochemical and biological parameters. Among the most critical factors are the biochemical composition of the feedstock—especially the carbon-to-nitrogen (C/N) ratio—along with operational conditions such as pH, hydraulic retention time (HRT), and the activity of microbial consortia that drive the sequential stages of anaerobic conversion: hydrolysis, acidogenesis, acetogenesis, and methanogenesis. Environmental temperature plays a decisive role in modulating microbial kinetics. In Mediterranean climates, thermophilic digestion (50–60 °C) is often favoured due to the elevated ambient temperatures, promoting faster reaction rates and more efficient pathogen inactivation. Nevertheless, thermophilic systems demand greater energy input and precise operational control compared to mesophilic processes, posing challenges in terms of energy balance and process stability [20,21].

A major bottleneck for biomethane production in the Mediterranean context is the heterogeneous nature of available biomass, particularly lignocellulosic residues derived from crop production and agro-industrial processing. These include straw, vineyard, and orchard pruning residues and fruit-processing by-products, which are inherently recalcitrant to microbial degradation. To address this, various pre-treatment technologies are being increasingly implemented to enhance the bioavailability of complex substrates. Mechanical comminution, steam explosion, thermal hydrolysis, and chemical conditioning (alkaline or acidic) are among the techniques employed to disrupt lignocellulosic structures, increase the accessible surface area, and facilitate enzymatic hydrolysis, ultimately improving methane yields [22,23].

Co-digestion strategies represent another key optimisation approach, particularly relevant in the Mediterranean agri-food sector. By combining livestock manure—an abundant resource that provides buffering capacity, essential micronutrients, and a natural microbial inoculum—with high-energy agro-industrial residues such as olive mill wastewater, citrus peels, and winery grape marc, it is possible to achieve a more balanced nutrient composition and stimulate synergistic microbial interactions. Co-digestion not only enhances process stability and methane production but also supports the integrated management of organic waste streams within agro-industrial clusters, thereby promoting industrial symbiosis and regional waste valorisation [24,25].

Beyond energy production, the anaerobic process generates a nutrient-rich digestate, which can be valorised as an organic fertiliser or soil amendment. The application of digestate in agriculture contributes to nutrient recycling and soil fertility improvement, aligning with circular bioeconomy principles. This dual role of AD—producing renewable biomethane while recycling nutrients—positions the technology as a cornerstone of sustainable energy and waste management strategies in Mediterranean agriculture. In this context, decentralised energy solutions and distributed resource recovery systems are increasingly vital to building climate-resilient and energy-independent farming communities, particularly in light of ongoing climate change challenges and evolving energy transition policies [3,26].

Raw biogas, as produced by anaerobic digestion, typically consists of 50–70% methane (CH_4) and 30–50% carbon dioxide (CO_2), along with trace amounts of hydrogen sulphide (H_2S), ammonia (NH_3), siloxanes, and water vapour. To transform this raw biogas into biomethane—a high-purity renewable gas suitable for grid injection or as a vehicle fuel—it is necessary to remove CO_2 and these contaminants. This upgrading process not only increases the methane concentration but also ensures compliance with pipeline specifications and protects end-use equipment from corrosion or fouling.

Several upgrading technologies are currently available, each with distinct operational characteristics and suitability depending on plant scale, feedstock variability, and economic considerations. The most widely applied methods include Pressure Swing Adsorption (PSA), water scrubbing, chemical absorption using amines, membrane separation, and cryogenic upgrading.

Pressure Swing Adsorption (PSA) systems operate based on the selective adsorption of gas components onto porous media under varying pressure conditions. PSA is particularly effective for plants with stable biogas flow rates and allows for reliable methane recovery [27,28]. Water scrubbing, a physical absorption process, exploits the greater solubility of CO_2 and H_2S in water relative to methane. It is widely applied in small- to medium-scale AD plants due to its relative simplicity, moderate operational costs, and minimal chemical use, making it a favourable option in decentralised Mediterranean agricultural contexts.

Chemical absorption, typically using amine solutions, provides high selectivity and methane purity by chemically binding CO_2. This technology is well-suited to large-scale facilities where the higher operational and maintenance costs are justified by economies of scale and the need for consistently high-purity biomethane [27,28].

Membrane separation technologies are gaining increasing traction for agricultural biogas upgrading. These systems rely on the differential permeability of gas molecules through polymeric or ceramic membranes, allowing for the selective separation of CO_2 from CH_4. Membrane systems are highly modular, energy-efficient, and easy to integrate into existing AD plants. Their compact footprint and scalability make them particularly attractive for decentralised, small- to medium-scale agricultural plants in the Mediterranean region, where space and economic efficiency are key considerations [29,30].

Cryogenic upgrading, which separates gas components through selective condensation and liquefaction at very low temperatures, achieves biomethane with exceptionally high purity. However, it is energy-intensive and economically viable primarily for large-scale or centralised operations, particularly those producing liquefied biomethane (Bio-LNG) for transport and storage applications.

Beyond these established technologies, several emerging approaches are expanding the potential for biomethane upgrading, particularly in distributed and small-scale systems. Biological methanation is an innovative process where hydrogen reacts with CO_2 under the action of methanogenic archaea, producing additional methane. This approach not only upgrades biogas but also integrates renewable electricity through power-to-gas applications, improving the overall energy system flexibility.

Additionally, compact, containerised upgrading solutions are being deployed in small cooperative networks and multifunctional farms. These modular systems offer scalable, plug-and-play solutions that require minimal infrastructure investment, enabling local energy production and consumption. Such systems are well suited to Mediterranean agricultural landscapes, where energy autonomy, circular resource management, and cost-effectiveness are essential for sustainable rural development [31,32].

Together, these advancements in upgrading technologies are facilitating the wider deployment of biomethane as a renewable energy carrier, supporting the transition towards decentralised energy systems and circular bioeconomies in Mediterranean agricultural regions.

3. Feedstocks in the Mediterranean Region

The Mediterranean region, characterised by favourable agro-climatic conditions and exceptional agricultural biodiversity, offers a diverse range of organic feedstocks suitable for biomethane production through anaerobic digestion (AD). The utilisation of these bioresources is consistent with the principles of the circular bioeconomy, fostering sustainable rural development in territories often challenged by water scarcity, energy dependency, and fragmented agricultural landscapes.

Among the most promising energy crops, cactus pear (*Opuntia ficus-indica*), a Crassulacean Acid Metabolism (CAM) species well adapted to arid environments, has been extensively studied. Comparetti et al. (2017) [14] demonstrated that the cladodes of *O. ficus-indica* serve as an efficient substrate for AD, yielding methane-rich biogas and nutrient-dense digestate suitable for soil improvement. Agro-industrial residues also provide a significant resource base. By-products from winemaking, such as grape marc, along with olive oil production residues like pomace and wastewater, and citrus fruit processing wastes such as peels, represent abundant and locally available feedstocks. Their integration into AD systems not only facilitates the reduction in organic waste but also contributes to decentralised energy production across rural and peri-urban areas [33,34].

Livestock manure continues to play a foundational role in Mediterranean AD systems, owing to its stable composition, buffering capacity, and balanced nutrient content. Studies by Greco et al. (2019a) and Attard et al. (2023) [13,17] have shown its critical role in island contexts, where the need for local energy autonomy is pronounced. Furthermore, co-digestion strategies that combine livestock manure with agro-industrial by-products such as tomato processing waste, olive pomace, and grape marc have proven effective in enhancing methane production. These combinations optimise the carbon-to-nitrogen (C/N) ratio of the feedstock mix and increase overall biodegradability, leading to more stable and productive AD processes [19,21].

In addition to these widely used feedstocks, emerging attention has been given to underutilised biomass streams. Pruning residues from orchards and vineyards, unsold fruit and vegetables from local markets, and biomass from invasive plant species such as *Ailanthus altissima* are increasingly being investigated as supplementary substrates for AD [35–37]. These resources are particularly well suited for small-scale or cooperative AD plants, which promote local energy self-sufficiency, reduce the environmental impact of organic waste, and diversify rural incomes. Such initiatives align with the objectives of the European Renewable Energy Directive and the Circular Economy Action Plan, both of which encourage the sustainable management of bioresources and the development of local energy systems.

Despite these advances, the number of studies specifically focusing on Mediterranean feedstocks for biomethane production remains limited when compared to the research output from Northern and Central Europe. While the scientific literature on the topic is expanding, further investigations are required to assess the availability, seasonal fluctuations, and local sustainability of these feedstocks. According to a bibliometric analysis by Zhou (2022) [38], Italy and Spain currently lead biomethane research in the region, but significant gaps remain in southern Mediterranean countries, where research efforts and technology deployment are still emerging.

Feedstock pretreatment represents a critical area for improving AD efficiency in the Mediterranean context. Various thermal processes, such as hot water or steam application, have been evaluated to enhance the biodegradability of lignocellulosic residues like pruning waste. Alkaline treatments, using sodium hydroxide (NaOH) or lime, have been applied to olive pomace and grape marc to disrupt lignocellulosic structures and improve microbial accessibility [39]. Biological pretreatment methods, such as the application of white-rot fungi, have also been explored to degrade lignin components, thus facilitating higher methane yields [40]. These pretreatment strategies contribute not only to increasing methane production but also to reducing the inhibitory effects of compounds such as limonene and polyphenols, which are commonly present in citrus peel and olive mill effluents [36,37]. Moreover, co-digestion with livestock manure or food waste further mitigates the presence of toxic substances and promotes the stability of microbial communities, improving overall process performance.

Digestate, the residual material from the anaerobic process, has shown substantial agronomic potential when applied to Mediterranean soils. Field experiments conducted in Sicily, Spain, and Tunisia have demonstrated that digestate enhances soil organic matter content, improves water retention capacity, and increases crop yields, particularly in degraded soils suffering from fertility loss [19]. These findings support the role of biomethane systems as a nexus between renewable energy production and sustainable agriculture, contributing to soil restoration and climate-resilient farming practices.

Table 2 provides a comprehensive summary of the principal feedstocks available in the Mediterranean region, outlining their agricultural or industrial origins, key biochemical characteristics, and representative countries where their use has been studied. The table is supported by data from the published literature and technical sources [14,41–44], reflecting both scientific research and applied case studies.

Table 2. Major Mediterranean feedstocks for biomethane production, including origin, biochemical composition, and countries of use based on key references.

Feedstock	Origin	Key Components	Countries	References
Olive pomace	Olive oil production	Lipids, cellulose	Italy, Spain, Greece	[41,42,44]
Citrus peel	Citrus processing	Sugars, pectin	Italy, Tunisia	[45,46]
Winery grape marc	Grape processing	Sugars, lignin	France, Italy, Spain	[43]
Cactus pear residues	Pruning and fruit industry	Mucilage, polysaccharides	Morocco, Tunisia	[14]
OFMSW	Municipal waste	Organic residues	All	[3,45]
Livestock manure	Dairy and pig farms	Organic matter, N	Spain, Italy	[13,17,44]
Agro-industrial wastewater	Canning, dairy, brewery	Mixed organic load	Greece, Turkey	[41]

Recent investigations have reported promising methane yields from the co-digestion of citrus peel and olive mill wastewater, two abundant agro-industrial by-products generated across the Mediterranean region. Methane production values typically range from 250 to 380 cubic meters of CH_4 per tonne of volatile solids (VSs). These substrates are characterised by a high content of readily biodegradable organic compounds, including simple sugars, organic acids, and essential oils, which contribute to substantial biogas production. However, despite their favourable biodegradability, these by-products also contain inhibitory substances—such as limonene in citrus peel and polyphenols in olive mill wastewater—that can negatively affect microbial activity during anaerobic digestion if not properly managed [46,47].

To mitigate these inhibitory effects, co-digestion strategies have been successfully employed. The addition of livestock manure or food waste to citrus peel and olive mill wastewater not only dilutes toxic compounds but also improves the overall nutrient balance of the feedstock mixture. This synergistic effect enhances microbial diversity and resilience, stabilising the anaerobic digestion process and improving methane yields [43,48]. Such integrated approaches reflect best practices for optimising methane production from challenging agro-industrial residues.

Beyond energy recovery, the process generates a nutrient-rich digestate, containing essential macro- and micronutrients with recognised agronomic value. When applied as a biofertiliser, this digestate contributes to soil fertility improvement, aligning with the European Union's sustainability goals for nutrient recycling and the reduction in dependency on synthetic chemical fertilisers. Field-scale experiments in Mediterranean areas have demonstrated that digestate derived from citrus and olive waste mixtures effectively enhances soil organic matter content, water retention capacity, and crop productivity. These benefits are particularly evident in degraded soils typical of many Mediterranean agricultural landscapes [44,49].

Recent co-digestion trials using citrus peel and olive mill wastewater have reported methane yields ranging between 250 and 380 m^3 CH_4 per tonne of volatile solids (VS). These studies confirm that co-digestion plays a critical role in mitigating the inhibitory effects of certain feedstocks, thereby enhancing overall biogas production efficiency [46–48].

In conclusion, while numerous Mediterranean feedstocks present significant potential for biomethane generation, additional research is necessary to fully assess their seasonal availability, scalability, and compatibility with various pretreatment technologies. Addressing these knowledge gaps will facilitate the broader deployment of biomethane systems, supporting climate-neutral energy strategies and fostering sustainable development pathways throughout the Mediterranean region.

4. Potential Biomethane Production in the Mediterranean Region

The Mediterranean region holds substantial and diversified potential for biomethane production, owing to the abundant availability of organic residues derived from agricultural, agro-industrial, municipal, and livestock activities. Agricultural by-products such as olive pomace and wastewater, grape marc, citrus peel, and tomato processing residues represent highly fermentable substrates, capable of yielding between 250 and 450 cubic meters of methane (CH_4) per tonne of volatile solids (VSs) [22,36]. In Greece and Italy alone, the olive oil industry annually generates over 2 and 5 million tonnes of olive pomace, respectively, while Spain surpasses 3 million tonnes [50]. Similarly, the winemaking sectors of France, Italy, and Greece collectively produce more than 6 million tonnes of grape marc each year, particularly concentrated in the September–October harvest period.

These lignocellulosic residues, although rich in fermentable carbohydrates, are often deficient in nitrogen, necessitating co-digestion with livestock manure or food waste to optimise nutrient balance and ensure process stability [23,43]. In Southern Spain and Portugal, integrated anaerobic digestion (AD) facilities have demonstrated that co-digestion of olive mill residues with pig slurry can increase methane yields by 15 to 20 percent [51]. Nevertheless, despite these positive examples, logistical constraints persist in fragmented rural landscapes, particularly in southern Italy and inland Greece, where more than 40 percent of the theoretical manure potential remains underutilised [7].

The Organic Fraction of Municipal Solid Waste (OFMSW) is also becoming a key biomethane resource in urban centres such as Barcelona, Rome, and Marseille. In these cities, biogas yields from OFMSW range between 100 and 250 cubic meters of CH_4 per ton, with enhanced performance observed when food service sector waste is included in the feedstock mix [3,52]. France and Italy have implemented dedicated biowaste collection schemes in major urban districts, achieving capture rates exceeding 70 percent and ensuring a stable feedstock supply for medium- and large-scale digesters. Portugal is actively expanding separate biowaste collection zones in Lisbon and Porto, aiming to boost biomethane output by 25 percent by 2030.

Among emerging bioenergy crops, cactus pear (*Opuntia ficus-indica*), extensively cultivated in Sicily, southern Spain, and Tunisia, is showing strong potential for biomethane production, especially in arid and semi-arid zones. Cladodes and fruit processing residues from this crop have demonstrated methane yields of 300 to 350 cubic meters of CH_4 per tonne of VS, with the additional advantage of low water and nutrient requirements [14,53]. Citrus peel from Calabria and Valencia, along with orange pulp from Moroccan juice industries, also represents a promising carbon source, although these residues require pre-treatment to reduce limonene content and avoid inhibition of methanogenic activity [46,54].

Municipal sewage sludge is another important substrate, particularly in Mediterranean cities such as Athens, Palermo, and Madrid, where it is increasingly co-digested with food industry by-products, including brewery and dairy residues [29]. Cross-sector integration models are gaining traction in France, where regional hubs collect food industry waste and municipal sewage sludge for joint anaerobic digestion and biogas upgrading, fostering synergies within the circular bioeconomy framework [55].

When considered in aggregate, the combined biomethane production potential from agro-industrial residues, OFMSW, livestock manure, cactus biomass, and sewage sludge exceeds 8 billion cubic meters of methane per year across the Mediterranean region. Realising this potential, however, will require coordinated improvements in biomass collection logistics, the adoption of effective pre-treatment strategies, and the deployment of advanced biogas upgrading technologies. If successfully implemented, this renewable capacity could significantly reduce regional dependence on fossil gas imports, contribute to national and EU decarbonisation targets, and strengthen rural bioeconomies through integrated biorefinery models in Italy, France, Spain, Greece, and Portugal.

Figure 2 illustrates the estimated annual biomethane production potential in billion cubic meters of methane per year, derived from organic feedstock availability in selected Mediterranean countries. Italy leads with a projected potential of 2.5 billion cubic meters of CH_4 per year, followed by Spain with 2.0 billion, France with 1.5 billion, Greece with 1.0 billion, and Portugal with 0.8 billion cubic meters of CH_4 per year. These figures reflect the combined contribution of agricultural residues, agro-industrial by-products, OFMSW, and livestock manure and highlight the importance of co-digestion strategies in maximising regional biomethane capacity.

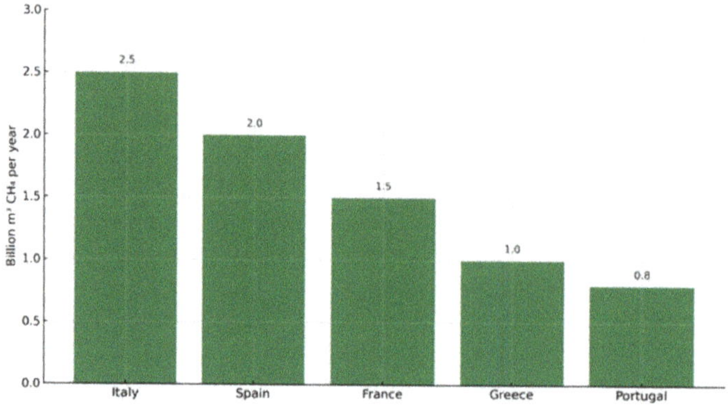

Figure 2. Estimated biomethane production potential in selected Mediterranean countries (Italy, Spain, France, Greece, and Portugal).

5. Policy and Economic Drivers

The policy and economic framework supporting biomethane production in the Mediterranean region is shaped by a complex interplay of European Union directives, national renewable energy strategies, and local incentive mechanisms. Collectively, these instruments aim to accelerate the deployment of renewable energy, promote the valorisation of organic waste, and strengthen the resilience of rural economies.

At the European Union level, the Renewable Energy Directive II (Directive (EU) 2018/2001) and its revision, RED III (Directive (EU) 2023/2413), have set ambitious renewable energy targets. RED III mandates that by 2030, at least 42.5 percent of the EU's final energy consumption must be derived from renewable sources, with a specific sub-target of producing 35 billion cubic meters of biomethane per year [35,56,57]. However, despite these policy goals, EU-wide biomethane production remains relatively modest, reaching just over 3.5 billion cubic meters per year as of 2022, underscoring both the challenge and the untapped potential for growth [58].

In response to this production gap, the EU launched the Biomethane Industrial Partnership (BIP) in 2022 to facilitate large-scale deployment. This initiative promotes investment, technological innovation, and regional collaboration, particularly in rural and agricultural areas where biomethane production can contribute to economic revitalisation and energy self-sufficiency. Complementary to these efforts, the EU Methane Strategy (2020) highlights biomethane as a crucial pathway for methane emission reductions, particularly through the anaerobic digestion of livestock manure and organic waste streams [59].

At the national level, Mediterranean countries are progressively incorporating biomethane into their broader climate and energy policies. Italy currently leads the Mediterranean region, with 44 operational biomethane plants as of early 2024 and over 200 additional projects in the development pipeline. The Italian Biometano Decree provides financial incentives, including feed-in premiums and support for grid injection and transport sector use of biomethane [60]. Spain, with 17 biomethane plants operational in 2023, anticipates significant growth through its Strategic Framework for Energy and Climate, which includes incentives for expanding gas grid infrastructure and facilitating biomethane injection [61].

In France, while not exclusively Mediterranean, biomethane production is well established, particularly in the southern regions bordering the Mediterranean coast. The country operates more than 400 biomethane plants and has set a national target of producing 7 terawatt-hours (TWh) of biomethane per year by 2030, supported by robust grid injection incentives [62]. Greece, although still in the early stages of development, has launched several pilot plants and feasibility studies supported by EU cohesion funds and national clean energy strategies [62]. Tunisia, through international cooperation initiatives, has also initiated pilot biomethane projects, focusing primarily on agricultural waste valorisation. Technical assistance and financing from agencies such as the German Agency for International Cooperation (GIZ) and the European Investment Bank have supported these initial efforts [62].

In this evolving policy environment, decision-support tools have become increasingly important for guiding both policy design and investment strategies. Asciuto et al. (2023) [8] developed an integrated assessment model for evaluating the feasibility of biogas plants in Mediterranean island contexts. This tool combines technical, environmental, and economic criteria, demonstrating how decentralised anaerobic digestion systems can align with the objectives of the circular bioeconomy.

Public–private partnerships, along with European funding programmes such as Horizon Europe, LIFE, and the Common Agricultural Policy (CAP), are playing a critical role in supporting the biomethane sector. These instruments help reduce investment risks and promote the deployment of small- and medium-scale AD systems, which are particularly suited to the fragmented agricultural landscapes and rural economies characteristic of the Mediterranean region [62].

6. Environmental and Agronomic Benefits

Biomethane production through anaerobic digestion (AD) represents not only a renewable energy pathway but also a key contributor to agronomic improvement and environmental sustainability in Mediterranean agricultural systems. As illustrated in Figure 3, AD supports climate-smart and circular agriculture by addressing both energy production and resource recycling challenges. By capturing methane emissions that would otherwise be released from the decomposition of unmanaged organic waste streams—including livestock manure, crop residues, and agro-industrial by-products—AD mitigates one of the most potent greenhouse gases. This reduction in methane emissions directly contributes to the climate neutrality targets set forth in the European Green Deal and the "Fit for 55" package, aligning biomethane production with broader strategies to decarbonise the agricultural and waste management sectors [3,63].

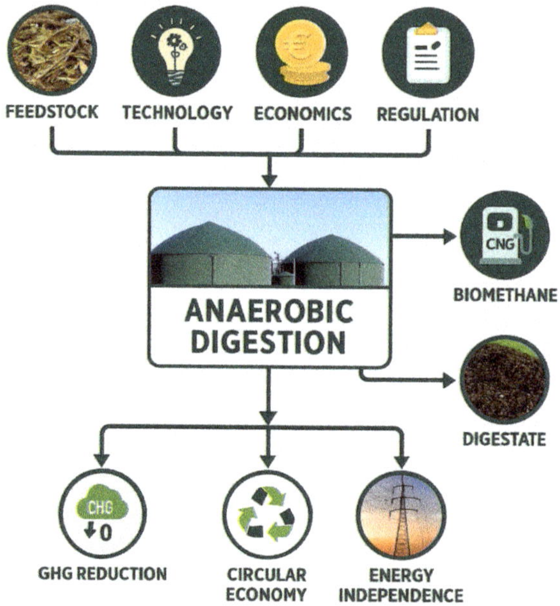

Figure 3. Agronomic and environmental benefits of anaerobic digestion and biomethane production in Mediterranean agriculture.

A central environmental and agronomic benefit of AD lies in the production of digestate, a nutrient-rich co-product in either liquid or solid form. Digestate contains essential plant nutrients, including readily available nitrogen in both ammonium and organic forms, phosphorus, potassium, micronutrients, and residual organic carbon. Its application to agricultural soils has been widely demonstrated to enhance fertility and structure, promote microbial activity, and increase water retention capacity. These benefits are particularly valuable in Mediterranean regions, where soils are frequently exposed to drought stress, nutrient depletion, and erosion risk.

Field studies conducted across southern Europe have consistently shown that the application of digestate can increase crop yields and reduce the reliance on synthetic mineral fertilisers, thereby contributing to nutrient circularity and lowering the environmental footprint of agricultural production. Notably, digestate from co-digestion processes often exhibits an improved carbon-to-nitrogen (C/N) balance and slower nutrient mineralisation rates compared to synthetic fertilisers, enhancing nitrogen-use efficiency and reducing nutrient losses. The reuse of digestate not only closes the nutrient loop but also decreases the dependency on fossil-based mineral fertilisers, in line with European Union objectives for reducing crop input costs and promoting more sustainable nutrient management practices.

In high-input production systems, such as Mediterranean greenhouse horticulture, digestate can be efficiently used in fertigation systems, supplying crops with both nutrients and irrigation water and thereby enhancing resource-use efficiency [16]. Additionally, composted digestate serves as a renewable alternative to peat in horticultural substrates. This practice not only improves substrate sustainability but also contributes to the protection of fragile peatland ecosystems, reducing carbon emissions associated with peat extraction. Research by Greco et al. (2020, 2021) [18,19] confirmed the agronomic suitability of compost and vermicompost derived from digestate for the cultivation of aromatic and medicinal plants such as sage (*Salvia officinalis*), with no adverse effects on crop yield or quality.

From a life cycle perspective, the environmental performance of biomethane production systems improves significantly when they are powered by on-site renewable energy sources. The integration of solar photovoltaic (PV) systems or biomass boilers into farm-scale AD plants enables on-site heat and electricity generation, reducing external energy inputs and lowering overall greenhouse gas (GHG) emissions. Farms adopting such diversified energy strategies benefit from enhanced energy autonomy, lower production costs, and improved environmental sustainability [32]. Moreover, co-digestion strategies involving a mix of livestock manure, food waste, and crop residues enhance the stability of the bioreactor, increase methane yields, and improve both the quality and consistency of the digestate produced.

Taken together, these environmental and agronomic synergies demonstrate the transformative potential of biomethane production for Mediterranean agriculture. Anaerobic digestion enables the integrated management of organic waste, renewable energy generation, and soil restoration, fully aligning with agroecological principles and bioeconomy objectives. This holistic approach reinforces the role of biomethane as a cornerstone technology in the transition toward more resilient and sustainable Mediterranean farming systems.

While the agronomic potential of digestate as a biofertiliser is widely recognised, its suitability for land application is closely linked to its chemical, biological, and physical quality. This consideration becomes particularly critical when the anaerobic digestion (AD) process involves the co-digestion of industrial effluents, such as olive mill wastewater, citrus peel residues, winery by-products, or municipal biowaste. These substrates may introduce a range of contaminants into the digestate, including polyphenols, limonene, heavy metals such as zinc (Zn), copper (Cu), lead (Pb), and cadmium (Cd), as well as residual antibiotics, microplastics, and pathogenic microorganisms [63].

Recent studies conducted in the Mediterranean context have drawn attention to these environmental risks. Pognani et al. (2019) [64], for example, analysed digestate derived from the co-digestion of food waste and agro-industrial effluents in northern Italy and observed elevated concentrations of copper and zinc, in some cases exceeding the thresholds established by Italian fertiliser legislation. The presence of winery residues in the feedstock was identified as a significant contributor to these elevated metal concentrations. Similarly, Vavouraki et al. (2020) [65] investigated digestate composition and reported considerable variability in polyphenol content depending on the proportion of olive mill pomace in the feedstock mixture. Their findings highlighted the need for post-treatment or dilution strategies to ensure the digestate could be safely applied to agricultural soils without phytotoxic effects. Conversely, research by Nayak and Bhushan (2019) [66] demonstrated that with appropriate composting and stabilisation techniques, digestate from Moroccan citrus and olive residues could meet the safety and quality standards set by EU Regulation 2019/1009, enabling its use in nutrient recycling.

In response to these concerns, Mediterranean countries have developed regulatory frameworks and quality assurance systems designed to safeguard soil and environmental health. In Italy, Legislative Decree 75/2010 classifies digestate into distinct categories based on feedstock origin and mandates compliance with strict limits for heavy metals, pathogens such as *E. coli* and *Salmonella*, and organic contaminants. Similar regulatory systems have been adopted in Spain and France, where digestate quality control is integrated into national fertiliser catalogues and biofertiliser certification programmes, including R.E.N.A.R.E. in Spain, the MAGRAMA guidelines, and ADEME protocols in France. These frameworks require routine analysis of digestate parameters such as pH, dry matter content, carbon-to-nitrogen (C/N) ratio, heavy metal concentrations, nitrogen content, and microbial load prior to its authorisation for agricultural use.

To further mitigate contamination risks and ensure environmental safety, best practices have been developed focusing on the segregation of waste sources, thorough characterisation of feedstocks prior to digestion, and the application of post-digestion treatments. Composting is frequently employed to stabilise digestate and reduce phytotoxicity, while techniques such as ammonia stripping, membrane filtration, and biochar addition are used to remove excess nutrients and contaminants. These approaches not only improve the chemical and microbiological stability of the digestate but also support its regulatory compliance, enhancing its value as a biofertiliser. Moreover, by promoting the safe and circular use of organic matter, these practices contribute to protecting soil health and water resources in line with the sustainability goals of the circular bioeconomy.

7. Biomethane in the Circular Bioeconomy

Anaerobic digestion (AD) plays a pivotal role within the circular bioeconomy (CBE), providing a sustainable solution for the valorisation of organic waste streams while simultaneously producing renewable energy and bio-based products. In the broader framework of the European Union's energy transition, AD is gaining strategic importance, particularly in light of recent geopolitical tensions. Energy dependency on third countries such as Russia, Algeria, and Libya, compounded by the ongoing instability in Ukraine, has underscored the vulnerability of fossil gas imports [3,55]. Against this backdrop, the production of biomethane through AD offers a pathway to enhance regional energy self-sufficiency, reduce exposure to external energy shocks, and strengthen the resilience of local economies.

By converting agricultural residues, livestock manure, and municipal biowaste into biomethane, AD systems provide a renewable substitute for fossil natural gas, supporting both energy transition goals and rural development objectives [67]. Equally important is the role of digestate, the nutrient-rich by-product of the AD process, which contributes to closing nutrient cycles and reducing reliance on synthetic chemical fertilisers [61]. Digestate application in agriculture improves soil fertility, enhances organic matter content, and promotes microbial biodiversity, all of which are critical factors in maintaining the health and productivity of Mediterranean agroecosystems. Recent studies have demonstrated the agronomic benefits of digestate and compost, particularly in enhancing soil structure and sustaining crop yields under Mediterranean climatic conditions [68,69].

A growing body of research highlights the potential of stabilised digestate and vermicompost as renewable alternatives to peat in soilless cultivation systems. This application is particularly relevant for the production of nutraceutical and aromatic crops such as sage (*Salvia officinalis*), oregano (*Origanum vulgare*), and rosemary (*Rosmarinus officinalis*), which are widely cultivated in Mediterranean regions [15,18,19]. The use of digestate-based substrates supports optimal plant growth and qualitative parameters while also reducing environmental pressures linked to peat extraction and habitat degradation. In this way, the integration of AD into agricultural and agro-industrial supply chains enhances environmental sustainability, fosters resource circularity, and contributes to the resilience of regional bioeconomies.

Figure 4 illustrates the integrated model of biomethane production within the circular bioeconomy. Agricultural, livestock, and food industry biowaste—including manure and slurry—are processed through AD systems, producing biomethane and digestate. The biomethane serves as a renewable energy carrier for grid injection or transport fuel, while the digestate is utilised as a biofertiliser or as a substitute for peat in horticultural substrates. This closed-loop approach simultaneously supports soil health, renewable energy generation, and the cultivation of aromatic crops, contributing to the reduction in external energy dependency and enhancing the sustainability of Mediterranean farming systems.

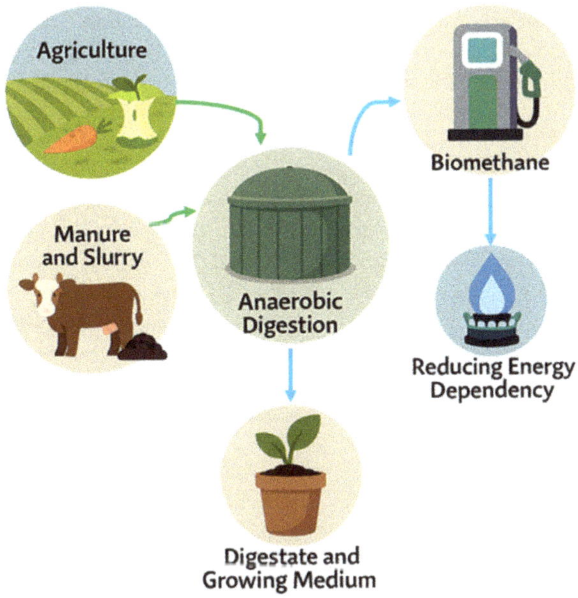

Figure 4. Circular bioeconomy model based on anaerobic digestion: valorisation of biowaste for energy, agriculture, and sustainability.

8. Challenges and Future Perspectives for Biomethane Development in the Mediterranean Circular Bioeconomy

Despite the clear environmental and socio-economic benefits associated with biomethane production, several technical, economic, and social challenges continue to limit its widespread adoption across the Mediterranean region. From a technical perspective, one of the primary constraints lies in the heterogeneity and seasonality of available feedstocks. Agro-industrial residues such as olive pomace and wastewater, citrus peel, grape marc, and livestock manure are highly variable in their composition. These feedstocks often exhibit fluctuations in moisture content, contain inhibitory compounds such as phenolics and limonene, and possess recalcitrant lignocellulosic fractions. Such characteristics necessitate the application of appropriate pre-treatment technologies to improve substrate hydrolysis and microbial accessibility, ensuring stable and efficient anaerobic digestion (AD) processes [22,23]. Moreover, sustaining a resilient and active microbial community capable of adapting to these variations and to fluctuating environmental conditions remains an essential research priority, particularly for decentralised and small-scale digesters operating in rural and island territories [67,70].

Infrastructure limitations further constrain the scalability of biomethane projects. Many peri-urban and remote Mediterranean areas lack sufficient gas upgrading facilities and grid injection points, limiting the capacity to valorise biogas as high-purity biomethane suitable for energy networks or transport applications. Addressing these gaps will require substantial investments in upgrading technologies and energy infrastructure.

Economic barriers also present significant hurdles. The development of AD plants, particularly those equipped with biomethane upgrading units, involves substantial capital investment. In addition, operational costs—including energy consumption, digestate handling, and system maintenance—are considerable, especially for small and medium-sized enterprises. While several Mediterranean countries have introduced supportive policies, such as Italy's Biometano Decree and Spain's Recovery, Transformation and Resilience Plan, the absence of a harmonised regulatory framework across the Mediterranean basin creates policy fragmentation. This disjointed policy environment discourages cross-border investment and impedes the development of a cohesive Mediterranean biomethane market [29]. Furthermore, the lack of long-term feed-in tariffs or stable biomethane pricing mechanisms creates financial uncertainty for project developers and investors, undermining market stability.

Social acceptance also represents a critical challenge. Public awareness of biomethane's role in sustainable energy and agricultural systems remains limited, and the use of digestate as a fertiliser in food-producing sectors faces moderate acceptance. Concerns persist regarding odour emissions, land application practices, and the proximity of AD plants to residential areas [71]. Overcoming these social barriers will require targeted stakeholder engagement strategies, transparent communication, and educational campaigns aimed at increasing public understanding and trust. The development of cooperative business models and multifunctional farm-based biomethane systems may also enhance community participation and foster socio-environmental acceptance.

Looking forward, the advancement of the Mediterranean biomethane sector will require coordinated multi-actor innovation strategies. Research priorities include optimising co-digestion practices by leveraging locally abundant and seasonal feedstocks, engineering microbial consortia with thermotolerant and inhibitor-resistant strains, and developing real-time process monitoring systems using Internet of Things (IoT) technologies and artificial intelligence-based diagnostics [30,31]. The deployment of modular and mobile AD plants represents a promising approach for off-grid or seasonal production systems, particularly suited to the fragmented agricultural landscapes of the Mediterranean.

Further integration of AD plants with on-site renewable electricity sources, such as solar photovoltaic (PV) systems, will enhance energy self-sufficiency and reduce carbon footprints. In parallel, the valorisation of digestate into high-value bio-based products—such as compost, biochar, or bioplastics—will contribute to the overall circularity and economic viability of biomethane systems [71–78].

Finally, fostering cross-border collaboration will be essential to harmonise technical standards, facilitate knowledge exchange, and stimulate innovation. International platforms such as PRIMA (Partnership for Research and Innovation in the Mediterranean Area), IEA Bioenergy, and Horizon Europe are pivotal in supporting joint research initiatives, sharing best practices, and building resilient innovation ecosystems for the Mediterranean biomethane sector.

Table 3 summarises the current status of biomethane-related policies, support mechanisms, and regulatory frameworks across five representative Mediterranean countries: Italy, Spain, France, Greece, and Tunisia. This comparative overview highlights the significant institutional and legislative disparities that characterise the region and influence the scale, pace, and efficiency of biomethane sector development.

Table 3. Overview of national policies and regulatory frameworks for biomethane in the Mediterranean region.

Country	Key Policies and Laws	Incentives and Support	Grid Injection Status	National Target (2030)
Italy	Biometano Decree 2018, PNIEC	Feed-in premiums, tax credits, CAP integration	Fully regulated by ARERA	5 billion m^3 biomethane/year
Spain	Climate Law 7/2021, PNIEC	Green gas incentives, infrastructure investment	Partial (pilot-based)	1.5–2 billion m^3/year
France	LTECV, PPE	15-year feed-in contracts, R&D grants	Fully regulated (GRDF)	10% of the gas grid by 2030
Greece	Law 4936/2022, REPowerEU plan	EU Cohesion Funds, pilot support	Not yet regulated	Target not defined
Tunisia	National Biogas Plan (draft), Decree-Law 2020-33	GIZ/EIB technical and financial support	Legal draft under development	~0.3 billion m^3 (draft target)

Italy represents one of the most advanced cases in the Mediterranean context, having established a comprehensive regulatory framework for biomethane. The Biometano Decree, originally introduced in 2018 and revised in 2022, supports the full biomethane value chain from upgrading to grid injection. Incentives include feed-in premiums, capital subsidies, and favourable tariffs, particularly targeting the transport sector. Furthermore, the Italian Regulatory Authority for Energy, Networks, and Environment (ARERA) has implemented detailed technical and safety standards for gas grid injection, positioning Italy among the few Mediterranean countries with an operational and regulated biomethane market. Italy has also set an ambitious national production target of 5 billion cubic meters of biomethane per year by 2030, directly contributing to the goals of the EU Biomethane Industrial Partnership [79].

France also benefits from a robust policy framework. Its Energy Transition for Green Growth Law (LTECV) and the Multi-Year Energy Programme (PPE) provide long-term policy continuity for renewable gas development. Biomethane producers are supported by 15-year guaranteed feed-in tariffs, while the national gas grid operator, GRDF, facilitates biomethane injection with traceability and certification tools. France has set a 2030 target of achieving a 10 percent renewable gas share in its national gas network, reflecting a strong policy commitment to decarbonisation [80].

In contrast, Spain's biomethane sector is still in an emerging phase. While the Climate Change and Energy Transition Law (7/2021) and the Integrated National Energy and Climate Plan (PNIEC) include biomethane in the national decarbonisation strategy, the country's biomethane infrastructure remains limited, with grid injection confined mainly to pilot projects. Current support consists of tax incentives and innovation grants, but the absence of long-term pricing mechanisms for biomethane hinders market certainty. The Institute for Energy Diversification and Saving (IDAE) has proposed a tentative production target of 1.5 to 2 billion cubic meters per year by 2030, although this objective has not yet been formalised into law [81].

Greece is at an even earlier stage of biomethane sector development. The National Climate Law 4936/2022 and EU-backed programmes such as REPowerEU acknowledge the potential of biomethane, but the country lacks a regulatory framework for gas grid injection. Current activities are largely confined to biogas-based electricity production, with no formal biomethane production targets established. Nevertheless, EU Cohesion Funds and Horizon Europe are beginning to support pilot initiatives and capacity-building programmes [82].

Tunisia, although not a member of the European Union, provides a relevant non-EU Mediterranean example. The country has drafted a National Biogas Plan and enacted Decree-Law 2020-33 to promote renewable energy development. Supported by international cooperation initiatives, including technical and financial assistance from the German Agency for International Cooperation (GIZ), the European Investment Bank (EIB), and the European Union, Tunisia has launched pilot agricultural anaerobic digestion projects in livestock and horticultural sectors. However, the legal and technical frameworks for gas grid injection are still under development, and a preliminary biomethane production target of approximately 0.3 billion cubic meters per year is envisioned in the long term.

This comparative analysis reveals clear differences in regulatory maturity, market incentives, and infrastructure accessibility across the Mediterranean region. Countries such as France and Italy demonstrate advanced regulatory environments and operational markets, whereas Greece and Tunisia are still in the foundational phases of policy and technical development. Key gaps persist in the harmonisation of grid injection standards, the availability of long-term feed-in mechanisms, and the definition of binding production targets. This heterogeneity represents both a constraint to regional market integration and an opportunity to prioritise policy harmonisation, capacity building, and cross-border cooperation. Initiatives such as Horizon Europe, PRIMA (Partnership for Research and Innovation in the Mediterranean Area), and the Green Deal Industrial Plan are well positioned to support these integration efforts.

Beyond policy discrepancies, the Mediterranean biomethane sector faces broader structural challenges. One critical barrier is the fragmentation of waste management systems and the lack of a unified circular bioeconomy strategy. In many Mediterranean countries, transitioning from linear waste disposal to multi-waste valorisation models requires not only supportive regulations but also cultural shifts and infrastructure upgrades. The integration of anaerobic digestion into a coherent circular economy approach remains incomplete.

Emerging technologies such as bio-electrochemical systems, which enable the simultaneous production of hydrogen and electricity from organic waste, offer complementary pathways to biomethane production. However, their large-scale deployment in Mediterranean rural contexts is limited by high capital costs and the need for specialised technical expertise. Additionally, the adoption of waste hierarchy-based decision frameworks and the digitalisation of waste valorisation processes, inspired by Industry 4.0 principles, are increasingly seen as key drivers for optimising resource recovery and improving the economic performance of biomethane systems.

Lessons from countries with advanced circular bioeconomies, such as Germany, illustrate that progress in biomethane production depends on systemic reforms that integrate environmental policies, innovation incentives, and market-driven mechanisms. Maximising the value of biowaste in the Mediterranean region will require a shift from waste treatment paradigms to integrated circular economy models, where biomethane functions not only as a renewable energy vector but also as a catalyst for regional economic development.

Looking forward, the future growth of biomethane in the Mediterranean will depend on coordinated efforts to harmonise policies, advance technological innovation, and mobilise regional investment strategies that simultaneously address energy security and ecological sustainability [78–86].

9. Conclusions

Anaerobic digestion (AD) represents a strategic technology at the nexus of sustainable agriculture, climate adaptation, and the circular bioeconomy (CBE), offering regionally tailored solutions to the environmental and energy challenges facing Mediterranean countries. In a context characterised by abundant biowaste streams, declining soil fertility, and persistent dependence on fossil gas imports, AD provides a multifunctional approach capable of converting agricultural and agro-industrial residues into two high-value products: biomethane, a renewable and locally produced energy vector, and digestate, a nutrient-rich amendment that contributes to the restoration of degraded soils and the closure of nutrient cycles.

The dual function of AD in generating renewable energy and regenerating soil health is fully aligned with the foundational principles of the CBE. Its integration into diversified and multifunctional farming systems has the potential to mitigate greenhouse gas emissions, improve soil organic matter content, and enhance water retention capacity—benefits that are particularly critical in Mediterranean regions such as Sicily, Spain, Malta, and Greece, where desertification and land degradation pose growing threats to agricultural sustainability. Moreover, the decentralised deployment of small- and medium-scale biomethane plants in rural and insular areas promotes local energy autonomy, supports rural economic diversification, and strengthens climate-smart infrastructure.

Empirical evidence from case studies in Italy, Spain, Tunisia, and other Mediterranean countries confirms the technical feasibility and socio-economic value of embedding AD systems within existing agricultural and waste management frameworks. These systems enable diverse utilisation pathways, including biomethane grid injection, vehicle biofuel production, on-site combined heat and power generation, and digestate reuse. Collectively, these applications create synergies that extend across the environmental, economic, and agronomic dimensions of Mediterranean rural systems.

Nevertheless, unlocking the full potential of biomethane in the region requires a paradigm shift from linear resource management models to circular bioeconomy approaches. Achieving this transition will depend on coordinated advances in innovation, policy harmonisation, and capacity building. Technological challenges—such as feedstock seasonality, lignocellulosic recalcitrance, and the high costs of biogas upgrading—demand further research into optimised co-digestion strategies, enhanced microbial consortia, and efficient pre-treatment technologies. Simultaneously, the adoption of digital monitoring tools, precision agriculture practices, and IoT-enabled management systems will be essential for improving process control and tailoring digestate application to site-specific soil and crop conditions.

Policy frameworks such as the Renewable Energy Directive III (RED III) and evolving national biogas strategies offer important regulatory momentum. However, scaling up biomethane deployment will require greater policy harmonisation across Mediterranean countries, the establishment of stable long-term investment incentives, and the promotion of stakeholder engagement at multiple governance levels. Social acceptance and knowledge transfer mechanisms, including training and cooperative platforms, will be critical for integrating AD into mainstream agricultural, waste management, and energy policies.

Ultimately, anaerobic digestion is not merely a waste management technology but a cornerstone of low-carbon, resilient, and regenerative Mediterranean agriculture. By positioning AD within a comprehensive circular bioeconomy framework, Mediterranean countries have the opportunity to simultaneously reduce organic waste, enhance food and energy security, mitigate climate risks, and revitalise rural economies. In doing so, they can lead the transition toward an agricultural future that is productive, environmentally sustainable, and socially inclusive.

Author Contributions: Conceptualisation, A.C. and C.G.; methodology, S.O., S.C., and F.S.; software, C.G. and S.O.; validation, A.C., C.G., and F.S.; formal analysis, S.C.; investigation, C.G. and A.C.; resources, S.C.; data curation, F.S.; writing—original draft preparation, A.C., C.G., and S.O.; writing—

review and editing, S.C. and F.S.; visualisation, S.O. All authors have read and agreed to the published version of the manuscript.

Funding: This research was financially supported by the Innovation Diffusion Plan within the first Food District Call—District Contract Program "Distretto del Cibo Bio Slow Pane e Olio"—CUP: J95B02000030007. Piano di diffusione delle innovazioni nell'ambito del primo Bando Distretti del cibo—Programma del Contratto di Distretto "Distretto del cibo Bio Slow Pane e Olio": J95B02000030007.

Data Availability Statement: The raw data supporting the conclusions of this article are available from the authors on request.

Conflicts of Interest: The authors declare no conflicts of interest.

References

1. Cramer, W.; Guiot, J.; Fader, M.; Garrabou, J.; Gattuso, J.P.; Iglesias, A.; Lange, M.A.; Lionello, P.; Llasat, M.C.; Paz, S.; et al. Climate change and interconnected risks to sustainable development in the Mediterranean. *Nat. Clim. Chang.* **2018**, *8*, 972–980. [CrossRef]
2. Lionello, P.; Scarascia, L. The relation between climate change in the Mediterranean region and global warming. *Reg. Environ. Chang.* **2018**, *18*, 1481–1493. [CrossRef]
3. Scarlat, N.; Dallemand, J.F.; Fahl, F. Biogas: Developments and perspectives in Europe. *Renew. Energy* **2018**, *129*, 457–472. [CrossRef]
4. Zupančič, M.; Možic, V.; Može, M.; Cimerman, F.; Golobič, I. Current status and review of waste-to-biogas conversion for selected European countries and worldwide. *Sustainability* **2022**, *14*, 1823. [CrossRef]
5. Comparetti, A.; Febo, P.; Greco, C.; Orlando, S.; Navickas, K.; Venslauskas, K. Sicilian potential biogas production. *J. Agric. Eng.* **2013**, *44*, e103. [CrossRef]
6. Comparetti, A.; Febo, P.; Greco, C.; Navickas, K.; Nekrosius, A.; Orlando, S.; Venslauskas, K. Assessment of organic waste management methods through energy balance. *Am. J. Appl. Sci.* **2014**, *11*, 1631–1644. [CrossRef]
7. Attard, G.; Comparetti, A.; Febo, P.; Greco, C.; Mammano, M.M.; Orlando, S. Case study of potential production of Renewable Energy Sources (RES) from livestock wastes in Mediterranean islands. *Chem. Eng. Trans.* **2017**, *58*, 553–558.
8. Asciuto, A.; Agosta, M.; Attard, G.; Comparetti, A.; Greco, C.; Mammano, M.M. Development of an Investment Decision Tool for Biogas Production from Biowaste in Mediterranean Islands. In *AIIA 2022: Biosystems Engineering Towards the Green Deal, Proceedings of the Conference of the Italian Society of Agricultural Engineering, Palermo, Italy, 19–22 September 2022*; Springer International Publishing: Cham, Switzerland, 2022; pp. 251–262.
9. Choudhary, A.; Kumar, A.; Kumar, S.; Govil, T.; Sani, R.K. Enhanced biogas production from pine litter codigestion with food waste, microbial community, kinetics, and technoeconomic feasibility. *J. Environ. Eng.* **2023**, *149*, 04022089. [CrossRef]
10. Kougias, P.G.; Angelidaki, I. Biogas and its opportunities—A review. *Front. Environ. Sci. Eng.* **2018**, *12*, 1–12. [CrossRef]
11. Comparetti, A.; Greco, C.; Navickas, K.; Venslauskas, K. Evaluation of potential biogas production in Sicily. In Proceedings of the 11th International Scientific Conference Engineering for Rural Development, Jelgava, Latvia, 24–25 May 2012.
12. Greco, C.; Comparetti, A.; Orlando, S.; Mammano, M.M. A contribution to environmental protection through the valorisation of kitchen biowaste. In *Safety, Health and Welfare in Agriculture and Agro-Food Systems*; SHWA 2020. Lecture Notes in Civil Engineering, (LNCE); Biocca, M., Cavallo, E., Cecchini, M., Failla, S., Romano, E., Eds.; Springer International Publishing: Cham, Switzerland, 2022; Volume 252, pp. 411–420.
13. Greco, C.; Attard, G.; Fenech, O.; Alessandro, A.; Comparetti, A. Manure as a potential source of renewable energy: The behaviour and characterisation of biofuels generated from three animal manure types when subjected to pyrolysis. *Riv. Studi Sulla Sostenibilità* **2019**, *IX* (Suppl. 2), 331–344.
14. Comparetti, A.; Febo, P.; Greco, C.; Mammano, M.M.; Orlando, S. Potential production of biogas from prinkly pear (*Opuntia ficus-indica* L.) in Sicilian uncultivated areas. *Chem. Eng. Trans.* **2017**, *58*, 559–564.
15. Greco, C.; Navickas, K.; Orlando, S.; Comparetti, A.; Venslauskas, K. Life Cycle Impact Assessment applied to cactus pear crop production for generating bioenergy and biofertiliser. *Riv. Studi Sulla Sostenibilità* **2019**, *IX* (Suppl. 2), 315–329.
16. Campiotti, C.A.; Bibbiani, C.; Greco, C. Renewable energy for greenhouse agriculture. *Qual. Access Success* **2019**, *20*, 152–156.
17. Attard, G.; Azzopardi, N.; Comparetti, A.; Greco, C.; Gruppetta, A.; Orlando, S. Potential bioenergy and biofertiliser production from livestock waste in Mediterranean Islands within circular bioeconomy. In *AIIA 2022: Biosystems Engineering Towards the Green Deal, Proceedings of the Conference of the Italian Society of Agricultural Engineering, Palermo, Italy, 19–22 September 2022*; Springer International Publishing: Cham, Switzerland, 2022; pp. 271–283.

18. Greco, C.; Comparetti, A.; Febo, P.; La Placa, G.; Mammano, M.M.; Orlando, S. Sustainable valorisation of biowaste for soilless cultivation of *Salvia officinalis* in a circular bioeconomy. *Agronomy* **2020**, *10*, 1158. [CrossRef]
19. Greco, C.; Comparetti, A.; Fascella, G.; Febo, P.; La Placa, G.; Saiano, F.; Mammano, M.M.; Orlando, S.; Laudicina, V.A. Effects of vermicompost, compost and digestate as commercial alternative peat-based substrates on qualitative parameters of *Salvia officinalis*. *Agronomy* **2021**, *11*, 98. [CrossRef]
20. Asunis, F.; De Gioannis, G.; Isipato, M.; Muntoni, A.; Polettini, A.; Pomi, R.; Rossi, A.; Spiga, D. Control of fermentation duration and pH to orient biochemicals and biofuels production from cheese whey. *Bioresour. Technol.* **2019**, *289*, 121722. [CrossRef]
21. Bolzonella, D.; Pavan, P.; Battistoni, P.; Cecchi, F. Anaerobic co-digestion of sludge with other organic wastes and phosphorus reclamation in wastewater treatment plants for biological nutrients removal. *Water Sci. Technol.* **2006**, *53*, 177–186. [CrossRef]
22. Bacenetti, J.; Fiala, M.; Baboun, S.H.; Demery, F.; Aburdeineh, I. Environmental impact assessment of electricity generation from biogas in Palestine. *Environ. Eng. Manag. J. (EEMJ)* **2016**, *15*, 1915–1922. [CrossRef]
23. Carrere, H.; Antonopoulou, G.; Affes, R.; Passos, F.; Battimelli, A.; Lyberatos, G.; Ferrer, I. Review of feedstock pretreatment strategies for improved anaerobic digestion: From lab-scale research to full-scale application. *Bioresour. Technol.* **2016**, *199*, 386–397. [CrossRef]
24. Tallou, A.; Salcedo, F.P.; Haouas, A.; Jamali, M.Y.; Atif, K.; Aziz, F.; Amir, S. Assessment of biogas and biofertilizer produced from anaerobic co-digestion of olive mill wastewater with municipal wastewater and cow dung. *Environ. Technol. Innov.* **2020**, *20*, 101152. [CrossRef]
25. Calabrò, P.S.; Pontoni, L.; Porqueddu, I.; Greco, R.; Pirozzi, F.; Malpei, F. Effect of the concentration of essential oil on orange peel waste biomethanization: Preliminary batch results. *Waste Manag.* **2016**, *48*, 440–447. [CrossRef] [PubMed]
26. Arfelli, F.; Cespi, D.; Ciacci, L.; Passarini, F. Application of life cycle assessment to high quality-soil conditioner production from biowaste. *Waste Manag.* **2023**, *172*, 216–225. [CrossRef]
27. Sun, Q.; Li, H.; Yan, J.; Liu, L.; Yu, Z.; Yu, X. Selection of appropriate biogas upgrading technology-a review of biogas cleaning, upgrading and utilisation. *Renew. Sustain. Energy Rev.* **2015**, *51*, 521–532. [CrossRef]
28. Petersson, A.; Wellinger, A. Biogas upgrading technologies–developments and innovations. *IEA Bioenergy* **2009**, *20*, 1–19.
29. Scarlat, N.; Prussi, M.; Padella, M. Quantification of the carbon intensity of electricity produced and used in Europe. *Appl. Energy* **2022**, *305*, 117901. [CrossRef]
30. Angelidaki, I.; Treu, L.; Tsapekos, P.; Luo, G.; Campanaro, S.; Wenzel, H.; Kougias, P.G. Biogas upgrading and utilization: Current status and perspectives. *Biotechnol. Adv.* **2018**, *36*, 452–466. [CrossRef] [PubMed]
31. Tian, G.; Yang, B.; Dong, M.; Zhu, R.; Yin, F.; Zhao, X.; Wang, Y.; Xiao, W.; Wang, Q.; Zhang, W.; et al. The effect of temperature on the microbial communities of peak biogas production in batch biogas reactors. *Renew. Energy* **2018**, *123*, 15–25. [CrossRef]
32. Rossi, E.; Pecorini, I.; Iannelli, R. Multilinear regression model for biogas production prediction from dry anaerobic digestion of OFMSW. *Sustainability* **2022**, *14*, 4393. [CrossRef]
33. Mata-Alvarez, J.; Dosta, J.; Romero-Güiza, M.S.; Fonoll, X.; Peces, M.; Astals, S. A critical review on anaerobic co-digestion achievements between 2010 and 2013. *Renew. Sustain. Energy Rev.* **2014**, *36*, 412–427. [CrossRef]
34. Riaño, B.; García-González, M.C. Influence of organic loading rate and hydraulic retention time on the performance of anaerobic digestion of olive-mill wastewater in a tubular plant. *Renew. Energy* **2014**, *68*, 625–632. [CrossRef]
35. EU RED III. Directive 2023/2413 on the Promotion of the Use of Energy from Renewable Sources. Available online: https://eur-lex.europa.eu/eli/dir/2023/2413/oj/eng (accessed on 1 June 2025).
36. Guo, D.S.; Ji, X.J.; Ren, L.J.; Li, G.L.; Yin, F.W.; Huang, H. Development of a real-time bioprocess monitoring method for docosahexaenoic acid production by *Schizochytrium* sp. *Bioresour. Technol.* **2016**, *216*, 422–427.
37. Fernández-Rodríguez, M.J.; Cubero-Cardoso, J.; de la Lama-Calvente, D.; Fernández-Prior, Á.; Rodríguez-Gutiérrez, G.; Borja, R. Anaerobic digestion of the residue (combination of wastewater and solid waste) from a new olive-oil manufacturing process based on an olive cold-pressing system: Kinetic approach and process performance. *Processes* **2022**, *10*, 2552. [CrossRef]
38. Zhou, Y. Transition towards carbon-neutral districts based on storage techniques and spatiotemporal energy sharing with electrification and hydrogenation. *Renew. Sustain. Energy Rev.* **2022**, *162*, 112444. [CrossRef]
39. Antonopoulou, I.; Spanopoulos, A.; Matsakas, L. Single cell oil and ethanol production by the oleaginous yeast Trichosporon fermentans utilizing dried sweet sorghum stalks. *Renew. Energy* **2020**, *146*, 1609–1617. [CrossRef]
40. Bouallagui, H.; Touhami, Y.; Ben Cheikh, R.; Hamdi, M. Bioreactor performance in anaerobic digestion of fruit and vegetable wastes. *Process Biochem.* **2005**, *40*, 989–995. [CrossRef]
41. Qin, Y.; Huang, L.; Jiang, Q.; Lu, T.; Xin, Y.; Zhen, Y.; Liu, J.; Shen, P. Anaerobic co-digestion of molasses vinasse and three kinds of manure: A comparative study of performance at different mixture ratio and organic loading rate. *J. Clean. Prod.* **2022**, *371*, 133631. [CrossRef]
42. Khalid, N.A.; Rajandas, H.; Parimannan, S.; Croft, L.J.; Loke, S.; Chong, C.S.; Bruce, N.C.; Yahya, A. Insights into microbial community structure and diversity in oil palm waste compost. *3 Biotech* **2019**, *9*, 1–11. [CrossRef]

43. Semeraro, B.; Summa, D.; Costa, S.; Zappaterra, F.; Tamburini, E. Bio-delignification of green waste (GW) in co-digestion with the organic fraction of municipal solid waste (OFMSW) to enhance biogas production. *Appl. Sci.* **2021**, *11*, 6061. [CrossRef]
44. Di Maria, F.; El-Hoz, M. A short review of comparative energy, economic and environmental assessment of different biogas-based power generation technologies. *Energy Procedia* **2018**, *148*, 846–851.
45. Drosg, B. *Process Monitoring in Biogas Plants*; IEA Bioenergy: Paris, France, 2013; pp. 1–38.
46. Zema, D.A.; Fòlino, A.; Zappia, G.; Calabrò, P.S.; Tamburino, V.; Zimbone, S.M. Anaerobic digestion of orange peel in a semi-continuous pilot plant: An environmentally sound way of citrus waste management in agro-ecosystems. *Sci. Total Environ.* **2018**, *630*, 401–408. [CrossRef] [PubMed]
47. García-González, M.C.; Hernández, D.; Molinuevo-Salces, B.; Riaño, B. Positive impact of biogas chain on GHG reduction. In *Improving Biogas Production: Technological Challenges, Alternative Sources, Future Developments*; Springer: Berlin/Heidelberg, Germany, 2019; pp. 217–242.
48. Capson-Tojo, G.; Moscoviz, R.; Astals, S.; Robles, Á.; Steyer, J.P. Unraveling the literature chaos around free ammonia inhibition in anaerobic digestion. *Renew. Sustain. Energy Rev.* **2020**, *117*, 109487. [CrossRef]
49. Chinea, L.; Slopiecka, K.; Bartocci, P.; Park, A.H.A.; Wang, S.; Jiang, D.; Fantozzi, F. Methane enrichment of biogas using carbon capture materials. *Fuel* **2023**, *334*, 126428. [CrossRef]
50. FAO (Agriculture Organization of the United Nations). Faostat: Crop Production Data. Production Data 2020. Available online: http://www.fao.org/faostat/ (accessed on 6 May 2025).
51. Deschamps, L.; Imatoukene, N.; Lemaire, J.; Mounkaila, M.; Filali, R.; Lopez, M.; Theoleyre, M.A. In-situ biogas upgrading by bio-methanation with an innovative membrane bioreactor combining sludge filtration and H2 injection. *Bioresour. Technol.* **2021**, *337*, 125444. [CrossRef]
52. Avena, L.G.; Almendrala, M.; Marron, E.J.; Obille, J.A. Biogas Production from the Co-and Tri-digestion of Pineapple Wastes with Food Wastes and Pig Manure. In *E3S Web of Conferences*; EDP Sciences: Les Ulis, France, 2024; Volume 521, p. 01004.
53. Antón-Herrero, R.; García-Delgado, C.; Alonso-Izquierdo, M.; Cuevas, J.; Carreras, N.; Mayans, B.; Camacho-Arévalo, R.; Eymar, E. New uses of treated urban waste digestates on stimulation of hydroponically grown tomato (*Solanum lycopersicon* L.). *Waste Biomass Valorization* **2021**, *12*, 1877–1889. [CrossRef]
54. Riaño, B.; García-González, M.C. On-farm treatment of swine manure based on solid-liquid separation and biological nitrification-denitrification of the liquid fraction. *J. Environ. Manag.* **2014**, *132*, 87–93. [CrossRef]
55. Kampman, N.; Busch, A.; Bertier, P.; Snippe, J.; Hangx, S.; Pipich, V.; Di, Z.; Rother, G.; Harrington, J.F.; Evans, J.P.; et al. Observational evidence confirms modelling of the long-term integrity of CO2-reservoir caprocks. *Nat. Commun.* **2016**, *7*, 12268. [CrossRef]
56. Directive (EU) 2018/2002 of the European Parliament and of the Council of 11 December 2018 Amending Directive 2012/27/EU on Energy Efficiency. Available online: https://eur-lex.europa.eu/eli/dir/2018/2002/oj/eng (accessed on 4 June 2025).
57. Nouicer, A.; Stampatori, D.; Iliopoulos, T. Renewable energy policy. In *The EU Green Deal*; European Commission: Brussels, Belgium, 2024.
58. Fiorentino, G.; Zucaro, A.; Cerbone, A.; Giocoli, A.; Motola, V.; Rinaldi, C.; Scalbi, S.; Ansanelli, G. The contribution of biogas to the electricity supply chain: An Italian life cycle assessment database. *Energies* **2024**, *17*, 3264. [CrossRef]
59. IDAE (Instituto para la Diversificación y Ahorro de la Energía). *Biometano y Energía Renovable en España 2023*; IDAE: Madrid, Spain, 2023.
60. ADEME. *Observatoire du Biométhane 2023*; Agence de la Transition Écologique: Paris, France, 2023.
61. Toufexis, C.; Makris, D.O.; Vlachokostas, C.; Michailidou, A.V.; Mertzanakis, C.; Vachtsiavanou, A. Bridging the Gap between Biowaste and Biomethane Production: A Systematic Review Meta-Analysis Methodological Approach. *Sustainability* **2024**, *16*, 6433. [CrossRef]
62. Tang, S.; Shao, N.; Zheng, C.; Yan, F.; Zhang, Z. Amino-functionalized sewage sludge-derived biochar as sustainable efficient adsorbent for Cu (II) removal. *Waste Manag.* **2019**, *90*, 17–28. [CrossRef]
63. Vavouraki, A.I.; Zakoura, M.V.; Dareioti, M.A.; Kornaros, M. Biodegradation of polyphenolic compounds from olive mill wastewaters (OMW) during two-stage anaerobic co-digestion of agro-industrial mixtures. *Waste Biomass Valorization* **2020**, *11*, 5783–5791. [CrossRef]
64. Nayak, A.; Bhushan, B. An overview of the recent trends on the waste valorization techniques for food wastes. *J. Environ. Manag.* **2019**, *233*, 352–370. [CrossRef]
65. European Commission. The European Green Deal. COM(2019) 640 Final. Brussels. 2019. Available online: https://eur-lex.europa.eu/legal-content/EN/TXT/?uri=CELEX:52019DC0640 (accessed on 13 May 2025).
66. European Commission. A European Green Deal—EU Strategy to Reduce Methane Emissions; COM(2020) 663 Final, Brussels. 2020. Available online: https://eur-lex.europa.eu/legal-content/EN/TXT/?uri=CELEX:52020DC0663 (accessed on 14 May 2025).

67. Borchers, M.; Thrän, D.; Chi, Y.; Dahmen, N.; Dittmeyer, R.; Dolch, T.; Dold, C.; Förster, J.; Herbst, M.; Heß, D.; et al. Scoping carbon dioxide removal options for Germany–What is their potential contribution to Net-Zero CO2? *Front. Clim.* **2022**, *4*, 810343. [CrossRef]
68. Kardung, M.; Cingiz, K.; Costenoble, O.; Delahaye, R.; Heijman, W.; Lovrić, M.; van Leeuwen, M.; M'Barek, R.; van Meijl, H.; Piotrowski, S.; et al. Development of the circular bioeconomy: Drivers and indicators. *Sustainability* **2021**, *13*, 413. [CrossRef]
69. European Biogas Association (EBA). Biomethane Industrial Partnership—Accelerating Biomethane Scale-up in the EU. EBA. 2022. Available online: https://energy.ec.europa.eu/topics/renewable-energy/bioenergy/biomethane_en#:~:text=The%20Biomethane%20Industrial%20Partnership%20(BIP,of%20its%20potential%20by%202050./ (accessed on 26 May 2025).
70. Bertolino, A.M.; Giganti, P.; dos Santos, D.D.; Falcone, P.M. A matter of energy injustice? comparative analysis of biogas development in Brazil and Italy. *Energy Res. Soc. Sci.* **2023**, *105*, 103278. [CrossRef]
71. Tambone, F.; Pradella, M.; Bedussi, F.; Adani, F. Moringa oleifera Lam. as an energy crop for biogas production in developing countries. *Biomass Convers. Biorefinery* **2020**, *10*, 1083–1089. [CrossRef]
72. Möller, K.; Müller, T. Effects of anaerobic digestion on digestate nutrient availability and crop growth: A review. *Eng. Life Sci.* **2012**, *12*, 242–257. [CrossRef]
73. Tamburini, G.; Pereira-Peixoto, M.H.; Borth, J.; Lotz, S.; Wintermantel, D.; Allan, M.J.; Dean, R.; Schwarz, J.M.; Knauer, A.; Albrecht, M.; et al. Fungicide and insecticide exposure adversely impacts bumblebees and pollination services under semi-field conditions. *Environ. Int.* **2021**, *157*, 106813. [CrossRef]
74. Fantozzi, F.; Buratti, C. Anaerobic digestion of mechanically treated OFMSW: Experimental data on biogas/methane production and residues characterization. *Bioresour. Technol.* **2011**, *102*, 8885–8892. [CrossRef]
75. Tamburini, M.; Pernetti, R.; Anelli, M.; Oddone, E.; Morandi, A.; Osuchowski, A.; Villani, S.; Montomoli, C.; Monti, M.C. Analysing the Impact on Health and Environment from Biogas Production Process and Biomass Combustion: A Scoping Review. *Int. J. Environ. Res. Public Health* **2023**, *20*, 5305. [CrossRef]
76. Alburquerque, J.A.; de la Fuente, C.; Ferrer-Costa, A.; Carrasco, L.; Cegarra, J.; Abad, M.; Bernal, M.P. Assessment of the fertiliser potential of digestates from farm and agroindustrial residues. *Biomass Bioenergy* **2012**, *40*, 181–189. [CrossRef]
77. Greco, C.; Navickas, K.; La Placa, G.; Mammano, M.M.; Agnello, A. Biowaste in a circular bioeconomy in Mediterranean area: A case study of compost and vermicompost as growing substrates alternative to peat. *Riv. Studi Sulla Sostenibilità* **2019**, *IX* (Suppl. 2), 345–362. [CrossRef]
78. De Gioannis, G.; Dell'Era, A.; Muntoni, A.; Pasquali, M.; Polettini, A.; Pomi, R.; Rossi, A.; Zonfa, T. Bio-electrochemical production of hydrogen and electricity from organic waste: Preliminary assessment. *Clean Technol. Environ. Policy* **2023**, *25*, 269–280. [CrossRef]
79. D'Adamo, I.; Mazzanti, M.; Morone, P.; Rosa, P. Assessing the relation between waste management policies and circular economy goals. *Waste Manag.* **2022**, *154*, 27–35. [CrossRef]
80. Hidalgo, D.; Martín-Marroquín, J.M.; Corona, F. A multi waste management concept as a basis towards a circular economy model. *Renew. Sustain. Energy Rev.* **2019**, *111*, 481–489. [CrossRef]
81. Luttenberger, L.R. Waste management challenges in transition to circular economy–case of Croatia. *J. Clean. Prod.* **2020**, *256*, 120495. [CrossRef]
82. Pires, A.; Martinho, G. Waste hierarchy index for circular economy in waste management. *Waste Manag.* **2019**, *95*, 298–305. [CrossRef] [PubMed]
83. Romero-Hernández, O.; Romero, S. Maximizing the value of waste: From waste management to the circular economy. *Thunderbird Int. Bus. Rev.* **2018**, *60*, 757–764. [CrossRef]
84. Chioatto, E.; Sospiro, P. Transition from waste management to circular economy: The European Union roadmap. *Environ. Dev. Sustain.* **2023**, *25*, 249–276. [CrossRef]
85. Nelles, M.; Gruenes, J.; Morscheck, G. Waste management in Germany–development to a sustainable circular economy? *Procedia Environ. Sci.* **2016**, *35*, 6–14. [CrossRef]
86. Afshari, H.; Gurtu, A.; Jaber, M.Y. Unlocking the potential of solid waste management with circular economy and Industry 4.0. *Comput. Ind. Eng.* **2024**, *195*, 110457. [CrossRef]

MDPI AG
Grosspeteranlage 5
4052 Basel
Switzerland
Tel.: +41 61 683 77 34

Agronomy Editorial Office
E-mail: agronomy@mdpi.com
www.mdpi.com/journal/agronomy

www.ingramcontent.com/pod-product-compliance
Lightning Source LLC
Chambersburg PA
CBHW061552210926
51850CB00009B/2079
9783725870769